# Study and Solutions Guide
# Trigonometry
## FOURTH EDITION
### Larson / Hostetler

# Dianna L. Zook

Indiana University
Purdue University at Fort Wayne, Indiana

**HOUGHTON MIFFLIN COMPANY**  Boston   New York

Sponsoring Editor: Christine B. Hoag
Senior Associate Editor: Maureen Brooks
Managing Editor: Catherine B. Cantin
Supervising Editor: Karen Carter
Associate Project Editor: Rachel D'Angelo Wimberly
Editorial Assistant: Caroline Lipscomb
Production Supervisor: Lisa Merrill
Art Supervisor: Gary Crespo
Associate Marketing Manager: Ros Kane
Marketing Assistant: Kate Burden Thomas

Printed in the United States of America.

International Standard Book Number: 0-669-41739-4

3456789-PO 01 00 99 98

# CONTENTS

# TO THE STUDENT

The *Study and Solutions Guide* for *Trigonometry* is a supplement to the text by Roland E. Larson and Robert P. Hostetler.

As a mathematics instructor, I often have students come to me with questions about assigned homework. When I ask to see their work, the reply often is "I didn't know where to start." The purpose of the *Study Guide* is to provide brief summaries of the topics covered in the textbook and enough detailed solutions to problems so that you will be able to work the remaining exercises.

A special thanks to Larson Texts, Inc. for typing this guide. Also I would like to thank my husband Edward L. Schlindwein for his support during the several months I worked on this project.

If you have any corrections or suggestions for improving this *Study Guide*, I would appreciate hearing from you.

Good luck with your study of trigonometry.

Dianna L. Zook
Indiana University,
Purdue University at
Fort Wayne, Indiana 46805

---

# STUDY STRATEGIES

- Attend all classes and come prepared. Have your homework completed. Bring the text, paper, pen or pencil, and a calculator (scientific or graphing) to each class.

- Read the section in the text that is to be covered before class. Make notes about any questions that you have and, if not answered during the lecture, ask them at the appropriate time.

- Participate in class. As mentioned above, ask questions. Also, do not be afraid to answer questions.

- Take notes on all definitions, concepts, rules, formulas and examples. After class, read your notes and fill in any gaps, or make notations of any questions that you have.

- DO THE HOMEWORK!!! You learn mathematics by doing it yourself. Allow at least two hours outside of each class for homework. Do not fall behind.

- Seek help when needed. Visit your instructor during office hours and come prepared with specific questions; check with your school's tutoring service; find a study partner in class; check additional books in the library for more examples—just do something before the problem becomes insurmountable.

- Do not cram for exams. Each chapter in the text contains a chapter review and this study guide contains a practice test at the end of each chapter. (The answers are at the back of the study guide.) Work these problems a few days before the exam and review any areas of weakness.

# PART I
## Solutions to Odd-Numbered Exercises

# C H A P T E R  P
## Prerequisites

# CHAPTER P
## Prerequisites

## Section P.1    Real Numbers

- You should know the following sets.

    (a) The set of real numbers includes the rational numbers and the irrational numbers.

    (b) The set of rational numbers includes all real numbers that can be written as the ratio $p/q$ of two integers, where $q \neq 0$.

    (c) The set of irrational numbers includes all real numbers which are not rational.

    (d) The set of integers: $\{\ldots, -3, -2, -1, 0, 1, 2, 3, \ldots\}$

    (e) The set of whole numbers: $\{0, 1, 2, 3, 4, \ldots\}$

    (f) The set of natural numbers: $\{1, 2, 3, 4, \ldots\}$

- The real number line is used to represent the real numbers.

- Know the inequality symbols.

    (a) $a < b$ means $a$ is less than $b$.

    (b) $a \leq b$ means $a$ is less than or equal to $b$.

    (c) $a > b$ means $a$ is greater than $b$.

    (d) $a \geq b$ means $a$ is greater than or equal to $b$.

- You should know that
$$|a| = \begin{cases} a, & \text{if } a \geq 0 \\ -a, & \text{if } a < 0. \end{cases}$$

- Know the properties of absolute value.

    (a) $|a| \geq 0$      (b) $|-a| = |a|$      (c) $|ab| = |a|\,|b|$      (d) $\left|\dfrac{a}{b}\right| = \dfrac{|a|}{|b|}$

- The distance between $a$ and $b$ on the real line is $|b - a| = |a - b|$.

- You should be able to identify the terms in an algebraic expression.

- You should know and be able to use the basic rules of algebra.

- Commutative Property

    (a) Addition: $a + b = b + a$          (b) Multiplication: $a \cdot b = b \cdot a$

- Associative Property

    (a) Addition: $(a + b) + c = a + (b + c)$      (b) Multiplication: $(ab)c = a(bc)$

- Identity Property

    (a) Addition: 0 is the identity; $a + 0 = 0 + a = a$.

    (b) Multiplication: 1 is the identity; $a \cdot 1 = 1 \cdot a = a$.

- Inverse Property

    (a) Addition: $-a$ is the inverse of $a$; $a + (-a) = -a + a = 0$.

    (b) Multiplication: $1/a$ is the inverse of $a$, $a \neq 0$; $a(1/a) = (1/a)a = 1$.

- Distributive Property

    (a) Left: $a(b + c) = ab + ac$          (b) Right: $(a + b)c = ac + bc$

*continued*

■ Properties of Negatives

    (a) $(-1)a = -a$

    (b) $-(-a) = a$

    (c) $(-a)b = a(-b) = -ab$

    (d) $(-a)(-b) = ab$

    (e) $-(a + b) = (-a) + (-b) = -a - b$

■ Properties of Zero

    (a) $a \pm 0 = a$

    (b) $a \cdot 0 = 0$

    (c) $0 \div a = 0/a = 0, a \neq 0$

    (d) If $ab = 0$, then $a = 0$ or $b = 0$.

    (e) $a/0$ is undefined.

■ Properties of Fractions $(b \neq 0, d \neq 0)$

    (a) Equivalent Fractions: $a/b = c/d$ if and only if $ad = bc$.

    (b) Rule of Signs: $-a/b = a/-b = -(a/b)$ and $-a/-b = a/b$

    (c) Equivalent Fractions: $a/b = ac/bc, c \neq 0$

    (d) Addition and Subtraction

       1. Like Denominators: $(a/b) \pm (c/b) = (a \pm c)/b$

       2. Unlike Denominators: $(a/b) \pm (c/d) = (ad \pm bc)/bd$

    (e) Multiplication: $(a/b) \cdot (c/d) = ac/bd$

    (f) Division: $(a/b) \div (c/d) = (a/b) \cdot (d/c) = ad/bc$ if $c \neq 0$.

■ Properties of Equality

    (a) If $a = b$, then $a + c = b + c$.

    (b) If $a = b$, then $ac = bc$.

    (c) If $a + c = b + c$, then $a = b$.

    (d) If $ac = bc$ and $c \neq 0$, then $a = b$.

## Solutions to Odd-Numbered Exercises

**1.** $-9, -\frac{7}{2}, 5, \frac{2}{3}, \sqrt{2}, 0, 1$

    (a) Natural numbers: 5, 1

    (b) Integers: $-9, 5, 0, 1$

    (c) Rational numbers: $-9, -\frac{7}{2}, 5, \frac{2}{3}, 0, 1$

    (d) Irrational numbers: $\sqrt{2}$

**3.** $2.01, \ 0.666\ldots, -13, 0.010110111\ldots$

    (a) Natural numbers: none

    (b) Integers: $-13$

    (c) Rational numbers: $2.01, 0.666\ldots, -13$

    (d) Irrational numbers: $0.010110111\ldots$

**5.** $-\pi, -\frac{1}{3}, \frac{6}{3}, \frac{1}{2}\sqrt{2}, -7.5$

    (a) Natural numbers: $\frac{6}{3}$ (since it equals 2)

    (b) Integers: $\frac{6}{3}$

    (c) Rational numbers: $-\frac{1}{3}, \frac{6}{3}, -7.5$

    (d) Irrational numbers: $-\pi, \frac{1}{2}\sqrt{2}$

**7.** $\frac{5}{8} - 0.625$

**9.** $\frac{41}{333} = 0.\overline{123}$

**11.** $-1 < 2.5$

**13.** $\frac{3}{2} < 7$

**15.** $-4 > -8$

**17.** $\frac{5}{6} > \frac{2}{3}$

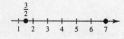

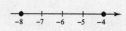

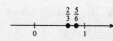

**19.** The inequality $x \leq 5$ is the set of all real numbers less than or equal to 5. The interval is unbounded.

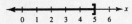

**21.** The inequality $x < 0$ is the set of all negative real numbers. The interval is unbounded.

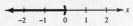

**23.** The inequality $x \geq 4$ is the set of all real numbers greater than or equal to 4. The interval is unbounded.

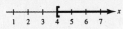

**25.** The inequality $-2 < x < 2$ is the set of all real numbers greater than $-2$ and less than 2. The interval is bounded.

**27.** The inequality $-1 \leq x < 0$ is the set of all negative real numbers greater than or equal to $-1$. The interval is bounded.

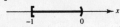

**29.** $\frac{127}{90} \approx 1.41111, \frac{584}{413} \approx 1.41404, \frac{7071}{5000} \approx 1.41420, \sqrt{2} \approx 1.41421, \frac{47}{33} \approx 1.42424$

**31.** $x < 0$

**33.** $y \geq 0$

**35.** $A \geq 30$

**37.** $|-10| = -(-10) = 10$

**39.** $|3 - \pi| = -(3 - \pi) = \pi - 3 \approx 0.1416$

**41.** $\dfrac{-5}{|-5|} = \dfrac{-5}{-(-5)} = \dfrac{-5}{5} = -1$

**43.** $-3|-3| = -3[-(-3)] = -9$

**45.** $-|16.25| + 20 = -16.25 + 20 = 3.75$

**47.** $|-3| > -|-3|$ since $3 > -3$.

**49.** $-5 = -|5|$ since $-5 = -5$.

**51.** $-|-2| = -|2|$ since $-2 = -2$.

**53.** $d(-1, 3) = |3 - (-1)| = |3 + 1| = 4$

**55.** $d\left(-\frac{5}{2}, 0\right) = \left|0 - \left(-\frac{5}{2}\right)\right| = \frac{5}{2}$

**57.** $d(126, 75) = |75 - 126| = 51$

**59.** $d\left(\frac{16}{5}, \frac{112}{75}\right) = \left|\frac{112}{75} - \frac{16}{5}\right| = \frac{128}{75}$

**61.** $d(x, 5) = |x - 5|$ and $d(x, 5) \leq 3$, thus $|x - 5| \leq 3$.

**63.** $d(7, 18) = |7 - 18| = 11$ miles

**65.** $d(y, 0) = |y - 0| = |y|$ and $d(y, 0) \geq 6$, thus $|y| \geq 6$.

**67.**

| Budgeted Expense, b | Actual Expense, a | $|a - b|$ | $0.05b$ |
|---|---|---|---|
| $112,700 | $113,356 | $656 | $5635 |

The actual expense difference is greater than $500 (but is less than 5% of the budget) so it does not pass the test.

**69.**

| Budgeted Expense, b | Actual Expense, a | $|a - b|$ | $0.05b$ |
|---|---|---|---|
| $37,640 | $37,335 | $305 | $1882 |

Since $305 < $500 and $305 < $1882, it passes the "budget variance test."

**71.** $|77.8 - 92.2| = \$14.4$ billion deficit for 1960

**73.** $|1031.3 - 1252.7| = \$221.4$ billion deficit for 1990

**75.** (a) $|u + v| \neq |u| + |v|$ if $u$ is positive and $v$ is negative or vice versa.

  (b) $|u + v| \leq |u| + |v|$

  They are equal when $u$ and $v$ have the same sign. If they differ in sign, $|u + v|$ is less than $|u| + |v|$.

**77.** $7x + 4$

Terms: $7x, 4$

**79.** $4x^3 + x - 5$

Terms: $4x^3, x, -5$

**81.** $4x - 6$

  (a) $4(-1) - 6 = -4 - 6 = -10$

  (b) $4(0) - 6 = 0 - 6 = -6$

**83.** $x^2 - 3x + 4$

  (a) $(-2)^2 - 3(-2) + 4 = 4 + 6 + 4 = 14$

  (b) $(2)^2 - 3(2) + 4 = 4 - 6 + 4 = 2$

**85.** $\dfrac{x + 1}{x - 1}$

  (a) $\dfrac{1 + 1}{1 - 1} = \dfrac{2}{0}$ is undefined.

  You cannot divide by zero.

  (b) $\dfrac{-1 + 1}{-1 - 1} = \dfrac{0}{-2} = 0$

**87.** $x + 9 = 9 + x$

Commutative (addition)

**89.** $\dfrac{1}{(h + 6)}(h + 6) = 1, h \neq -6$

Inverse (multiplication)

**91.** $2(x + 3) = 2x + 6$

Distributive Property

**93.** $1 \cdot (1 + x) = 1 + x$

Identity (multiplication)

**95.** $x(3y) = (x \cdot 3)y$   Associative (multiplication)

  $= (3x)y$   Commutative (multiplication)

**97.** $\dfrac{81 - (90 - 9)}{5} = \dfrac{81 - 81}{5} = \dfrac{0}{5} = 0$

**99.** $\dfrac{8 - 8}{-9 + (6 + 3)} = \dfrac{0}{-9 + 9} = \dfrac{0}{0}$ which is undefined.

**101.** $(4 - 7)(-2) = (-3)(-2) = 6$

**103.** $\frac{3}{16} + \frac{5}{16} = \frac{8}{16} = \frac{1}{2}$

**105.** $\dfrac{5}{8} - \dfrac{5}{12} + \dfrac{1}{6} = \dfrac{15}{24} - \dfrac{10}{24} + \dfrac{4}{24} = \dfrac{9}{24} = \dfrac{3}{8}$

**107.** $\dfrac{4}{5} \cdot \dfrac{1}{2} \cdot \dfrac{3}{4} \cdot \ = \dfrac{3}{10}$

**109.** $12 \div \frac{1}{4} = 12 \cdot \frac{4}{1} = 12 \cdot 4 = 48$

**111.** $-3 + \frac{3}{7} \approx -2.57$

**113.** $\dfrac{11.46 - 5.37}{3.91} \approx 1.56$

**115.**

| $n$ | 1 | 0.5 | 0.01 | 0.0001 | 0.000001 |
|-----|---|-----|------|--------|----------|
| $5/n$ | 5 | 10 | 500 | 50,000 | 5,000,000 |

**117.**

| $n$ | 1 | 10 | 100 | 10,000 | 100,000 |
|-----|---|-----|------|--------|----------|
| $5/n$ | 5 | 0.5 | 0.05 | 0.0005 | 0.00005 |

# Section P.2   Solving Equations

- You should know how to solve linear equations.
  $$ax + b = 0$$

- An identity is an equation whose solution consists of every real number in its domain.

- To solve an equation you can:

  (a) Add or subtract the same quantity from both sides.

  (b) Multiply or divide both sides by the same nonzero quantity.

- To solve an equation that can be simplified to a linear equation:

  (a) Remove all symbols of grouping and all fractions.

  (b) Combine like terms.

  (c) Solve by algebra.

  (d) Check the answer.

- A "solution" that does not satisfy the original equation is called an extraneous solution.

- You should be able to solve a quadratic equation by factoring, if possible.

- You should be able to solve a quadratic equation of the form $u^2 = d$ by extracting square roots.

- You should be able to solve a quadratic equation by completing the square.

- You should know and be able to use the Quadratic Formula: For $ax^2 + bx + c = 0, a \neq 0$,

  $$x = \frac{-b \pm \sqrt{b^2 - 4ac}}{2a}.$$

- You should be able to solve polynomials of higher degree by factoring.

- For equations involving radicals or fractional powers, raise both sides to the same power.

- For equations with fractions, multiply both sides by the least common denominator to clear the fractions.

- For equations involving absolute value, remember that the expression inside the absolute value can be positive or negative.

- Always check for extraneous solutions.

## Solutions to Odd-Numbered Exercises

**1.** $5x - 3 = 3x + 5$

(a) $5(0) - 3 \stackrel{?}{=} 3(0) + 5$

$-3 \neq 5$

$x = 0$ *is not* a solution.

(b) $5(-5) - 3 \stackrel{?}{=} 3(-5) + 5$

$-28 \neq -10$

$x = -5$ *is not* a solution.

(c) $5(4) - 3 \stackrel{?}{=} 3(4) + 5$

$17 = 17$

$x = 4$ *is* a solution.

(d) $5(10) - 3 \stackrel{?}{=} 3(10) + 5$

$47 \neq 35$

$x = 10$ *is not* a solution.

**3.** $3x^2 + 2x - 5 = 2x^2 - 2$

   (a) $3(-3) + 2(-3) - 5 \stackrel{?}{=} 2(-3)^2 - 2$

                       $16 = 16$

     $x = -3$ *is* a solution.

   (c) $3(4)^2 + 2(4) - 5 \stackrel{?}{=} 2(4)^2 - 2$

                     $51 \neq 30$

     $x = 4$ *is not* a solution.

   (b) $3(1)^2 + 2(1) - 5 \stackrel{?}{=} 2(1)^2 - 2$

                       $0 = 0$

     $x = 1$ *is* a solution.

   (d) $3(-5)^2 + 2(-5) - 5 \stackrel{?}{=} 2(-5)^2 - 2$

                     $60 \neq 48$

     $x = -5$ *is not* a solution.

**5.** $\dfrac{5}{2x} - \dfrac{4}{x} = 3$

   (a) $\dfrac{5}{2(-1/2)} - \dfrac{5}{(-1/2)} \stackrel{?}{=} 3$

               $3 = 3$

     $x = -\frac{1}{2}$ *is* a solution.

   (c) $\dfrac{5}{2(0)} - \dfrac{4}{0}$ is undefined.

     $x = 0$ *is not* a solution.

   (b) $\dfrac{5}{2(4)} - \dfrac{4}{4} \stackrel{?}{=} 3$

           $-\dfrac{3}{8} \neq 3$

     $x = 4$ *is not* a solution.

   (d) $\dfrac{5}{2(1/4)} - \dfrac{4}{1/4} \stackrel{?}{=} 3$

            $-6 \neq 3$

     $x = \frac{1}{4}$ *is not* a solution.

**7.** $2(x - 1) = 2x - 2$ is an *identity* by the Distributive Property. It is true for all real values of $x$.

**9.** $-6(x - 3) + 5 = -2x + 10$ is *conditional*. There are real values of $x$ for which the equation is not true.

**11.** $x^2 - 8x + 5 = (x - 4)^2 - 11$ is an *identity* since $(x - 4)^2 - 11 = x^2 - 8x + 16 - 11 = x^2 - 8x + 5$.

**13.** (a) Equivalent equations are derived from the substitution principle and simplification techniques. They have the same solution(s).

     $2x + 3 = 8$ and $2x = 5$ are equivalent equations.

   (b) Equivalent equations are produced by removing symbols of grouping, adding or subtracting the same quantity from both sides of the equation, multiplying or dividing both sides of the equation by the same nonzero quantity, or by interchanging the two sides of the equation.

**15.** $2(x + 5) - 7 = 3(x - 2)$

    $2x + 10 - 7 = 3x - 6$

       $2x + 3 = 3x - 6$

          $-x = -9$

            $x = 9$

**17.**      $\dfrac{5x}{4} + \dfrac{1}{2} = x - \dfrac{1}{2}$

    $4\left(\dfrac{5x}{4}\right) + 4\left(\dfrac{1}{2}\right) = 4(x) - 4\left(\dfrac{1}{2}\right)$

          $5x + 2 = 4x - 2$

               $x = -4$

**19.**
$$0.25x + 0.75(10 - x) = 3$$
$$4(0.25x) + 4(0.75)(10 - x) = 4(3)$$
$$x + 3(10 - x) = 12$$
$$x + 30 - 3x = 12$$
$$-2x = -18$$
$$x = 9$$

**21.** $x + 8 = 2(x - 2) - x$
$$x + 8 = 2x - 4 - x$$
$$x + 8 = x - 4$$
$$8 = -4$$
Contradiction: no solution

**23.**
$$\frac{100 - 4u}{3} = \frac{5u + 6}{4} + 6$$
$$12\left(\frac{100 - 4u}{3}\right) = 12\left(\frac{5u + 6}{4}\right) + 12(6)$$
$$4(100 - 4u) = 3(5u + 6) + 72$$
$$400 - 16u = 15u + 18 + 72$$
$$-31u = -310$$
$$u = 10$$

**25.**
$$\frac{5x - 4}{5x + 4} = \frac{2}{3}$$
$$3(5x - 4) = 2(5x + 4)$$
$$15x - 12 = 10x + 8$$
$$5x = 20$$
$$x = 4$$

**27.** $10 - \dfrac{13}{x} = 4 + \dfrac{5}{x}$
$$\frac{10x - 13}{x} = \frac{4x + 5}{x}$$
$$10x - 13 = 4x + 5$$
$$6x = 18$$
$$x = 3$$

**29.**
$$\frac{1}{x - 3} + \frac{1}{x + 3} = \frac{10}{x^2 - 9}$$
$$\frac{(x + 3) + (x - 3)}{x^2 - 9} = \frac{10}{x^2 - 9}$$
$$2x = 10$$
$$x = 5$$

**31.** $\dfrac{x}{x + 4} + \dfrac{4}{x + 4} + 2 = 0$
$$\frac{x + 4}{x + 4} + 2 = 0$$
$$1 + 2 = 0$$
$$3 = 0$$
Contradiction : no solution

**33.**
$$\frac{7}{2x + 1} - \frac{8x}{2x - 1} = -4$$
$$7(2x - 1) - 8x(2x + 1) = -4(2x + 1)(2x - 1)$$
$$14x - 7 - 16x^2 - 8x = -16x^2 + 4$$
$$6x = 11$$
$$x = \frac{11}{6}$$

**35.**
$$(x + 2)^2 + 5 = (x + 3)^2$$
$$x^2 + 4x + 4 + 5 = x^2 + 6x + 9$$
$$4x + 9 = 6x + 9$$
$$-2x = 0$$
$$x = 0$$

**37.**
$$(x + 2)^2 - x^2 = 4(x + 1)$$
$$x^2 + 4x + 4 - x^2 = 4x + 4$$
$$4 = 4$$
The equation is an identity; every real number is a solution.

**39.** $4 - 2(x - 2b) = ax + 3$

$4 - 2x + 4b = ax + 3$

$1 + 4b = ax + 2x$

$1 + 4b = x(a + 2)$

$\dfrac{1 + 4b}{a + 2} = x, \ a \neq -2$

**41.** (a)

| $x$ | $-1$ | 0 | 1 | 2 | 3 | 4 |
|---|---|---|---|---|---|---|
| $3.2x - 5.8$ | $-9$ | $-5.8$ | $-2.6$ | 0.6 | 3.8 | 7 |

(b) Since the sign changes from negative at 1 to positive at 2, the root is somewhere between 1 and 2.
$1 < x < 2$

(c)

| $x$ | 1.5 | 1.6 | 1.7 | 1.8 | 1.9 | 2 |
|---|---|---|---|---|---|---|
| $3.2x - 5.8$ | $-1$ | $-0.68$ | $-0.36$ | $-0.04$ | 0.28 | 0.6 |

(d) Since the sign changes from negative at 1.8 to positive at 1.9, the root is somewhere between 1.8 and 1.9.
$1.8 < x < 1.9$.

To improve accuracy, evaluate the expression in this interval and determine where the sign changes.

**43.**  $6x^2 + 3x = 0$

$3x(2x + 1) = 0$

$3x = 0 \quad \text{or} \quad 2x + 1 = 0$

$x = 0 \quad \text{or} \qquad x = -\frac{1}{2}$

**45.**  $x^2 - 2x - 8 = 0$

$(x - 4)(x + 2) = 0$

$x - 4 = 0 \quad \text{or} \quad x + 2 = 0$

$x = 4 \quad \text{or} \qquad x = -2$

**47.**  $3 + 5x - 2x^2 = 0$

$(3 - x)(1 + 2x) = 0$

$3 - x = 0 \quad \text{or} \quad 1 + 2x = 0$

$x = 3 \quad \text{or} \qquad x = -\frac{1}{2}$

**49.**  $\qquad 2x^2 = 19x + 33$

$2x^2 - 19x - 33 = 0$

$(2x + 3)(x - 11) = 0$

$2x + 3 = 0 \quad \text{or} \quad x - 11 = 0$

$x = -\frac{3}{2} \quad \text{or} \qquad x = 11$

**51.**  $\qquad 2x^4 - 18x^2 = 0$

$2x^2(x^2 - 9) = 0$

$2x^2(x + 3)(x - 3) = 0$

$2x^2 = 0 \quad \text{or} \quad x + 3 = 0 \qquad \text{or} \quad x - 3 = 0$

$x = 0 \quad \text{or} \qquad x = -3 \quad \text{or} \qquad x = 3$

**53.**  $x^3 - 2x^2 - 3x = 0$

$x(x^2 - 2x - 3) = 0$

$x(x + 1)(x - 3) = 0$

$x = 0 \quad \text{or} \quad x + 1 = 0 \qquad \text{or} \quad x - 3 = 0$

$x = 0 \quad \text{or} \qquad x = -1 \quad \text{or} \qquad x = 3$

**55.**  $2x^4 - 15x^3 + 18x^2 = 0$

$x^2(2x^2 - 15x + 18) = 0$

$x^2(2x - 3)(x - 6) = 0$

$x^2 = 0 \quad \text{or} \quad 2x - 3 = 0 \quad \text{or} \quad x - 6 = 0$

$x = 0 \quad \text{or} \qquad x = \frac{3}{2} \quad \text{or} \qquad x = 6$

**57.** $x^2 = 16$

$x = \pm 4$

$\quad = \pm 4.00$

**59.** $3x^2 = 36$

$x^2 = 12$

$x = \pm 2\sqrt{3}$

$\approx \pm 3.46$

**61.** $(x - 12)^2 = 18$

$x - 12 = \pm 3\sqrt{2}$

$x = 12 \pm 3\sqrt{2}$

$x \approx 16.24 \quad \text{or} \quad \approx 7.76$

**63.** $(x + 2)^2 = 12$

$x + 2 = \pm 2\sqrt{3}$

$x = -2 \pm 2\sqrt{3}$

$x \approx 1.46 \quad \text{or} \quad x \approx -5.46$

**65.** $\quad x^2 - 2x = 0$

$x^2 - 2x + 1^2 = 0 + 1$

$x^2 - 2x + 1 = 1$

$(x - 1)^2 = 1$

$x - 1 = \pm\sqrt{1}$

$x = 1 \pm 1$

$x = 0 \quad \text{or} \quad x = 2$

**67.** $x^2 + 6x + 2 = 0$

$x^2 + 6x = -2$

$x^2 + 6x + 3^2 = -2 + 3^2$

$(x + 3)^2 = 7$

$x + 3 = \pm\sqrt{7}$

$x = -3 \pm\sqrt{7}$

**69.** $\quad 8 + 4x - x^2 = 0$

$-x^2 + 4x + 8 = 0$

$x^2 - 4x - 8 = 0$

$x^2 - 4x = 8$

$x^2 - 4x + 2^2 = 8 + 2^2$

$(x - 2)^2 = 12$

$x - 2 = \pm\sqrt{12}$

$x = 2 \pm 2\sqrt{3}$

**71.** $2x^2 + x - 1 = 0$

$x = \dfrac{-b \pm \sqrt{b^2 - 4ac}}{2a}$

$\quad = \dfrac{-1 \pm \sqrt{1^2 - 4(2)(-1)}}{2(2)}$

$\quad = \dfrac{-1 \pm 3}{4} = \dfrac{1}{2}, -1$

**73.** $x^2 + 8x - 4 = 0$

$x = \dfrac{-b \pm \sqrt{b^2 - 4ac}}{2a}$

$\quad = \dfrac{-8 \pm \sqrt{8^2 - 4(1)(-4)}}{2(1)}$

$\quad = \dfrac{-8 \pm 4\sqrt{5}}{2}$

$\quad = -4 \pm 2\sqrt{5}$

**75.** $\quad 12x - 9x^2 = -3$

$-9x^2 + 12x + 3 = 0$

$x = \dfrac{-b \pm \sqrt{b^2 - 4ac}}{2a}$

$\quad = \dfrac{-12 \pm \sqrt{12^2 - 4(-9)(3)}}{2(-9)}$

$\quad = \dfrac{-12 \pm 6\sqrt{7}}{-18} = \dfrac{2}{3} \pm \dfrac{\sqrt{7}}{3}$

**77.** $3x + x^2 - 1 = 0$

$x^2 + 3x - 1 = 0$

$x = \dfrac{-b \pm \sqrt{b^2 - 4ac}}{2a}$

$\quad = \dfrac{-3 \pm \sqrt{3^2 - 4(1)(-1)}}{2(1)}$

$\quad = \dfrac{-3 \pm \sqrt{13}}{2}$

$\quad = -\dfrac{3}{2} \pm \dfrac{\sqrt{13}}{2}$

**79.**
$$28x - 49x^2 = 4$$
$$-49x^2 + 28x - 4 = 0$$
$$x = \frac{-b \pm \sqrt{b^2 - 4ac}}{2a}$$
$$= \frac{-28 \pm \sqrt{28^2 - 4(-49)(-4)}}{2(-49)}$$
$$= \frac{-28 \pm 0}{-98} = \frac{2}{7}$$

**81.**
$$8t = 5 + 2t^2$$
$$-2t^2 + 8t - 5 = 0$$
$$t = \frac{-b \pm \sqrt{b^2 - 4ac}}{2a}$$
$$= \frac{-8 \pm \sqrt{8^2 - 4(-2)(-5)}}{2(-2)}$$
$$= \frac{-8 \pm 2\sqrt{6}}{-4} = 2 \pm \frac{\sqrt{6}}{2}$$

**83.** False. The product must equal zero to use the Zero-Factor Property.

**85.** (a) $ax^2 + bx = 0$
$$x(ax + b) = 0$$
$$x = 0 \quad \text{or} \quad x = -\frac{b}{a}$$

(b) $ax^2 - ax = 0$
$$ax(x - 1) = 0$$
$$x = 0 \quad \text{or} \quad x = 1$$

**87.**
$$x^4 - 4x^2 + 3 = 0$$
$$(x^2 - 3)(x^2 - 1) = 0$$
$$\left(x + \sqrt{3}\right)\left(x - \sqrt{3}\right)(x + 1)(x - 1) = 0$$
$$x + \sqrt{3} = 0 \Rightarrow x = -\sqrt{3}$$
$$x - \sqrt{3} = 0 \Rightarrow x = \sqrt{3}$$
$$x + 1 = 0 \quad \Rightarrow \quad x = -1$$
$$x - 1 = 0 \quad \Rightarrow \quad x = 1$$

**89.**
$$\frac{1}{t^2} + \frac{8}{t} + 15 = 0$$
$$1 + 8t + 15t^2 = 0$$
$$(1 + 3t)(1 + 5t) = 0$$
$$1 + 3t = 0 \quad \Rightarrow \quad t = -\frac{1}{3}$$
$$1 + 5t = 0 \quad \Rightarrow \quad t = -\frac{1}{5}$$

**91.**
$$2x + 9\sqrt{x} - 5 = 0$$
$$(2\sqrt{x} - 1)(\sqrt{x} + 5) = 0$$
$$\sqrt{x} = \frac{1}{2} \Rightarrow x = \frac{1}{4}$$
$$\left(\sqrt{x} = -5 \text{ is not a solution.}\right)$$

**93.**
$$3x^{1/3} + 2x^{2/3} = 5$$
$$2x^{2/3} + 3x^{1/3} - 5 = 0$$
$$(2x^{1/3} + 5)(x^{1/3} - 1) = 0$$
$$2x^{1/3} + 5 = 0 \quad \Rightarrow \quad x^{1/3} = -\frac{5}{2} \quad \Rightarrow \quad x = \left(-\frac{5}{2}\right)^3 = -\frac{125}{8}$$
$$x^{1/3} - 1 = 0 \quad \Rightarrow \quad x^{1/3} = 1 \quad \Rightarrow \quad x = (1)^3 = 1$$

**95.** $\sqrt{x - 10} - 4 = 0$

$\sqrt{x - 10} = 4$

$x - 10 = 16$

$x = 26$

**97.** $\sqrt[3]{2x + 5} + 3 = 0$

$\sqrt[3]{2x + 5} = -3$

$2x + 5 = -27$

$2x = -32$

$x = -16$

**99.** $x = \sqrt{11x - 30}$

$x^2 = 11x - 30$

$x^2 - 11x + 30 = 0$

$(x - 5)(x - 6) = 0$

$x - 5 = 0 \implies x = 5$

$x - 6 = 0 \implies x = 6$

**101.** $\sqrt{x + 1} - 3x = 1$

$\sqrt{x + 1} = 3x + 1$

$x + 1 = 9x^2 + 6x + 1$

$0 = 9x^2 + 5x$

$0 = x(9x + 5)$

$x = 0$

$9x + 5 = 0 \implies x = -\frac{5}{9},$ extraneous

**103.** $\sqrt{x} - \sqrt{x - 5} = 1$

$\sqrt{x} = 1 + \sqrt{x - 5}$

$\left(\sqrt{x}\right)^2 = \left(1 + \sqrt{x - 5}\right)^2$

$x = 1 + 2\sqrt{x - 5} + x - 5$

$4 = 2\sqrt{x - 5}$

$2 = \sqrt{x - 5}$

$4 = x - 5$

$9 = x$

**105.** $2\sqrt{x + 1} - \sqrt{2x + 3} = 1$

$2\sqrt{x + 1} = \sqrt{2x + 3} + 1$

$\left(2\sqrt{x + 1}\right)^2 = \left(\sqrt{2x + 3} + 1\right)^2$

$4(x + 1) = 2x + 3 + 2\sqrt{2x + 3} + 1$

$4x + 4 = 2x + 4 + 2\sqrt{2x + 3}$

$2x = 2\sqrt{2x + 3}$

$x = \sqrt{2x + 3}$

$x^2 = 2x + 3$

$x^2 - 2x - 3 = 0$

$(x - 3)(x + 1) = 0$

$x - 3 = 0 \implies x = 3$

$x + 1 = 0 \implies x = -1,$ extraneous

**107.** $(x - 5)^{2/3} = 16$

$x - 5 = \pm 16^{3/2}$

$x - 5 = \pm 64$

$x = 69, -59$

**109.** $37.55 = 40 - \sqrt{0.01x + 1}$

$\sqrt{0.01x + 1} = 2.45$

$0.01x + 1 = 6.0025$

$0.01x = 5.0025$

$x = 500.25$

Rounding $x$ to the nearest whole unit yields $x \approx 500$ units.

**111.** $S = \pi r\sqrt{r^2 + h^2}$

$S^2 = \pi^2 r^2 (r^2 + h^2)$

$S^2 = \pi^2 r^4 + \pi^2 r^2 h^2$

$\dfrac{S^2 - \pi^2 r^4}{\pi^2 r^2} = h^2$

$h = \dfrac{\sqrt{S^2 - \pi^2 r^4}}{\pi r}$

**113.**
$$20 + \sqrt{20 - a} = b$$
$$\sqrt{20 - a} = b - 20$$
$$20 - a = b^2 - 40b + 400$$
$$-a = b^2 - 40b + 380$$
$$a = -b^2 + 40b - 380$$

This formula gives the relationship between $a$ and $b$. From the original equation we know that $a \le 20$ and $b \ge 20$. Choose a $b$ value, where $b \ge 20$ and then solve for $a$, keeping in mind that $a \le 20$.

Some possibilities are:

$b = 20, \quad a = 20$

$b = 21, \quad a = 19$

$b = 22, \quad a = 16$

$b = 23, \quad a = 11$

$b = 24, \quad a = 4$   $\leftarrow \left(\begin{array}{l}\text{This is the one given}\\\text{in your textbook.}\end{array}\right)$

$b = 25, \quad a = -5$

**115.**
$$|x + 1| = 2$$
$$x + 1 = 2 \;\Rightarrow\; x = 1$$
$$-(x + 1) = 2 \;\Rightarrow\; x = -3$$

**117.**
$$|2x - 1| = 5$$
$$2x - 1 = 5 \;\Rightarrow\; x = 3$$
$$-(2x - 1) = 5 \;\Rightarrow\; x = -2$$

**119.**
$$|x^2 + 6x| = 3x + 18$$

$$x^2 + 6x = 3x + 18$$
$$x^2 + 3x - 18 = 0$$
$$(x + 6)(x - 3) = 0$$
$$x + 6 = 0 \;\Rightarrow\; x = -6$$
$$x - 3 = 0 \;\Rightarrow\; x = 3$$

or

$$-(x^2 + 6x) = 3x + 18$$
$$-x^2 - 6x = 3x + 18$$
$$0 = x^2 + 9x + 18$$
$$0 = (x + 3)(x + 6)$$
$$x + 3 = 0 \;\Rightarrow\; x = -3$$
$$x + 6 = 0 \;\Rightarrow\; x = -6$$

The solutions are $x = -6$ and $x = \pm 3$.

**121.**
$$[x - (-3)](x - 5) = 0$$
$$(x + 3)(x - 5) = 0$$
$$x^2 - 2x - 15 = 0$$

This equation has $x = -3$ and $x = 5$ as solutions, as does any nonzero multiple of this equation.

**123.**
$$4x + 3y = 100$$
$$3y = 100 - 4x$$
$$y = \tfrac{1}{3}(100 - 4x)$$
$$\text{Area} = \text{length} \cdot \text{width}$$
$$350 = (2x)(y)$$
$$350 = 2x\left[\tfrac{1}{3}(100 - 4x)\right]$$
$$350 = \tfrac{2}{3}x(100 - 4x)$$
$$1050 = 2x(100 - 4x)$$
$$1050 = 200x - 8x^2$$
$$8x^2 - 200x + 1050 = 0$$
$$2(4x^2 - 100x + 525) = 0$$
$$2(2x - 35)(2x - 15) = 0$$
$$2x - 35 = 0 \quad \text{or} \quad 2x - 15 = 0$$

$$x = \tfrac{35}{2} \qquad\qquad x = \tfrac{15}{2}$$
$$y = 10 \qquad\qquad y = \tfrac{70}{3}$$

There are two different possible dimensions:

$x = \frac{35}{2} = 17.5$ meters and $y = 10$ meters or $x = \frac{15}{2} = 7.5$ meters and $y = \frac{70}{3} = 23\tfrac{1}{3}$ meters

# Section P.3   Graphs and Equations

- You should be able to plot points.
- You should know that the distance between $(x_1, y_1)$ and $(x_2, y_2)$ in the plane is

  $$d = \sqrt{(x_2 - x_1)^2 + (y_2 - y_1)^2}.$$

- You should know that the midpoint of the line segment joining $(x_1, y_1)$ and $(x_2, y_2)$ is

  $$\left( \frac{x_1 + x_2}{2}, \frac{y_1 + y_2}{2} \right).$$

- You should be able to use the point-plotting method of graphing.
- You should be able to find $x$- and $y$-intercepts.
  - (a) To find the $x$-intercepts, let $y = 0$ and solve for $x$.
  - (b) To find the $y$-intercepts, let $x = 0$ and solve for $y$.
- You should be able to test for symmetry.
  - (a) To test for $x$-axis symmetry, replace $y$ with $-y$.
  - (b) To test for $y$-axis symmetry, replace $x$ with $-x$.
  - (c) To test for origin symmetry, replace $x$ with $-x$ and $y$ with $-y$.
- You should know the standard equation of a circle with center $(h, k)$ and radius $r$:

  $$(x - h)^2 + (y - k)^2 = r^2$$

**Solutions to Odd-Numbered Exercises**

1.

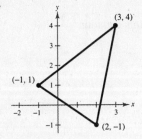

3. $A$: $(2, 6)$,   $B$: $(-6, -2)$,   $C$: $(4, -4)$,   $D$: $(-3, 2)$

5. $(-3, 4)$

7. $(-5, -5)$

9. $x > 0$ and $y < 0$ in Quadrant IV.

11. $(x, -y)$ is in the second Quadrant means that $(x, y)$ is in Quadrant III.

13. On the $x$-axis, $y = 0$.
    On the $y$-axis, $x = 0$.

15.

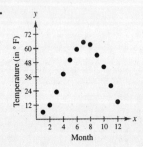

17. (a) The distance between $(0, 2)$ and $(4, 2)$ is 4.
       The distance between $(4, 2)$ and $(4, 5)$ is 3.
       the distance between $(0, 2)$ and $(4, 5)$ is

$$\sqrt{(4 - 0)^2 + (5 - 2)^2} = \sqrt{16 + 9}$$
$$= \sqrt{25} = 5.$$

    (b) $4^2 + 3^2 = 16 + 9 = 25 = 5^2$

**19. (a)** The distance between $(-1, 1)$ and $(9, 1)$ is 10.
The distance between $(9, 1)$ and $(9, 4)$ is 3.
The distance between $(-1, 1)$ and $(9, 4)$ is

$$\sqrt{(9 - (-1))^2 + (4 - 1)^2} = \sqrt{100 + 9} = \sqrt{109}.$$

**(b)** $10^2 + 3^2 = 109 = \left(\sqrt{109}\right)^2$

**21. (a)**

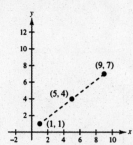

**(b)** $d = \sqrt{(9 - 1)^2 + (7 - 1)^2}$
$= \sqrt{64 + 36} = 10$

**(c)** $\left(\dfrac{9 + 1}{2}, \dfrac{7 + 1}{2}\right) = (5, 4)$

**23. (a)**

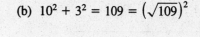

**(b)** $d = \sqrt{(5 + 1)^2 + (4 - 2)^2}$
$= \sqrt{36 + 4} = 2\sqrt{10}$

**(c)** $\left(\dfrac{-1 + 5}{2}, \dfrac{2 + 4}{2}\right) = (2, 3)$

**25. (a)**

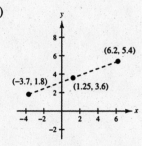

**(b)** $d = \sqrt{(6.2 + 3.7)^2 + (5.4 - 1.8)^2}$
$= \sqrt{98.01 + 12.96}$
$= \sqrt{110.97}$

**(c)** $\left(\dfrac{6.2 - 3.7}{2}, \dfrac{5.4 + 1.8}{2}\right) = (1.25, 3.6)$

**27.** $\left(\dfrac{1991 + 1995}{2}, \dfrac{\$520{,}000 + \$740{,}000}{2}\right) = (1993, \$630{,}000)$

**29.**

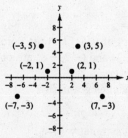

The points are reflected through the $y$-axis.

**31.** $y = \sqrt{x + 4}$

**(a)** $(0, 2)$: $2 \overset{?}{=} \sqrt{0 + 4}$
$2 = 2 \checkmark$

Yes, the point *is* on the graph.

**(b)** $(5, 3)$: $3 \overset{?}{=} \sqrt{5 + 4}$
$3 = \sqrt{9} \checkmark$

Yes, the point *is* on the graph.

**33.** $x^2y - x^2 + 4y = 0$

**(a)** $\left(1, \frac{1}{5}\right)$: $(1)^2\left(\frac{1}{5}\right) - (1)^2 + 4\left(\frac{1}{5}\right) \overset{?}{=} 0$
$\frac{1}{5} - 1 + \frac{4}{5} = 0 \checkmark$

Yes, the point *is* on the graph.

**(b)** $\left(2, \frac{1}{2}\right)$: $(2)^2\left(\frac{1}{2}\right) - (2)^2 + 4\left(\frac{1}{2}\right) \overset{?}{=} 0$
$2 - 4 + 2 = 0 \checkmark$

Yes, the point *is* on the graph.

**35.** $y = \frac{3}{2}x - 1$

| $x$ | $-2$ | $0$ | $\frac{2}{3}$ | $1$ | $2$ |
|---|---|---|---|---|---|
| $y$ | $-4$ | $-1$ | $0$ | $\frac{1}{2}$ | $2$ |

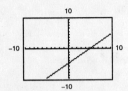

**37.**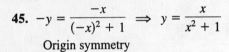

$y = x - 5$

Intercepts: $(5, 0)$, $(0, -5)$

**39.**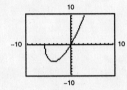

$y = x\sqrt{x + 6}$

Intercepts: $(0, 0)$, $(-6, 0)$

**41.** $(-x)^2 - y = 0 \implies x^2 - y = 0$
  $y$-axis symmetry

**43.** $y = \sqrt{9 - (-x)^2} \implies y = \sqrt{9 - x^2}$
  $y$-axis symmetry

**45.** $-y = \dfrac{-x}{(-x)^2 + 1} \implies y = \dfrac{x}{x^2 + 1}$
  Origin symmetry

**47.** $y$-axis symmetry

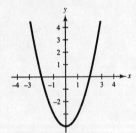

**49.** Origin symmetry

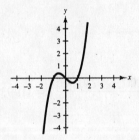

**51.** $y = 1 - x$ has intercepts $(1, 0)$ and $(0, 1)$.
Matches graph (c).

**53.** $y = \sqrt{9 - x^2}$ has intercepts $(\pm 3, 0)$ and $(0, 3)$.
Matches graph (f).

**55.** $y = x^3 - x + 1$ has a $y$-intercept of $(0, 1)$ and the points $(1, 1)$ and $(-2, -5)$ are on the graph.
Matches graph (b).

**57.** $y = -3x + 2$
No symmetry

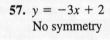

**59.** $y = x^2 - 3x$
No symmetry

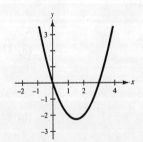

**61.** $y = x^3 + 2$
No symmetry

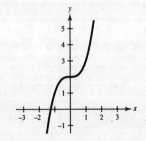

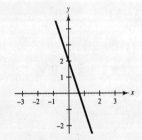

**63.** $y = \sqrt{x - 3}$

No symmetry

Domain: $x \geq 3$

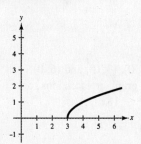

**65.** $y = |x - 2|$

No symmetry

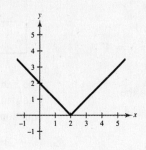

**67.** $x = y^2 - 1$

$x$-axis symmetry

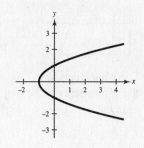

**69.** $y = \frac{5}{2}x + 5$

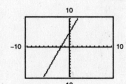  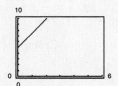

The standard setting gives a more complete graph.

**71.** $y = 4x^2 - 25$

Range/Window

| | |
|---|---|
| Xmin = -5 |
| Xmax = 5 |
| Xscl = 1 |
| Ymin = -30 |
| Ymax = 30 |
| Yscl = 10 |

**73.** $y = |x| + |x + 10|$

Range/Window

| | |
|---|---|
| Xmin = -10 |
| Xmax = 20 |
| Xscl = 5 |
| Ymin = -5 |
| Ymax = 30 |
| Yscl = 5 |

**75.** $y = 12 - 4x$

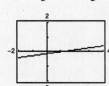

$0 = 12 - 4x$

$4x = 12$

$x = 3$

The $x$-intercept is $(3, 0)$.

The $x$-intercept of the graph is a solution to the equation $0 = 12 - 4x$.

**77.** $y = \dfrac{x + 2}{3} - \dfrac{x - 1}{5} - 1$

$0 = \dfrac{x + 2}{3} - \dfrac{x - 1}{5} - 1$

$0 = 5(x + 2) - 3(x - 1) - 15$

$0 = 5x + 10 - 3x + 3 - 15$

$0 = 2x - 2$

$x = 1$

The $x$-intercept is $(1, 0)$.

The $x$-intercept of the graph is a solution to the equation $0 = \dfrac{x + 2}{3} - \dfrac{x - 1}{5} - 1$.

**79.** $y = 1 - (x - 2)^2$

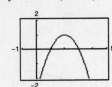

$$0 = 1 - (x - 2)^2$$
$$(x - 2)^2 = 1$$
$$x - 2 = \pm 1$$
$$x = 2 \pm 1$$
$$x = 3 \text{ or } x = 1$$

The $x$-intercepts are $(1, 0)$ and $(3, 0)$.
The $x$-intercepts of the graph are solutions to the equation $0 = 1 - (x - 2)^2$.

**81.** $y = x^3 - 9x^2 + 18x$

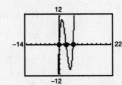

$$0 = x^3 - 9x^2 + 18x$$
$$0 = x(x - 3)(x - 6)$$
$$x = 0, x = 3, x = 6$$

The $x$-intercepts are $(0, 0)$, $(3, 0)$, and $(6, 0)$.
The $x$-intercepts of the graph are solutions to the equation $0 = x^3 - 9x^2 + 18x$.

**83.** $x^2 + y^2 = 3^2$
$x^2 + y^2 = 9$

**85.** $(x - 2)^2 + [y - (-1)]^2 = 4^2$
$(x - 2)^2 + (y + 1)^2 = 16$

**87.** $r = \sqrt{(0 - (-1))^2 + (0 - 2)^2} = \sqrt{1 + 4} = \sqrt{5}$
$[x - (-1)]^2 + (y - 2)^2 = (\sqrt{5})^2$
$(x + 1)^2 + (y - 2)^2 = 5$

**89.** $r = \frac{1}{2}\sqrt{(6 - 0)^2 + (8 - 0)^2} = 5$

$$\text{Center} = \left(\frac{0 + 6}{2}, \frac{0 + 8}{2}\right) = (3, 4)$$

$$(x - 3)^2 + (y - 4)^2 = 25$$

**91.** $y = 225{,}000 - 20{,}000t, \ 0 \le t \le 8$

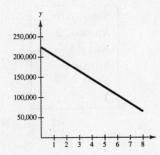

**95.** (a)

| Year | 1920 | 1930 | 1940 | 1950 | 1960 | 1970 | 1980 | 1990 |
|---|---|---|---|---|---|---|---|---|
| Life Expectancy | 54.1 | 59.7 | 62.9 | 68.2 | 69.7 | 70.8 | 73.7 | 75.4 |
| Model | 52.8 | 58.7 | 63.3 | 66.9 | 69.9 | 72.4 | 74.6 | 76.4 |

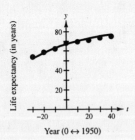

(b) When $t = 48$, $y \approx 77.7$ years.

(c) When $t = 50$, $y \approx 78.0$ years.

# Section P.4    Lines in the Plane and Slope

You should know the following important facts about lines.

- The graph of $y = mx + b$ is a straight line. It is called a linear equation.

- The slope of the line through $(x_1, y_1)$ and $(x_2, y_2)$ is

$$m = \frac{y_2 - y_1}{x_2 - x_1}.$$

- (a) If $m > 0$, the line rises from left to right.

  (b) If $m = 0$, the line is horizontal.

  (c) If $m < 0$, the line falls from left to right.

  (d) If $m$ is undefined, the line is vertical.

- Equations of Lines

  (a) Slope-Intercept: $y = mx + b$

  (b) Point-Slope: $y - y_1 = m(x - x_1)$

  (c) Two-Point: $y - y_1 = \dfrac{y_2 - y_1}{x_2 - x_1}(x - x_1)$

  (d) General: $Ax + By + C = 0$

  (e) Vertical: $x = a$

  (f) Horizontal: $y = b$

- Given two distinct nonvertical lines

$$L_1: y = m_1 x + b_1 \quad \text{and} \quad L_2: y = m_2 x + b_2$$

  (a) $L_1$ is parallel to $L_2$ if and only if $m_1 = m_2$ and $b_1 \neq b_2$.

  (b) $L_1$ is perpendicular to $L_2$ if and only if $m_1 = -1/m_2$.

## Solutions to Odd-Numbered Exercises

1. (a) $m = \frac{2}{3}$. Since the slope is positive, the line rises. Matches $L_2$.

   (b) $m$ is undefined. The line is vertical. Matches $L_3$.

   (c) $m = -2$. The line falls. Matches $L_1$.

5. Slope $= \dfrac{\text{rise}}{\text{run}} = \dfrac{8}{5}$

7. Slope $= \dfrac{\text{rise}}{\text{run}} = \dfrac{0}{1} = 0$

3.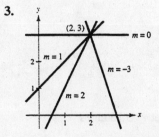

9. Slope $= \dfrac{\text{rise}}{\text{run}} = \dfrac{-8}{2} = -4$

**11.**

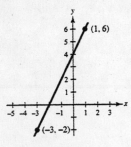

$$\text{slope} = \frac{6+2}{1+3} = 2$$

**13.**

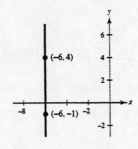

Slope is undefined.

**15.**

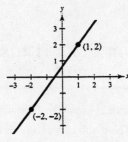

$$\text{slope} = \frac{2+2}{1+2} = \frac{4}{3}$$

**17.** Since $m = 0$, $y$ does not change. Three points are $(0, 1)$, $(3, 1)$, and $(-1, 1)$.

**19.** Since $m = 1$, $y$ increases by 1 for every one unit increase in $x$. Three points are $(6, -5)$, $(7, -4)$, and $(8, -3)$.

**21.** Since $m$ is undefined, $x$ does not change. Three points are $(-8, 0)$, $(-8, 2)$, and $(-8, 3)$.

**23.** Slope of $L_1$: $m = \dfrac{9+1}{5-0} = 2$

Slope of $L_2$: $m = \dfrac{1-3}{4-0} = -\dfrac{1}{2}$

$L_1$ and $L_2$ are perpendicular.

**25.** Slope of $L_1$: $m = \dfrac{0-6}{-6-3} = \dfrac{2}{3}$

Slope of $L_2$: $m = \dfrac{\frac{7}{3}+1}{5-0} = \dfrac{2}{3}$

$L_1$ and $L_2$ are parallel.

**27.** Yes, any pair of points on a line can be used to calculate the slope of the line. The rate of change remains the same on a line.

**29.** (a) $m = 135$. The sales are increasing 135 units per year.

(b) $m = 0$. There is no change in sales.

(c) $m = -40$. The sales are decreasing 40 units per year.

**31.** (a) The slope is negative and steep in 1989 to 1990.

(b) The slope is positive and steep in 1988 to 1989.

**33.** Slope $= \dfrac{\text{rise}}{\text{run}}$

$-\dfrac{12}{100} = -\dfrac{2000}{y}$

$-12y = -200,000$

$y = 16,666\frac{2}{3}$ feet $\approx 3.16$ miles

**35.** $5x - y + 3 = 0$

$\qquad y = 5x + 3$

Slope: $m = 5$

$y$-intercept: $(0, 3)$

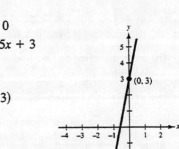

**37.** $5x - 2 = 0$

$$x = \frac{2}{5}$$

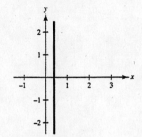

Slope:  undefined

No $y$-intercept

**39.** $7x + 6y - 30 = 0$

$$y = -\frac{7}{6}x + 5$$

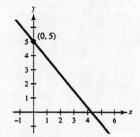

Slope:  $m = -\frac{7}{6}$

$y$-intercept:  $(0, 5)$

**41.** $y + 1 = \dfrac{5 + 1}{-5 - 5}(x - 5)$

$$y = -\frac{3}{5}(x - 5) - 1$$

$$y = -\frac{3}{5}x + 2 \implies 3x + 5y - 10 = 0$$

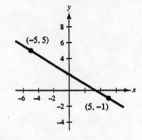

**43.** $y - \dfrac{1}{2} = \dfrac{\frac{5}{4} - \frac{1}{2}}{\frac{1}{2} - 2}(x - 2)$

$$y = -\frac{1}{2}(x - 2) + \frac{1}{2}$$

$$y = -\frac{1}{2}x + \frac{3}{2} \implies x + 2y - 3 = 0$$

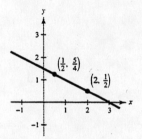

**45.** Since both points have $x = -8$, the slope is undefined.

$$x = -8 \implies x + 8 = 0$$

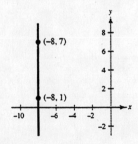

**47.** $y - 0.6 = \dfrac{-0.6 - 0.6}{-2 - 1}(x - 1)$

$$y = 0.4(x - 1) + 0.6$$

$$y = 0.4x + 0.2 \implies 2x - 5y + 1 = 0$$

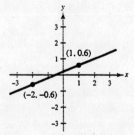

**49.** $y + 2 = 3(x - 0)$

$y = 3x - 2 \implies 3x - y - 2 = 0$

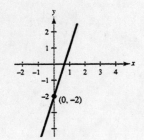

**51.** $y - 6 = -2(x + 3)$

$y = -2x \implies 2x + y = 0$

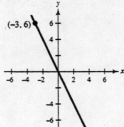

**53.** $y - 0 = -\frac{1}{3}(x - 4)$

$y = -\frac{1}{3}x + \frac{4}{3} \implies x + 3y - 4 = 0$

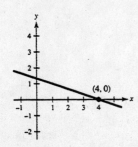

**55.** $x = 6$

$x - 6 = 0$

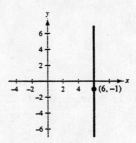

**57.** $y - \frac{5}{2} = \frac{4}{3}(x - 4)$

$y = \frac{4}{3}x - \frac{17}{6} \implies 8x - 6y - 17 = 0$

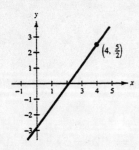

**59.** $\dfrac{x}{2} + \dfrac{y}{3} = 1$

$3x + 2y - 6 = 0$

**61.** $\dfrac{x}{-1/6} + \dfrac{y}{-2/3} = 1$

$6x + \dfrac{3}{2}y = -1$

$12x + 3y + 2 = 0$

**63.** $\dfrac{x}{a} + \dfrac{y}{a} = 1, \ a \neq 0$

$x + y = a$

$1 + 2 = a$

$3 = a$

$x + y = 3$

$x + y - 3 = 0$

**65.** $4x - 2y = 3$

$\qquad y = 2x - \frac{3}{2}$

slope: $m = 2$

(a) $y - 1 = 2(x - 2)$

$\qquad y = 2x - 3 \implies 2x - y - 3 = 0$

(b) $y - 1 = -\frac{1}{2}(x - 2)$

$\qquad y = -\frac{1}{2}x + 2 \implies x + 2y - 4 = 0$

**67.** $3x + 4y = 7$

$\qquad y = -\frac{3}{4}x + \frac{7}{4}$

slope: $m = -\frac{3}{4}$

(a) $y - 4 = -\frac{3}{4}(x + 6)$

$\qquad y = -\frac{3}{4}x - \frac{1}{2} \implies 3x + 4y + 2 = 0$

(b) $y - 4 = \frac{4}{3}(x + 6)$

$\qquad y = \frac{4}{3}x + 12 \implies 4x - 3y + 36 = 0$

**69.** $y = -3$

slope: $m = 0$

(a) $y = 0$

(b) $x = -1 \implies x + 1 = 0$

**71.** $L_1: y = \frac{1}{3}x - 2$

$L_2: y = \frac{1}{3}x + 3$

The lines are parallel.

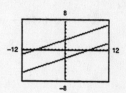

**73.** $L_1: y = \frac{1}{2}x - 3$

$L_2: y = -\frac{1}{2}x + 1$

Neither parallel nor perpendicular

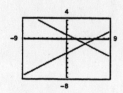

**75.** $L_1: y = \frac{2}{3}x - 3$

$L_2: y = -\frac{3}{2}x + 2$

The lines are perpendicular.

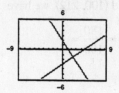

**77.** $y = 0.5x - 3$

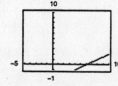 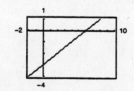

The second setting shows the $x$- and $y$-intercepts more clearly.

**79.** (a) $y = 2x$    (b) $y = -2x$    (c) $y = \frac{1}{2}x$

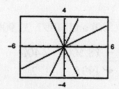

(b) and (c) are perpendicular.

**81.** (a) $y = -\frac{1}{2}x$    (b) $y = -\frac{1}{2}x + 3$    (c) $y = 2x - 4$

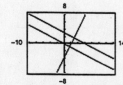

(a) and (b) are parallel.

(c) is perpendicular to (a) and (b).

**83.** $(6, 2540), m = 125$

$V - 2540 = 125(t - 6)$

$V - 2540 = 125t - 750$

$\qquad\quad V = 125t + 1790$

**85.** The slope is $m = -20$. This represents the decrease in the amount of the loan each week. Matches graph (b).

**87.** The slope is $m = 0.32$. This represents the increase in travel cost for each mile driven. Matches graph (a).

**89.** Set the distance between $(4, -1)$ and $(x, y)$ equal to the distance between $(-2, 3)$ and $(x, y)$.

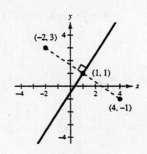

$$\sqrt{(x - 4)^2 + [y - (-1)]^2} = \sqrt{[x - (-2)]^2 + (y - 3)^2}$$

$$(x - 4)^2 + (y + 1)^2 = (x + 2)^2 + (y - 3)^2$$

$$x^2 - 8x + 16 + y^2 + 2y + 1 = x^2 + 4x + 4 + y^2 - 6y + 9$$

$$-8x + 2y + 17 = 4x - 6y + 13$$

$$0 = 12x - 8y - 4$$

$$0 = 4(3x - 2y - 1)$$

$$0 = 3x - 2y - 1$$

This line is the perpendicular bisector of the line segment connecting $(4, -1)$ and $(-2, 3)$.

**91.** Using the points $(0, 32)$ and $(100, 212)$, we have

$$m = \frac{212 - 32}{100 - 0} = \frac{180}{100} = \frac{9}{5}$$

$$F - 32 = \frac{9}{5}(C - 0)$$

$$F = \frac{9}{5}C + 32.$$

**93.** Using the points $(1995, 28{,}500)$ and $(1997, 32{,}900)$, we have

$$m = \frac{32{,}900 - 28{,}500}{1997 - 1995} = \frac{4400}{2} = 2200$$

$$S - 28{,}500 = 2200(t - 1995)$$

$$S = 2200t - 4{,}360{,}500$$

When $t = 2000$ we have $S = 2200(2000) - 4{,}360{,}500$ or $\$39{,}500$.

**95.** Using the points $(0, 875)$ and $(5, 0)$, where the first coordinate represents the year $t$ and the second coordinate represents the value $V$, we have

$$m = \frac{0 - 875}{5 - 0} = -175$$

$$V = -175t + 875, \ 0 \le t \le 5.$$

**97.** Sale price = List price − 15% of the list price

$$S = L - 0.15L$$

$$S = 0.85L$$

**99.** (a) $C = 36,500 + 5.25t + 11.50t$

$\qquad = 16.75t + 36,500$

(b) $R = 27t$

(c) $P = R - C$

$\qquad = 27t - (16.75t + 36,500)$

$\qquad = 10.25t - 36,500$

(d) $\quad 0 = 10.25t - 36,500$

$\qquad 36,500 = 10.25t$

$\qquad\qquad t \approx 3561$ hours

**101.** (a)

(b) $y = 2(15 + 2x) + 2(10 + 2x)$

$\qquad = 8x + 50$

(c)

(d) Since $m = 8$, each 1 meter increase in $x$ will increase $y$ by 8 meters.

**103.** $C = 120 + 0.26x$

**105.** Two approximate points on this line are $(6, 710)$ and $(10, 1075)$.

$$m = \frac{1075 - 710}{10 - 6} \approx 91$$

$$y - 710 = 91(x - 6)$$

$$y = 91x + 164$$

This answer may vary depending on the points used.

**107.** $y = 8 - 3x$ is a linear equation with slope $m = -3$. Matches graph (d).

**109.** $y = \frac{1}{2}x^2 + 2x + 1$ is a quadratic equation. Its graph is a parabola. Matches graph (a).

# Section P.5    Functions

- Given a set or an equation, you should be able to determine if it represents a function.
- Given a function, you should be able to do the following.
  - (a) Find the domain.
  - (b) Evaluate it at specific values.

**Solutions to Odd-Numbered Exercises**

1. Yes, it does represent a function. Each domain value is matched with only one range value.

3. No, it does not represent a function. The domain values are each matched with three range values.

5. Yes, it does represent a function. Each input value is matched with only one output value.

7. No, it does not represent a function. The input values of 10 and 7 are each matched with two output values.

9. (a) Each element of $A$ is matched with exactly one element of $B$, so it does represent a function.

   (b) The element 1 in $A$ is matched with two elements, $-2$ and 1 of $B$, so it does not represent a function.

   (c) Each element of $A$ is matched with exactly one element of $B$, so it does represent a function.

   (d) The element 2 in $A$ is not matched with an element of $B$, so it does not represent a function.

11. Each is a function. For each year there corresponds one and only one circulation.

13. $x^2 + y^2 = 4 \implies y = \pm\sqrt{4 - x^2}$
    No, $y$ *is not* a function of $x$.

15. $x^2 + y = 4 \implies y = 4 - x^2$
    Yes, $y$ *is* a function of $x$.

17. $2x + 3y = 4 \implies y = \frac{1}{3}(4 - 2x)$
    Yes, $y$ *is* a function of $x$.

19. $y^2 = x^2 - 1 \implies y = \pm\sqrt{x^2 - 1}$
    No, $y$ *is not* a function of $x$.

21. $y = |4 - x|$
    Yes, $y$ is a function of $x$.

23. $f(s) = \dfrac{1}{s + 1}$

   (a) $f(4) = \dfrac{1}{(4) + 1} = \dfrac{1}{5}$

   (b) $f(0) = \dfrac{1}{(0) + 1} = 1$

   (c) $f(4x) = \dfrac{1}{(4x) + 1} = \dfrac{1}{4x + 1}$

   (d) $f(x + c) = \dfrac{1}{(x + c) + 1} = \dfrac{1}{x + c + 1}$

25. $f(x) = 2x - 3$

   (a) $f(1) = 2(1) - 3 = -1$

   (b) $f(-3) = 2(-3) - 3 = -9$

   (c) $f(x - 1) = 2(x - 1) - 3 = 2x - 5$

27. $h(t) = t^2 - 2t$

   (a) $h(2) = 2^2 - 2(2) = 0$

   (b) $h(1.5) = (1.5)^2 - 2(1.5) = -0.75$

   (c) $h(x + 2) = (x + 2)^2 - 2(x + 2) = x^2 + 2x$

29. $f(y) = 3 - \sqrt{y}$

   (a) $f(4) = 3 - \sqrt{4} = 1$

   (b) $f(0.25) = 3 - \sqrt{0.25} = 2.5$

   (c) $f(4x^2) = 3 - \sqrt{4x^2} = 3 - 2|x|$

**31.** $q(x) = \dfrac{1}{x^2 - 9}$

(a) $q(0) = \dfrac{1}{0^2 - 9} = -\dfrac{1}{9}$

(b) $q(3) = \dfrac{1}{3^2 - 9}$ is undefined.

(c) $q(y + 3) = \dfrac{1}{(y + 3)^2 - 9} = \dfrac{1}{y^2 + 6y}$

**33.** $f(x) = \dfrac{|x|}{x}$

(a) $f(2) = \dfrac{|2|}{2} = 1$

(b) $f(-2) = \dfrac{|-2|}{-2} = -1$

(c) $f(x - 1) = \dfrac{|x - 1|}{x - 1}$

**35.** $f(x) = \begin{cases} 2x + 1, & x < 0 \\ 2x + 2, & x \geq 0 \end{cases}$

(a) $f(-1) = 2(-1) + 1 = -1$

(b) $f(0) = 2(0) + 2 = 2$

(c) $f(2) = 2(2) + 2 = 6$

**37.** $f(x) = x^2 - 3$

| $x$ | $-2$ | $-1$ | $0$ | $1$ | $2$ |
|-----|------|------|-----|-----|-----|
| $f(x)$ | $1$ | $-2$ | $-3$ | $-2$ | $1$ |

**39.** $h(t) = \frac{1}{2}|t + 3|$

| $t$ | $-5$ | $-4$ | $-3$ | $-2$ | $-1$ |
|-----|------|------|------|------|------|
| $h(t)$ | $1$ | $\frac{1}{2}$ | $0$ | $\frac{1}{2}$ | $1$ |

**41.** $f(x) = \begin{cases} -\frac{1}{2}x + 4, & x \leq 0 \\ (x - 2)^2, & x > 0 \end{cases}$

| $x$ | $-2$ | $-1$ | $0$ | $1$ | $2$ |
|-----|------|------|-----|-----|-----|
| $f(x)$ | $5$ | $\frac{9}{2}$ | $4$ | $1$ | $0$ |

**43.** $15 - 3x = 0$

$\quad 3x = 15$

$\quad x = 5$

**45.** $x^2 - 9 = 0$

$\quad x^2 = 9$

$\quad x = \pm 3$

**47.** $\qquad f(x) = g(x)$

$\qquad\quad x^2 = x + 2$

$\quad x^2 - x - 2 = 0$

$(x + 1)(x - 2) = 0$

$x = -1 \ \text{ or } \ x = 2$

**49.** $\qquad f(x) = g(x)$

$\sqrt{3x} + 1 = x + 1$

$\qquad \sqrt{3x} = x$

$\qquad\quad 3x = x^2$

$\qquad\quad\ 0 = x^2 - 3x$

$\qquad\quad\ 0 = x(x - 3)$

$x = 0 \ \text{ or } \ x = 3$

**51.** $f(x) = 5x^2 + 2x - 1$

Since $f(x)$ is a polynomial, the domain is all real numbers $x$.

**53.** $h(t) = \dfrac{4}{t}$

Domain: All real numbers except $t = 0$

**55.** $g(y) = \sqrt{y - 10}$

Domain: $y - 10 \geq 0$

$y \geq 10$

**57.** $f(x) = \sqrt[4]{1 - x^2}$

Domain: $1 - x^2 \geq 0$

$-x^2 \geq -1$

$x^2 \leq 1$

$x^2 - 1 \leq 0$

$-1 \leq x \leq 1$  (See Section 1.8.)

**59.** $g(x) = \dfrac{1}{x} - \dfrac{1}{x + 2}$

Domain:  All real numbers except
$x = 0, \ x = -2$

**61.** $f(x) = x^2$

$\{(-2, 4), (-1, 1), (0, 0), (1, 1), (2, 4)\}$

**63.** $f(x) = \sqrt{x + 2}$

$\{(-2, 0), (-1, 1), (0, \sqrt{2}), (1, \sqrt{3}), (2, 2)\}$

**65.** The domain is the set of inputs of the function and the range is the set of corresponding outputs.

**67.** By plotting the points, we have a parabola, so $g(x) = cx^2$. Since $(-4, -32)$ is on the graph, we have $-32 = c(-4)^2 \implies c = -2$. Thus, $g(x) = -2x^2$.

**69.** Since the function is undefined at 0, we have $r(x) = c/x$. Since $(-8, -4)$ is on the graph, we have $-4 = c/-8 \implies c = 32$. Thus, $r(x) = 32/x$.

**71.**

$$f(x) = x^2 - x + 1$$

$$f(2 + h) = (2 + h)^2 - (2 + h) + 1$$

$$= 4 + 4h + h^2 - 2 - h + 1$$

$$= h^2 + 3h + 3$$

$$f(2) = (2)^2 - 2 + 1 = 3$$

$$f(2 + h) - f(2) = h^2 + 3h$$

$$\cdot \frac{f(2 + h) - f(2)}{h} = h + 3, \ h \neq 0$$

**73.** $f(x) = x^3$

$$f(x + c) = (x + c)^3 = x^3 + 3x^2c + 3xc^2 + c^3$$

$$\frac{f(x + c) - f(x)}{c} = \frac{(x^3 + 3x^2c + 3xc^2 + c^3) - x^3}{c}$$

$$= \frac{c(3x^2 + 3xc + c^2)}{c}$$

$$= 3x^2 + 3xc + c^2, \ c \neq 0$$

**75.** $g(x) = 3x - 1$

$$\frac{g(x) - g(3)}{x - 3} = \frac{(3x - 1) - 8}{x - 3} = \frac{3x - 9}{x - 3} = \frac{3(x - 3)}{x - 3} = 3, \ x \neq 3$$

**77.** $A = \pi r^2, \ C = 2\pi r$

$$r = \frac{C}{2\pi}$$

$$A = \pi \left(\frac{C}{2\pi}\right)^2 = \frac{C^2}{4\pi}$$

**79.** (a)

| Height, $x$ | Width | Volume, $V$ |
|:-:|:-:|:-:|
| 1 | $24 - 2(1)$ | $1[24 - 2(1)]^2 = 484$ |
| 2 | $24 - 2(2)$ | $2[24 - 2(2)]^2 = 800$ |
| 3 | $24 - 2(3)$ | $3[24 - 2(3)]^2 = 972$ |
| 4 | $24 - 2(4)$ | $4[24 - 2(4)]^2 = 1024$ |
| 5 | $24 - 2(5)$ | $5[24 - 2(5)]^2 = 980$ |
| 6 | $24 - 2(6)$ | $6[24 - 2(6)]^2 = 864$ |

The volume is maximum when $x = 4$.

(b)

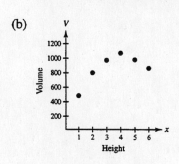

(c) $V = x(24 - 2x)^2$

Domain: $0 < x < 12$

$V$ is a function of $x$.

**81.** $A = \dfrac{1}{2}bh = \dfrac{1}{2}xy$

Since $(0, y)$, $(2, 1)$, and $(x, 0)$ all lie on the same line, the slopes between any pair are equal.

$$\frac{1 - y}{2 - 0} = \frac{0 - 1}{x - 2}$$

$$\frac{1 - y}{2} = \frac{-1}{x - 2}$$

$$y = \frac{2}{x - 2} + 1$$

$$y = \frac{x}{x - 2}$$

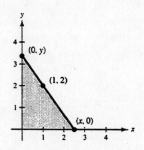

Therefore,

$$A = \frac{1}{2}x\left(\frac{x}{x - 2}\right) = \frac{x^2}{2(x - 2)}.$$

The domain of $A$ includes $x$-values such that $x^2/[2(x - 2)] > 0$. Using methods of Section 1.8 we find that the domain is $x > 2$.

**83.** $V = l \cdot w \cdot h = x \cdot y \cdot x = x^2y$ where $4x + y = 108$.

Thus, $y = 108 - 4x$ and $V = x^2(108 - 4x) = 108x^2 - 4x^3$ where $0 < x < 27$.

**85.** (a) Cost = variable costs + fixed costs

$$C = 12.30x + 98,000$$

(b) Revenue = price per unit × number of units

$$R = 17.98x$$

(c) Profit = Revenue − Cost

$$P = 17.98x - (12.30x + 98,000)$$

$$P = 5.68x - 98,000$$

**87.** (a) $R = n(\text{rate}) = n[8.00 - 0.05(n - 80)]$, $n \ge 80$

$$R = 12.00n - 0.05n^2 = 12n - \frac{n^2}{20} = \frac{240n - n^2}{20}, \ n \ge 80$$

(b)

| $n$ | 90 | 100 | 110 | 120 | 130 | 140 | 150 |
|---|---|---|---|---|---|---|---|
| $R(n)$ | \$675 | \$700 | \$715 | \$720 | \$715 | \$700 | \$675 |

The revenue is maximum when 120 people take the trip.

**89.** (a)

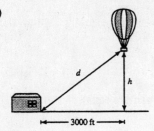

(b) $(3000)^2 + h^2 = d^2$

$$h = \sqrt{d^2 - (3000)^2}$$

Domain: $[3000, \infty)$

(since both $d \ge 0$ and $d^2 - (3000)^2 \ge 0$)

**91.** $\dfrac{t}{3} + \dfrac{t}{5} = 1$

$$15\left(\frac{t}{3} + \frac{t}{5}\right) = 15(1)$$

$$5t + 3t = 15$$

$$8t = 15$$

$$t = \frac{15}{8}$$

**93.** $\dfrac{3}{x(x + 1)} - \dfrac{4}{x} = \dfrac{1}{x + 1}$

$$x(x + 1)\left[\frac{3}{x(x + 1)} - \frac{4}{x}\right] = x(x + 1)\left(\frac{1}{x + 1}\right)$$

$$3 - 4(x + 1) = x$$

$$3 - 4x - 4 = x$$

$$-1 = 5x$$

$$-\frac{1}{5} = x$$

# Section P.6    Analyzing Graphs of Functions

- You should be able to determine the domain and range of a function from its graph.
- You should be able to use the vertical line test for functions.
- You should be able to determine when a function is constant, increasing, or decreasing.
- You should know that $f$ is
  - (a) odd if $f(-x) = -f(x)$.
  - (b) even if $f(-x) = f(x)$.

**Solutions to Odd-Numbered Exercises**

**1.** $f(x) = 1 - x^2$

Domain: All real numbers

Range: $(-\infty, 1]$

**3.** $f(x) = \sqrt{x^2 - 1}$

Domain: $(-\infty, -1] \cup [1, \infty)$

Range: $[0, \infty)$

**5.** $h(x) = \sqrt{16 - x^2}$

Domain: $[-4, 4]$

Range: $[0, 4]$

**7.** $y = \frac{1}{2}x^2$

A vertical line intersects the graph just once, so $y$ is a function of $x$.

**9.** $x - y^2 = 1 \implies y = \pm\sqrt{x - 1}$

$y$ is not a function of $x$.

**11.** $x^2 = 2xy - 1$

A vertical line intersects the graph just once, so $y$ is a function of $x$.

**13.** Yes, the graph in Exercise 9 does represent $x$ as a function of $y$. For each $y$-value there corresponds one and only one $x$-value.

**15.** $f(x) = -0.2x^2 + 3x + 32$

The second setting shows the most complete graph.

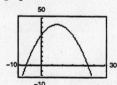

**17.** $f(x) = 4x^3 - x^4$

The first setting shows the most complete graph.

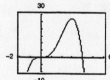

**19.** $f(x) = \frac{3}{2}x$

(a) $f$ is increasing on $(-\infty, \infty)$.

(b) Since $f(-x) = -f(x)$, $f$ is odd.

**21.** $f(x) = x^3 - 3x^2 + 2$

(a) $f$ is increasing on $(-\infty, 0)$ and $(2, \infty)$.

   $f$ is decreasing on $(0, 2)$.

(b) $f(-x) \neq -f(x)$

   $f(-x) \neq f(x)$

   $f$ is neither odd nor even.

**23.** $f(x) = 3x^4 - 6x^2$

(a)

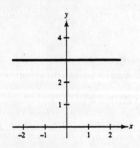

(b) Increasing on $(-1, 0)$ and $(1, \infty)$

Decreasing on $(-\infty, -1)$ and $(0, 1)$

(c) Since $f(-x) = f(x), f$ is even.

**25.** $f(x) = x\sqrt{x + 3}$

(a)

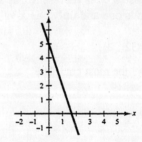

(b) Increasing on $(-2, \infty)$

Decreasing on $(-3, -2)$

(c) $f(-x) \neq -f(x)$

$f(-x) \neq f(x)$

$f$ is neither odd nor even.

**27.** $f(-x) = (-x)^6 - 2(-x)^2 + 3$

$\quad = x^6 - 2x^2 + 3$

$\quad = f(x)$

$f$ is even.

**29.** $g(-x) = (-x)^3 - 5(-x)$

$\quad = -x^3 + 5x$

$\quad = -g(x)$

$g$ is odd.

**31.** $f(-t) = (-t)^2 + 2(-t) - 3$

$\quad = t^2 - 2t - 3$

$\quad \neq f(t) \neq -f(t)$

$f$ is neither even nor odd.

**33.** $\left(-\frac{3}{2}, 4\right)$

(a) If $f$ is even, another point is $\left(\frac{3}{2}, 4\right)$.

(b) If $f$ is odd, another point is $\left(\frac{3}{2}, -4\right)$.

**35.** $f(x) = 3$, even

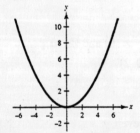

**37.** $f(x) = 5 - 3x$, neither even nor odd

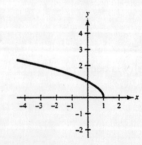

**39.** $g(s) = \dfrac{s^2}{4}$, even

**41.** $f(x) = \sqrt{1 - x}$, neither even nor odd

**43.** $g(t) = \sqrt[3]{t - 1}$, neither even nor odd

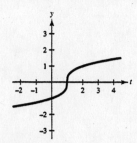

**45.** $f(x) = \begin{cases} x + 3, & x \le 0 \\ 3, & 0 < x \le 2 \\ 2x - 1, & x > 2 \end{cases}$

Neither even nor odd

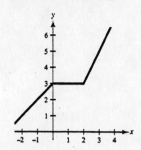

**47.** $f(x) = 4 - x$

$f(x) \ge 0$ on $(-\infty, 4]$.

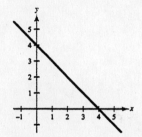

**49.** $f(x) = x^2 - 9$

$f(x) \ge 0$ on $(-\infty, -3]$ and $[3, \infty)$.

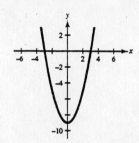

**51.** $f(x) = 1 - x^4$

$f(x) \ge 0$ on $[-1, 1]$.

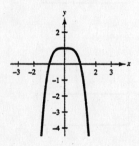

**53.** $f(x) = x^2 + 1$

$f(x) \ge 0$ on $(-\infty, \infty)$.

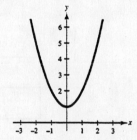

**55.** $f(x) = -5$, $f(x) < 0$ for all $x$.

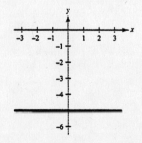

**57.** $f(x) = \begin{cases} 2x + 3, & x < 0 \\ 3 - x, & x \geq 0 \end{cases}$

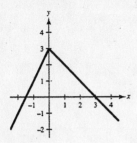

**59.** $f(x) = \begin{cases} x^2 + 5, & x \leq 1 \\ -x^2 + 4x + 3, & x > 1 \end{cases}$

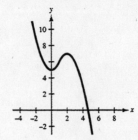

**61.** $f(x) = |x + 3|$

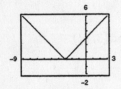

Domain:  All real numbers or $(-\infty, \infty)$

Range:  $[0, \infty)$

**63.** $s(x) = 2\left(\frac{1}{4}x - \left[\!\left[\frac{1}{4}x\right]\!\right]\right)$

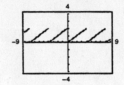

Domain: $(-\infty, \infty)$

Range: $[0, 2)$

Sawtooth pattern

**65.** (a) $y = x$

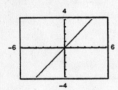

(b) $y = x^2$

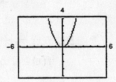

(c) $y = x^3$

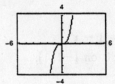

(d) $y = x^4$

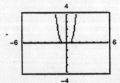

(e) $y = x^5$

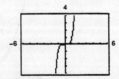

(f) $y = x^6$

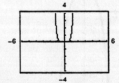

All the graphs pass through the origin. The graphs of the odd powers of $x$ are symmetric to the origin and the graphs of the even powers are symmetric to the $y$-axis. As the powers increase, the graphs become narrower in the interval $-1 < x < 1$.

**67.** (a) $C_2(t) = 0.65 - 0.4\left[\!\left[-(t - 1)\right]\!\right]$ is the appropriate model since the cost does not increase until after the next minute of conversation has started.

(b)

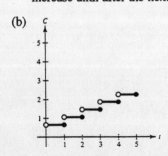

$C = 0.65 + 0.40(18) = \$7.85$

**69.**  $P = R - C = xp - C = x(100 - 0.0001x) - (350,000 + 30x)$

$= -0.0001x^2 + 70x - 350,000, \ 0 \le x$

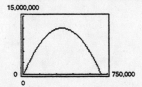

This function is maximized when $x = 350,000$ units.

**71.**  $h = $ top $-$ bottom

$= (-x^2 + 4x - 1) - 2$

$= -x^2 + 4x - 3$

**73.**  $h = $ top $-$ bottom

$= (4x - x^2) - 2x$

$= 2x - x^2$

**75.**  $L = $ right $-$ left

$= \frac{1}{2}y^2 - 0$

$= \frac{1}{2}y^2$

**77.**  $L = $ right $-$ left

$= 4 - y^2$

**79.**  $y = -87.49 + 16.28t - 4.82t^2 - 1.20t^3$

(a) Domain: $-4 \le t \le 3$

(b)

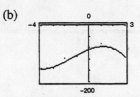

(c) Most accurate in 1986

Least accurate in 1990

(d) The balance would continue to decrease.

**81.**  (a) For average salaries of college professors, a scale of $10,000 would be appropriate.

(b) For the population of the United States, use a scale of 50,000,000.

(c) For the percent of the civilian workforce that is unemployed, use a scale of 1%.

**83.**  $f(x) = a_{2n+1}x^{2n+1} + a_{2n-1}x^{2n-1} + \cdots + a_3x^3 + a_1x$

$f(-x) = a_{2n+1}(-x)^{2n+1} + a_{2n-1}(-x)^{2n-1} + \cdots + a_3(-x)^3 + a_1(-x)$

$= -a_{2n+1}x^{2n+1} - a_{2n-1}x^{2n-1} - \cdots - a_3x^3 - a_1x = -f(x)$

Therefore, $f(x)$ is odd.

**85.**  $x^2 - 10x = 0$

$x(x - 10) = 0$

$x = 0 \quad \text{or} \quad x = 10$

**87.**  $x^3 + x = 0$

$x(x^2 + 1) = 0$

$x = 0 \quad \text{or} \quad x^2 + 1 = 0$

$x^2 = -1$

$x = \pm\sqrt{-1} = \pm i$

# Section P.7   Translations and Combinations

- You should know the basic types of transformations.

  Let $y = f(x)$ and let $c$ be a positive real number.

  | | |
  |---|---|
  | 1. $h(x) = f(x) + c$ | Vertical shift $c$ units upward |
  | 2. $h(x) = f(x) - c$ | Vertical shift $c$ units downward |
  | 3. $h(x) = f(x - c)$ | Horizontal shift $c$ units to the right |
  | 4. $h(x) = f(x + c)$ | Horizontal shift $c$ units to the left |
  | 5. $h(x) = -f(x)$ | Reflection in the $x$-axis |
  | 6. $h(x) = f(-x)$ | Reflection in the $y$-axis |
  | 7. $h(x) = cf(x), c > 1$ | Vertical stretch |
  | 8. $h(x) = cf(x), 0 < c < 1$ | Vertical shrink |

- Given two functions, $f$ and $g$, you should be able to form the following functions (if defined):

  1. Sum: $(f + g)(x) = f(x) + g(x)$
  2. Difference: $(f - g)(x) = f(x) - g(x)$
  3. Product: $(fg)(x) = f(x)g(x)$
  4. Quotient: $(f/g)(x) = f(x)/g(x), g(x) \neq 0$
  5. Composition of $f$ with $g$: $(f \circ g)(x) = f(g(x))$
  6. Composition of $g$ with $f$: $(g \circ f)(x) = g(f(x))$

**Solutions to Odd-Numbered Exercises**

**1.** (a) $f(x) = x^3 + c$   (b) $f(x) = (x - c)^3$

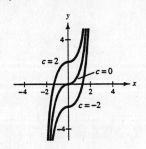
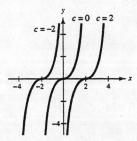

**3.** (a) $f(x) = |x + c|$   (b) $f(x) = |x - c|$   (c) $f(x) = |x + 4| + c$

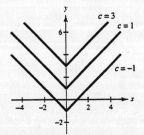
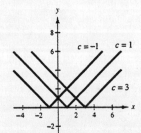
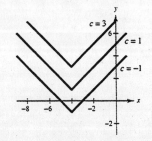

**5.** (a) $y = f(x) + 2$        (b) $y = -f(x)$        (c) $y = f(x - 2)$

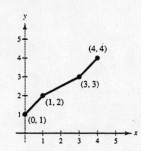        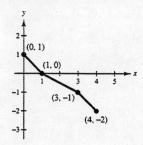        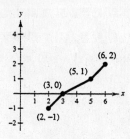

(d) $y = f(x + 3)$        (e) $y = f(2x)$        (f) $y = f(-x)$

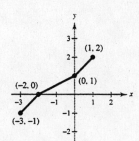        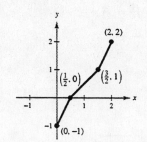        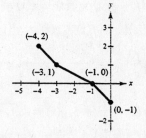

**7.** (a) Vertical shift one unit downward

$$y = x^2 - 1$$

(b) Vertical shift one unit upward, horizontal shift one unit to the left, and a reflection in the *x*-axis

$$y = 1 - (x + 1)^2$$

**9.** Horizontal shift two units to the right of $y = x^3$

$$y = (x - 2)^3$$

**11.** Reflection in the *x*-axis of $y = x^2$

$$y = -x^2$$

**13.** Reflection in the *x*-axis and a vertical shift one unit upward of $y = \sqrt{x}$

$$y = 1 - \sqrt{x}$$

**15.**

| $x$ | 0 | 1 | 2 | 3 |
|-----|---|---|---|---|
| $f(x)$ | 2 | 3 | 1 | 2 |
| $g(x)$ | $-1$ | 0 | $\frac{1}{2}$ | 0 |
| $h(x) = f(x) + g(x)$ | 1 | 3 | $\frac{3}{2}$ | 2 |

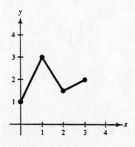

**17.**

| $x$ | $-2$ | $-1$ | 0 | 1 | 2 | 3 |
|-----|------|------|---|---|---|---|
| $f(x)$ | 2 | 1 | 0 | 1 | 2 | 3 |
| $g(x)$ | 4 | 3 | 2 | 1 | 0 | 1 |
| $h(x) = f(x) + g(x)$ | 6 | 4 | 2 | 2 | 2 | 4 |

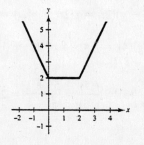

**19.** $f(x) = x + 1, g(x) = x - 1$

$(f + g)(x) = f(x) + g(x) = (x + 1) + (x - 1) = 2x$

$(f - g)(x) = f(x) - g(x) = (x + 1) - (x - 1) = 2$

$(fg)(x) = f(x) \cdot g(x) = (x + 1)(x - 1) = x^2 - 1$

$\left(\dfrac{f}{g}\right)(x) = \dfrac{f(x)}{g(x)} = \dfrac{x + 1}{x - 1}, \ x \neq 1$

**21.** $f(x) = x^2, g(x) = 1 - x$

$(f + g)(x) = f(x) + g(x) = x^2 + (1 - x) = x^2 - x + 1$

$(f - g)(x) = f(x) - g(x) = x^2 - (1 - x) = x^2 + x - 1$

$(fg)(x) = f(x) \cdot g(x) = x^2(1 - x) = x^2 - x^3$

$\left(\dfrac{f}{g}\right)(x) = \dfrac{f(x)}{g(x)} = \dfrac{x^2}{1 - x}, x \neq 1$

**23.** $f(x) = x^2 + 5, g(x) = \sqrt{1 - x}$

$(f + g)(x) = f(x) + g(x) = (x^2 + 5) + \sqrt{1 - x}$

$(f - g)(x) = f(x) - g(x) = (x^2 + 5) - \sqrt{1 - x}$

$(fg)(x) = f(x) \cdot g(x) = (x^2 + 5)\sqrt{1 - x}$

$\left(\dfrac{f}{g}\right)(x) = \dfrac{f(x)}{g(x)} = \dfrac{x^2 + 5}{\sqrt{1 - x}}, \ x < 1$

**25.** $f(x) = \dfrac{1}{x}, g(x) = \dfrac{1}{x^2}$

$(f + g)(x) = f(x) + g(x) = \dfrac{1}{x} + \dfrac{1}{x^2} = \dfrac{x + 1}{x^2}$

$(f - g)(x) = f(x) - g(x) = \dfrac{1}{x} - \dfrac{1}{x^2} = \dfrac{x - 1}{x^2}$

$(fg)(x) = f(x) \cdot g(x) = \dfrac{1}{x}\left(\dfrac{1}{x^2}\right) = \dfrac{1}{x^3}$

$\left(\dfrac{f}{g}\right)(x) = \dfrac{f(x)}{g(x)} = \dfrac{1/x}{1/x^2} = \dfrac{x^2}{x} = x, \ x \neq 0$

**27.** $(f + g)(3) = f(3) + g(3) = (3^2 + 1) + (3 - 4) = 9$

**29.** $(f - g)(0) = f(0) - g(0) = [0^2 + 1] - (0 - 4) = 5$

**31.** $(f - g)(2t) = f(2t) - g(2t) = [(2t)^2 + 1] - (2t - 4) = 4t^2 - 2t + 5$

**33.** $(fg)(4) = f(4)g(4) = (4^2 + 1)(4 - 4) = 0$

**35.** $\left(\dfrac{f}{g}\right)(5) = \dfrac{f(5)}{g(5)} = \dfrac{5^2 + 1}{5 - 4} = 26$

**37.** $\left(\dfrac{f}{g}\right)(-1) - g(3) = \dfrac{f(-1)}{g(-1)} - g(3)$

$$= \dfrac{(-1)^2 + 1}{-1 - 4} - (3 - 4)$$

$$= -\dfrac{2}{5} + 1 = \dfrac{3}{5}$$

**39.** $f(x) = \frac{1}{2}x, \; g(x) = x - 1, \; (f + g)(x) = \frac{3}{2}x - 1$

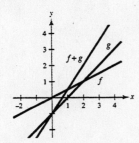

**41.** $f(x) = x^2, \; g(x) = -2x, \; (f + g)(x) = x^2 - 2x$

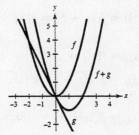

**43.** $f(x) = 3x, \; g(x) = -\dfrac{x^3}{10}, \; (f + g)(x) = 3x - \dfrac{x^3}{10}$

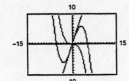

For $0 \le x \le 2$, $f(x)$ contributes most to the magnitude.
For $x > 6$, $g(x)$ contributes most to the magnitude.

**45.** $T(x) = R(x) + B(x) = \frac{3}{4}x + \frac{1}{15}x^2$

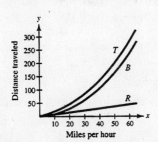

**47.**

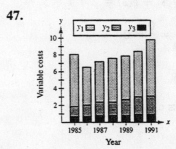

**49.** (a) $T$ is a function of $t$ since for each time $t$ there corresponds one and only one temperature $T$.

(b) $T(4) = 60°$
$T(15) = 72°$

(c) $H(t) = T(t - 1)$; All the temperature changes would be one hour later.

(d) $H(t) = T(t) - 1$; The temperature would be decreased by one degree.

**51.** $f(x) = x^2, g(x) = x - 1$

(a) $(f \circ g)(x) = f(g(x)) = f(x - 1) = (x - 1)^2$

(b) $(g \circ f)(x) = g(f(x)) = g(x^2) = x^2 - 1$

(c) $(f \circ f)(x) = f(f(x)) = f(x^2) = (x^2)^2 = x^4$

**53.** $f(x) = 3x + 5, g(x) = 5 - x$

(a) $(f \circ g)(x) = f(g(x)) = f(5 - x) = 3(5 - x) + 5 = 20 - 3x$

(b) $(g \circ f)(x) = g(f(x)) = g(3x + 5) = 5 - (3x + 5) = -3x$

(c) $(f \circ f)(x) = f(f(x)) = f(3x + 5) = 3(3x + 5) + 5 = 9x + 20$

**55.** (a) $(f \circ g)(x) = f(g(x)) = f(x^2) = \sqrt{x^2 + 4}$

(b) $(g \circ f)(x) = g(f(x)) = g\left(\sqrt{x + 4}\right) = \left(\sqrt{x + 4}\right)^2 = x + 4, \ x \geq 4$

**57.** (a) $(f \circ g)(x) = f(g(x)) = f(3x + 1) = \frac{1}{3}(3x + 1) - 3 = x - \frac{8}{3}$

(b) $(g \circ f)(x) = g(f(x)) = g\left(\frac{1}{3}x - 3\right) = 3\left(\frac{1}{3}x - 3\right) + 1 = x - 8$

**59.** (a) $(f \circ g)(x) = f(g(x)) = f\left(\sqrt{x}\right) = (x^{1/2})^{1/2} = x^{1/4} = \sqrt[4]{x}$

(b) Since $f(x) = g(x), (g \circ f)(x) = (f \circ g)(x) = \sqrt[4]{x}$.

**61.** (a) $(f \circ g)(x) = f(g(x)) = f(x + 6) = |x + 6|$

(b) $(g \circ f)(x) = g(f(x)) = g(|x|) = |x| + 6$

**63.** (a) $(f + g)(3) = f(3) + g(3) = 2 + 1 = 3$

(b) $\left(\frac{f}{g}\right)(2) = \frac{f(2)}{g(2)} = \frac{0}{2} = 0$

**65.** (a) $(f \circ g)(2) = f(g(2)) = f(2) = 0$

(b) $(g \circ f)(2) = g(f(2)) = g(0) = 4$

**67.** $g(x) = f(x) + 2$

Vertical shift 2 units upward

**69.** $g(x) = f(-x)$

Reflection in the $y$-axis

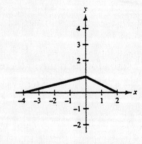

**71.** Let $f(x) = x^2$ and $g(x) = 2x + 1$, then $(f \circ g)(x) = h(x)$. This is not a unique solution.

For example, if $f(x) = (x + 1)^2$ and $g(x) = 2x$, then $(f \circ g)(x) = h(x)$ as well.

**73.** Let $f(x) = \sqrt[3]{x}$ and $g(x) = x^2 - 4$, then $(f \circ g)(x) = h(x)$.
This answer is not unique. Other possibilities may be:

$$f(x) = \sqrt[3]{x - 4} \text{ and } g(x) = x^2$$
$$\text{or } f(x) = \sqrt[3]{-x} \text{ and } g(x) = 4 - x^2$$
$$\text{or } f(x) = \sqrt[9]{x} \text{ and } g(x) = (4 - x^2)^3$$

**75.** Let $f(x) = 1/x$ and $g(x) = x + 2$, then $(f \circ g)(x) = h(x)$. Again, this is not a unique solution. Other possibilities may be:

$$f(x) = \frac{1}{x + 2} \text{ and } g(x) = x$$

$$\text{or } f(x) = \frac{1}{x + 1} \text{ and } g(x) = x + 1$$

$$\text{or } f(x) = \frac{1}{x^2 + 2} \text{ and } g(x) = \sqrt{x}$$

**77.** (a) The domain of $f(x) = \sqrt{x}$ is $x \geq 0$.

(b) The domain of $g(x) = x^2 + 1$ is all real numbers.

(c) $(f \circ g)(x) = f(g(x)) = f(x^2 + 1) = \sqrt{x^2 + 1}$

The domain of $f \circ g$ is all real numbers.

**79.** (a) The domain of $f(x) = \dfrac{3}{x^2 - 1}$ is all real numbers except $x = \pm 1$.

(b) The domain of $g(x) = x + 1$ is all real numbers.

(c) $(f \circ g)(x) = f(g(x)) = f(x + 1) = \dfrac{3}{(x + 1)^2 - 1} = \dfrac{3}{x^2 + 2x} = \dfrac{3}{x(x + 2)}$

This domain of $f \circ g$ is all real numbers except $x = 0$ and $x = -2$.

**81.** $f(x) = 3x - 4$

$$\frac{f(x + h) - f(x)}{h} = \frac{[3(x + h) - 4] - (3x - 4)}{h}$$

$$= \frac{3x + 3h - 4 - 3x + 4}{h}$$

$$= \frac{3h}{h}$$

$$= 3$$

**83.** $f(x) = \dfrac{4}{x}$

$$\frac{f(x+h) - f(x)}{h} = \frac{\dfrac{4}{x+h} - \dfrac{4}{x}}{h} = \frac{\dfrac{4x - 4(x+h)}{x(x+h)}}{\dfrac{h}{1}}$$

$$= \frac{4x - 4x - 4h}{x(x+h)} \cdot \frac{1}{h}$$

$$= \frac{-4h}{x(x+h)} \cdot \frac{1}{h}$$

$$= \frac{-4}{x(x+h)}$$

**85.** (a) $r(x) = \dfrac{x}{2}$

(b) $A(r) = \pi r^2$

(c) $(A \circ r)(x) = A(r(x)) = A\left(\dfrac{x}{2}\right) = \pi\left(\dfrac{x}{2}\right)^2$

$(A \circ r)(x)$ represents the area of the circular base of the tank on the square foundation with side length $y$.

**87.** $(C \circ x)(t) = C(x(t))$

$$= 60(50t) + 750$$

$$= 3000t + 750$$

$(C \circ x)(t)$ represents the cost after $t$ production hours.

**89.** (a) $R = p - 1200$

(b) $S = p - 0.08p = 0.92p$

(c) $(R \circ S)(p) = R(S(p)) = R(0.92p) = 0.92p - 1200$

$(S \circ R)(p) = S(R(p)) = S(p - 1200) = 0.92(p - 1200)$

$R \circ S$ represents taking a discount of 8% of the retail price and then receiving a $1200 rebate. $S \circ R$ represents taking the $1200 rebate first and then receiving an 8% discount on the difference.

(d) $(R \circ S)(18,400) = 0.92(18,400) - 1200 = \$15,728$

$(S \circ R)(18,400) = 0.92(18,400 - 1200) = \$15,824$

$R \circ S$ is a better deal. $S \circ R$ takes an 8% discount on a smaller amount.

**91.** Let $f(x)$ be an odd function, $g(x)$ be an even function and define $h(x) = f(x)g(x)$. Then

$$h(-x) = f(-x)g(-x)$$

$$= [-f(x)]g(x) \qquad \text{Since } f \text{ is odd and } g \text{ is even.}$$

$$= -f(x)g(x)$$

$$= -h(x)$$

Thus, $h$ is odd.

# Section P.8    Inverse Functions

■ Two functions $f$ and $g$ are inverses of each other if $f(g(x)) = x$ for every $x$ in the domain of $g$ and $g(f(x)) = x$ for every $x$ in the domain of $f$.

■ Be able to find the inverse of a function, if it exists.

1. Replace $f(x)$ with $y$.

2. Interchange $x$ and $y$.

3. Solve for $y$. If this equation represents $y$ as a function of $x$, then you have found $f^{-1}(x)$. If this equation does not represent $y$ as a function of $x$, then $f$ does not have an inverse function.

■ A function $f$ has an inverse function if and only if no **horizontal** line crosses the graph of $f$ at more than one point.

**Solutions to Odd-Numbered Exercises**

**1.** The inverse is a line through $(-1, 0)$.
Matches graph (c).

**3.** The inverse is half a parabola starting at $(1, 0)$.
Matches graph (a).

**5.** $f^{-1}(x) = \dfrac{x}{8} = \dfrac{1}{8}x$

$f(f^{-1}(x)) = f\left(\dfrac{x}{8}\right) = 8\left(\dfrac{x}{8}\right) = x$

$f^{-1}(f(x)) = f^{-1}(8x) = \dfrac{8x}{8} = x$

**7.** $f^{-1}(x) = x - 10$

$f(f^{-1}(x)) = f(x - 10) = (x - 10) + 10 = x$

$f^{-1}(f(x)) = f^{-1}(x + 10) = (x + 10) - 10 = x$

**9.** $f^{-1}(x) = x^3$

$f(f^{-1}(x)) = f(x^3) = \sqrt[3]{x^3} = x$

$f^{-1}(f(x)) = f^{-1}(\sqrt[3]{x}) = (\sqrt[3]{x})^3 = x$

**11.** (a) $f(g(x)) = f\left(\dfrac{x}{2}\right) = 2\left(\dfrac{x}{2}\right) = x$

$g(f(x)) = g(2x) = \dfrac{2x}{2} = x$

**13.** (a) $f(g(x)) = f\left(\dfrac{x-1}{5}\right) = 5\left(\dfrac{x-1}{5}\right) + 1 = x$

$g(f(x)) = g(5x + 1) = \dfrac{(5x + 1) - 1}{5} = x$

(b)

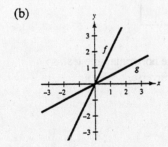

(b)

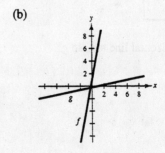

**15. (a)** $f(g(x)) = f(\sqrt[3]{x}) = (\sqrt[3]{x})^3 = x$

$g(f(x)) = g(x^3) = \sqrt[3]{x^3} = x$

**(b)**

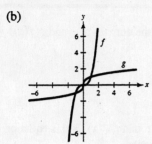

**17. (a)** $f(g(x)) = f(x^2 + 4), \ x \geq 0$

$= \sqrt{(x^2 + 4) - 4} = x$

$g(f(x)) = g(\sqrt{x - 4})$

$= (\sqrt{x - 4})^2 + 4 = x$

**(b)**

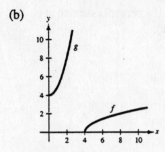

**19. (a)** $f(g(x)) = f(\sqrt{9 - x}), \ x \leq 9$

$= 9 - (\sqrt{9 - x})^2 = x$

$g(f(x)) = g(9 - x^2), \ x \geq 0$

$= \sqrt{9 - (9 - x^2)} = x$

**(b)**

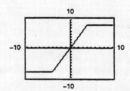

**21.** No, $\{(-2, -1), (1, 0), (2, 1), (1, 2), (-2, 3), (-6, 4)\}$ does not represent a function.

**23.** Since no horizontal line crosses the graph of $f$ at more than one point, $f$ **has** an inverse.

**25.** Since some horizontal lines cross the graph of $f$ twice, $f$ does **not** have an inverse.

**27.** $g(x) = \dfrac{4 - x}{6}$

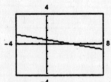

$g$ passes the horizontal line test, so $g$ **has** an inverse.

**29.** $h(x) = |x + 4| - |x - 4|$

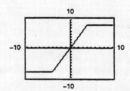

$h$ does not pass the horizontal line test, so $h$ does **not** have an inverse.

**31.** $f(x) = -2x\sqrt{16 - x^2}$

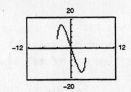

$f$ does not pass the horizontal line test, so $f$ does **not** have an inverse.

**33.**  $f(x) = 2x - 3$

$y = 2x - 3$

$x = 2y - 3$

$y = \dfrac{x + 3}{2}$

$f^{-1}(x) = \dfrac{x + 3}{2}$

**35.**  $f(x) = x^5$

$y = x^5$

$x = y^5$

$y = \sqrt[5]{x}$

$f^{-1}(x) = \sqrt[5]{x}$

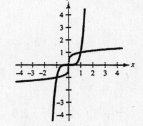

**37.**  $f(x) = \sqrt{x}$

$y = \sqrt{x}$

$x = \sqrt{y}$

$y = x^2$

$f^{-1}(x) = x^2,\ x \geq 0$

**39.**  $f(x) = \sqrt{4 - x^2},\ 0 \leq x \leq 2$

$y = \sqrt{4 - x^2}$

$x = \sqrt{4 - y^2}$

$f^{-1}(x) = \sqrt{4 - x^2},\ 0 \leq x \leq 2$

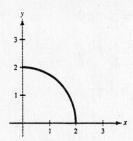

**41.**  $f(x) = \sqrt[3]{x - 1}$

$y = \sqrt[3]{x - 1}$

$x = \sqrt[3]{y - 1}$

$x^3 = y - 1$

$y = x^3 + 1$

$f^{-1}(x) = x^3 + 1$

**43.** $f(x) = x^4$

$y = x^4$

$x = y^4$

$y = \pm\sqrt[4]{x}$

This does not represent $y$ as a function of $x$.
$f$ does not have an inverse.

**45.** $g(x) = \dfrac{x}{8}$

$y = \dfrac{x}{8}$

$x = \dfrac{y}{8}$

$y = 8x$

This is a function of $x$, so $g$ has an inverse.
$g^{-1}(x) = 8x$

**47.** $p(x) = -4$

$y = -4$

Since $y = -4$ for all $x$, the graph is a horizontal line and fails the horizontal line test. $p$ does not have an inverse.

**49.** $f(x) = (x + 3)^2, \ x \geq -3 \ \Rightarrow \ y \geq 0$

$y = (x + 3)^2, \ x \geq -3, \ y \geq 0$

$x = (y + 3)^2, \ y \geq -3, \ x \geq 0$

$\sqrt{x} = y + 3, \ y \geq -3, \ x \geq 0$

$y = \sqrt{x} - 3, \ x \geq 0, \ y \geq -3$

This is a function of $x$, so $f$ has an inverse.
$f^{-1}(x) = \sqrt{x} - 3, \ x \geq 0$

**51.** $h(x) = \dfrac{1}{x}$

$y = \dfrac{1}{x}$

$xy = 1$

$y = \dfrac{1}{x}$

This is a function of $x$, so $h$ has an inverse.
$h^{-1}(x) = \dfrac{1}{x}$

**53.** $f(x) = \sqrt{2x + 3} \ \Rightarrow \ x \geq -\dfrac{3}{2}, \ y \geq 0$

$y = \sqrt{2x + 3}, \ x \geq -\dfrac{3}{2}, \ y \geq 0$

$x = \sqrt{2y + 3}, \ y \geq -\dfrac{3}{2}, \ x \geq 0$

$x^2 = 2y + 3, \ x \geq 0, \ y \geq -\dfrac{3}{2}$

$y = \dfrac{x^2 - 3}{2}, \ x \geq 0, \ y \geq -\dfrac{3}{2}$

This is a function of $x$, so $f$ has an inverse.

$f^{-1}(x) = \dfrac{x^2 - 3}{2}, \ x \geq 0$

**55.** $g(x) = x^2 - x^4$

The graph fails the horizontal line test, so $g$ does not have an inverse.

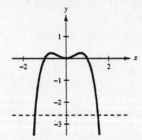

**57.** $f(x) = 25 - x^2, \ x \leq 0 \ \Rightarrow \ y \leq 25$

$y = 25 - x^2, \ x \leq 0, \ y \leq 25$

$x = 25 - y^2, \ y \leq 0, \ x \leq 25$

$y^2 = 25 - x, \ x \leq 25, \ y \leq 0$

$y = -\sqrt{25 - x}, \ x \leq 25, \ y \leq 0$

This is a function of $x$, so $f$ has an inverse.

$f^{-1}(x) = -\sqrt{25 - x}, \ x \leq 25$

**59.** If we let $f(x) = (x - 2)^2$, $x \geq 2$, then $f$ has an inverse. [Note: we could also let $x \leq 2$.]

$$f(x) = (x - 2)^2, \ x \geq 2 \ \Rightarrow \ y \geq 0$$
$$y = (x - 2)^2, \ x \geq 2, \ y \geq 0$$
$$x = (y - 2)^2, \ x \geq 0, \ y \geq 2$$
$$\sqrt{x} = y - 2, \ x \geq 0, \ y \geq 2$$
$$\sqrt{x} + 2 = y, \ x \geq 0, \ y \geq 2$$

Thus, $f^{-1}(x) = \sqrt{x} + 2$, $x \geq 0$.

**61.** If we let $f(x) = |x + 2|$, $x \geq -2$, then $f$ has an inverse. [Note: we could also let $x \leq -2$.]

$$f(x) = |x + 2|, \ x \geq -2$$
$$f(x) = x + 2 \text{ when } x \geq -2.$$
$$y = x + 2, \ x \geq -2, \ y \geq 0$$
$$x = y + 2, \ x \geq 0, \ y \geq -2$$
$$x - 2 = y, \ x \geq 0, \ y \geq -2$$

Thus, $f^{-1}(x) = x - 2$, $x \geq 0$.

**63.**

| $x$ | $f(x)$ |
|-----|--------|
| $-2$ | $-4$ |
| $-1$ | $-2$ |
| $1$ | $2$ |
| $3$ | $3$ |

| $x$ | $f^{-1}(x)$ |
|-----|-------------|
| $-4$ | $-2$ |
| $-2$ | $-1$ |
| $2$ | $1$ |
| $3$ | $3$ |

**65.** False, $f(x) = x^2$ is even and does not have an inverse.

**67.** True

In Exercises 69, 71, and 73, $f(x) = \frac{1}{8}x - 3$, $f^{-1}(x) = 8(x + 3)$, $g(x) = x^3$, $g^{-1}(x) = \sqrt[3]{x}$.

**69.** $(f^{-1} \circ g^{-1})(1) = f^{-1}(g^{-1}(1)) = f^{-1}(\sqrt[3]{1}) = 8(\sqrt[3]{1} + 3) = 32$

**71.** $(f^{-1} \circ f^{-1})(6) = f^{-1}(f^{-1}(6)) = f^{-1}(8[6 + 3]) = 8[8(6 + 3) + 3] = 600$

**73.** $(f \circ g)(x) = f(g(x)) = f(x^3) = \frac{1}{8}x^3 - 3$

$$y = \frac{1}{8}x^3 - 3$$
$$x = \frac{1}{8}y^3 - 3$$
$$x + 3 = \frac{1}{8}y^3$$
$$8(x + 3) = y^3$$
$$\sqrt[3]{8(x + 3)} = y$$
$$(f \circ g)^{-1}(x) = 2\sqrt[3]{x + 3}$$

In Exercises 75 and 77, $f(x) = x + 4$, $f^{-1}(x) = x - 4$, $g(x) = 2x - 5$, $g^{-1}(x) = \dfrac{x + 5}{2}$.

**75.** $(g^{-1} \circ f^{-1})(x) = g^{-1}(f^{-1}(x)) = g^{-1}(x - 4) = \dfrac{(x - 4) + 5}{2} = \dfrac{x + 1}{2}$

**77.** $(f \circ g)(x) = f(g(x)) = f(2x - 5) = (2x - 5) + 4 = 2x - 1$

$$(f \circ g)^{-1}(x) = \dfrac{x + 1}{2}$$

Note: Comparing Exercises 75 and 77, we see that $(f \circ g)^{-1}(x) = (g^{-1} \circ f^{-1})(x)$.

**79.** (a)    $y = 8 + 0.75x$

   $x = 8 + 0.75y$

   $x - 8 = 0.75y$

   $\dfrac{x - 8}{0.75} = y$

   $f^{-1}(x) = \dfrac{x - 8}{0.75}$

   (b)  $x$ = hourly wage

   $y$ = number of units produced

   (c)  $y = \dfrac{22.25 - 8}{0.75} = 19$ units

**81.** (a)    $y = 0.03x^2 + 254.50, \ 0 < x < 100$

   $x = 0.03y^2 + 254.50$

   $x - 254.50 = 0.03y^2$

   $\dfrac{x - 254.50}{0.03} = y^2$

   $\sqrt{\dfrac{x - 254.50}{0.03}} = y, \ 254.5 < x < 545.5$

   $f^{-1}(x) = \sqrt{\dfrac{x - 254.50}{0.03}}$

   $x$ = temperature in degrees Fahrenheit

   $y$ = percent load for a diesel engine

(b)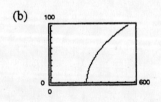

(c)  $0.03x^2 + 254.50 < 500$

   $0.03x^2 < 245.5$

   $x^2 < 8183\tfrac{1}{3}$

   $x < 90.46$

   Thus, $0 < x < 90.46$.

**83.** (a) Yes, since no $y$-value is paired with two different $x$-values, $f^{-1}$ does exist.

(b) $f^{-1}$ yields the year for a given average fuel consumption.

(c) $f^{-1}(19.95) = 8$

**85.** $x^2 = 64$

   $x = \pm\sqrt{64} = \pm 8$

**87.** $4x^2 - 12x + 9 = 0$

   $(2x - 3)^2 = 0$

   $2x - 3 = 0$

   $x = \tfrac{3}{2}$

**89.** $x^2 - 6x + 4 = 0$

$x^2 - 6x = -4$

$x^2 - 6x + 9 = -4 + 9$

$(x - 3)^2 = 5$

$x - 3 = \pm\sqrt{5}$

$x = 3 \pm \sqrt{5}$

**91.** $50 + 5x = 3x^2$

$0 = 3x^2 - 5x - 50$

$0 = (3x + 10)(x - 5)$

$3x + 10 = 0 \implies x = -\frac{10}{3}$

$x - 5 = 0 \implies x = 5$

**93.** Let $2n =$ first positive even integer. Then $2n + 2 =$ next positive even integer.

$2n(2n + 2) = 288$

$4n^2 + 4n - 288 = 0$

$4(n^2 + n - 72) = 0$

$4(n + 9)(n - 8) = 0$

$n + 9 = 0 \implies n = -9$    Not a solution since the integers are positive.

$n - 8 = 0 \implies n = 8$

Thus, $2n = 16$ and $2n + 2 = 18$.

**95.**

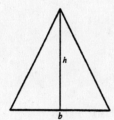

Given $b = h$ and $A = 10$ sq ft:

$A = \frac{1}{2}bh$

$10 = \frac{1}{2}bb$

$20 = b^2$

$\sqrt{20} = b$

$2\sqrt{5} = b$

Thus, $b = h = 2\sqrt{5}$ feet.

# ❑ Review Exercises for Chapter P

**Solutions to Odd-Numbered Exercises**

**1.** $\{11, -14, -\frac{8}{9}, \frac{5}{2}, \sqrt{6}, 0.4\}$

(a) Natural numbers: 11

(b) Integers: $11, -14$

(c) Rational numbers: $11, -14, -\frac{8}{9}, \frac{5}{2}, 0.4$

(d) Irrational numbers: $\sqrt{6}$

**3.** $x \leq 7$    The set consists of all real numbers less than or equal to 7.

**5.** $d(x, 7) = |x - 7|$ and $d(x, 7) \geq 4$, thus $|x - 7| \geq 4$.

**7.** $2x + (3x - 10) = (2x + 3x) - 10$

Illustrates the Associative Property of Addition.

**9.** $(t + 4)(2t) = (2t)(t + 4)$

Illustrates the Commutative Property of Multiplication.

**11.** $3x^2 + 7x = x^2 + 4$

(a) $x = 0$

$3(0)^2 + 7(0) \stackrel{?}{=} (0)^2 + 4$

$0 \neq 4$

No, $x = 0$ is not a solution.

(b) $x = -4$

$3(-4)^2 + 7(-4) \stackrel{?}{=} (-4)^2 + 4$

$20 = 20$

Yes, $x = -4$ is a solution.

(c) $x = \frac{1}{2}$

$3\left(\frac{1}{2}\right)^2 + 7\left(\frac{1}{2}\right) \stackrel{?}{=} \left(\frac{1}{2}\right)^2 + 4$

$\frac{17}{4} = \frac{17}{4}$

Yes, $x = \frac{1}{2}$ is a solution.

(d) $x = -1$

$3(-1)^2 + 7(-1) \stackrel{?}{=} (-1)^2 + 4$

$-4 \neq 5$

No, $x = -1$ is not a solution.

**13.** $4(x + 3) - 3 = 2(4 - 3x) - 4$

$4x + 12 - 3 = 8 - 6x - 4$

$4x + 9 = -6x + 4$

$10x = -5$

$x = -\frac{1}{2}$

**15.** $3\left(1 - \dfrac{1}{5t}\right) = 0$

$1 - \dfrac{1}{5t} = 0$

$1 = \dfrac{1}{5t}$

$5t = 1$

$t = \dfrac{1}{5}$

**17.** $(x + 4)^2 = 18$

$x + 4 = \pm\sqrt{18}$

$x = -4 \pm 3\sqrt{2}$

**19.** Let $x =$ the number of farmers in the group.

Cost per farmer $= \dfrac{48,000}{x}$

If two more farmers join the group, the cost per farmer will be $48,000/(x + 2)$. Since this new cost is \$4000 less than the original cost,

$$\frac{48,000}{x} - 4000 = \frac{48,000}{x + 2}$$

$$48,000(x + 2) - 4000x(x + 2) = 48,000x$$

$$12(x + 2) - x(x + 2) = 12x \qquad \text{Divide both sides by 4000.}$$

$$12x + 24 - x^2 - 2x = 12x$$

$$0 = x^2 + 2x - 24$$

$$0 = (x + 6)(x - 4)$$

$$x = -6, \text{ extraneous} \quad \text{or} \quad x = 4$$

$$x = 4 \text{ farmers.}$$

**21.** $y = -\frac{1}{2}x + 2$

| $x$ | $-2$ | $0$ | $2$ | $3$ | $4$ |
|---|---|---|---|---|---|
| $y$ | $3$ | $2$ | $1$ | $\frac{1}{2}$ | $0$ |

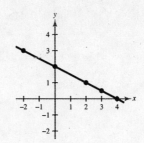

**23.** (a)

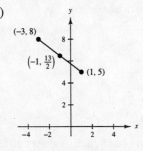

(b) $d = \sqrt{(1 - (-3))^2 + (5 - 8)^2}$

$\qquad = \sqrt{4^2 + (-3)^2}$

$\qquad = \sqrt{16 + 9}$

$\qquad = \sqrt{25}$

$\qquad = 5$

(c) $\left(\dfrac{-3 + 1}{2}, \dfrac{8 + 5}{2}\right) = \left(-1, \dfrac{13}{2}\right)$

**25.** $y - 2x - 3 = 0$

$\quad y = 2x + 3$

Line with $x$-intercept $\left(-\frac{3}{2}, 0\right)$ and $y$-intercept $(0, 3)$

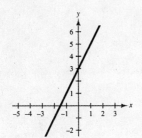

**27.** $y = \sqrt{5 - x}$

Domain: $(-\infty, 5]$

| $x$ | $5$ | $4$ | $1$ | $-4$ |
|---|---|---|---|---|
| $y$ | $0$ | $1$ | $2$ | $3$ |

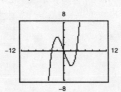

**29.** $y = \frac{1}{4}x^4 - 2x^2$

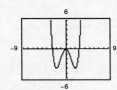

Intercepts: $(0, 0), \left(\pm 2\sqrt{2}, 0\right)$

$y$-axis symmetry

**31.** $y = \frac{1}{4}x^3 - 3x$

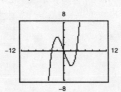

Intercepts: $(0, 0), \left(\pm 2\sqrt{3}, 0\right)$

Origin symmetry

**33.** $y = 10x^3 - 21x^2$

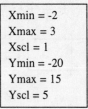

```
Xmin = -2
Xmax = 3
Xscl = 1
Ymin = -20
Ymax = 15
Yscl = 5
```

**35.** $(x - 3)^2 + (y + 1)^2 = 9$

$(x - 3)^2 + [y - (-1)]^2 = 3^2$

Center: $(3, -1)$

Radius: $r = 3$

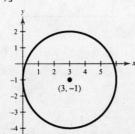

**37.**    $m = \dfrac{6 - 1}{14 - 2} = \dfrac{5}{12}$

$y - 1 = \dfrac{5}{12}(x - 2)$

$12y - 12 = 5x - 10$

$0 = 5x - 12y + 2$

**39.** $y - (-5) = \frac{3}{2}(x - 0)$

$\quad\quad 2y + 10 = 3x$

$\quad\quad\quad\quad 0 = 3x - 2y - 10$

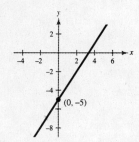

**41.** $\quad\quad\quad y - 0 = -\frac{2}{3}(x - 3)$

$\quad\quad\quad\quad 3y = -2x + 6$

$\quad 2x + 3y - 6 = 0$

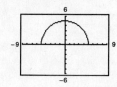

**43.** $(6, 12{,}500)\ m = 850$

$\quad y - 12{,}500 = 850(t - 6)$

$\quad\quad\quad\quad y = 850t - 5100 + 12{,}500$

$\quad\quad\quad\quad y = 850t + 7400$

**45.** $16x - y^4 = 0$

$\quad\quad\quad y^4 = 16x$

$\quad\quad\quad\ y = \pm 2\sqrt[4]{x}$

$y$ is **not** a function of $x$. Some $x$ values correspond to two $y$ values.

**47.** $y = \sqrt{1 - x}$

Each $x$ value, $x \le 1$, corresponds to only one $y$ value so $y$ **is** a function of $x$.

**49.** $g(x) = x^{4/3}$

    (a) $g(8) = 8^{4/3} = \left(\sqrt[3]{8}\right)^4 = 16$

    (b) $g(t + 1) = (t + 1)^{4/3}$

    (c) $\dfrac{g(8) - g(1)}{8 - 1} = \dfrac{16 - 1}{7} = \dfrac{15}{7}$

    (d) $g(-x) = (-x)^{4/3} = x^{4/3}$

**51.** $f(x) = \sqrt{25 - x^2}$

Domain: $[-5, 5]$

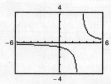

**53.** $g(s) = \dfrac{5}{3s - 9}$

Domain: All real numbers except $s = 3$

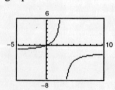

**55.** $h(x) = \dfrac{x}{x^2 - x - 6} = \dfrac{x}{(x + 2)(x - 3)}$

Domain: All real numbers except $x = -2, 3$

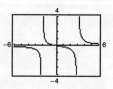

**57.** $f(x) = \dfrac{3x}{2(3 - x)}$

The second setting shows the most complete graph.

**59.** $y = 4x^3 - 12x^2 + 8x$

$\quad\quad 0 = 4x^3 - 12x^2 + 8x$

$\quad\quad 0 = 4x(x^2 - 3x + 2)$

$\quad\quad 0 = 4x(x - 1)(x - 2)$

$\quad\quad x = 0,\ x = 1,$

$\quad\quad\quad\quad\quad$ or $x = 2$

$x$-intercepts: $(0, 0),\ (1, 0),\ (2, 0)$

**61.** $y = \dfrac{1}{x} + \dfrac{1}{x+1} - 2$

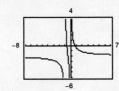

$x$-intercepts: $\left(\pm\dfrac{\sqrt{2}}{2}, 0\right)$

$0 = \dfrac{1}{x} + \dfrac{1}{x+1} - 2$

$2 = \dfrac{1}{x} + \dfrac{1}{x+1}$

$2x(x+1) = (x+1) + x$

$2x^2 + 2x = 2x + 1$

$2x^2 = 1$

$x^2 = \dfrac{1}{2}$

$x = \pm\sqrt{\dfrac{1}{2}} = \pm\dfrac{\sqrt{2}}{2}$

**63.** (a) $f(x) = \sqrt{x} + c$

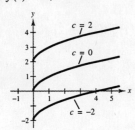

(b) $f(x) = \sqrt{x - c}$

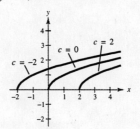

**65.** (a) $\begin{aligned} f(x) &= \tfrac{1}{2}x - 3 \\ y &= \tfrac{1}{2}x - 3 \\ x &= \tfrac{1}{2}y - 3 \\ 2x &= y - 6 \\ 2x + 6 &= y \\ f^{-1}(x) &= 2x + 6 \end{aligned}$

(b)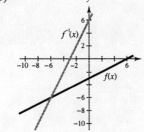

(c) $\begin{aligned} f^{-1}(f(x)) &= f^{-1}\left(\tfrac{1}{2}x - 3\right) \\ &= 2\left(\tfrac{1}{2}x - 3\right) + 6 \\ &= x - 6 + 6 \\ &= x \\ f(f^{-1}(x)) &= f(2x + 6) \\ &= \tfrac{1}{2}(2x + 6) - 3 \\ &= x + 3 - 3 \\ &= x \end{aligned}$

**67.** (a) $\begin{aligned} f(x) &= \sqrt{x + 1} \\ y &= \sqrt{x + 1} \\ x &= \sqrt{y + 1} \\ x^2 &= y + 1 \\ x^2 - 1 &= y \\ f^{-1}(x) &= x^2 - 1, x \geq 0 \end{aligned}$

Note: The inverse must have a restricted domain.

(b)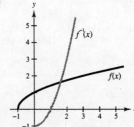

(c) $\begin{aligned} f^{-1}(f(x)) &= f^{-1}\left(\sqrt{x + 1}\right) \\ &= \left(\sqrt{x + 1}\right)^2 - 1 \\ &= x + 1 - 1 \\ &= x \\ f(f^{-1}(x)) &= f(x^2 - 1) \\ &= \sqrt{(x^2 - 1) + 1} \\ &= \sqrt{x^2} = x \text{ for } x \geq 0 \end{aligned}$

**69.** $f(x) = 2(x - 4)^2$ is increasing on $[4, \infty)$.

Let $f(x) = 2(x - 4)^2$, $x \geq 4$ and $y \geq 0$

$$y = 2(x - 4)^2$$

$$x = 2(y - 4)^2, x \geq 0, y \geq 4$$

$$\frac{x}{2} = (y - 4)^2$$

$$\sqrt{\frac{x}{2}} = y - 4$$

$$\sqrt{\frac{x}{2}} + 4 = y$$

$$f^{-1}(x) = \sqrt{\frac{x}{2}} + 4, x \geq 0$$

**71.** $f(x) = \sqrt{x^2 - 4}$ is increasing on $[2, \infty)$.

Let $f(x) = \sqrt{x^2 - 4}$, $x \geq 2, y \geq 0$

$$y = \sqrt{x^2 - 4}$$

$$x = \sqrt{y^2 - 4}$$

$$x^2 = y^2 - 4$$

$$x^2 + 4 = y^2$$

$$f^{-1}(x) = \sqrt{x^2 + 4}, x \geq 0$$

**73.** $(f - g)(4) = f(4) - g(4)$

$$= [3 - 2(4)] - \sqrt{4}$$

$$= -5 - 2$$

$$= -7$$

**75.** $(fh)(1) = f(1) \cdot h(1)$

$$= [3 - 2(1)][3(1)^2 + 2]$$

$$= (1)(5)$$

$$= 5$$

**77.** $(h \circ g)(7) = h(g(7))$

$$= h(\sqrt{7})$$

$$= 3(\sqrt{7})^2 + 2$$

$$= 23$$

**79.** $g^{-1}(x) = x^2, x \geq 0$

$$g^{-1}(3) = 3^2$$

$$= 9$$

# ❑ Practice Test for Chapter P

1. Evaluate $\dfrac{|-42| - 20}{15 - |-4|}$.

2. The distance between $x$ and 7 is no more than 4. Use absolute value notation to describe this expression.

3. (a) Plot the points $(-3, 7)$ and $(5, -1)$,

   (b) find the distance between the points, and

   (c) find the midpoint of the line segment joining the points.

4. Solve $5x + 4 = 7x - 8$.

5. Solve $\dfrac{x}{3} - 5 = \dfrac{x}{5} + 1$.

6. Solve $\dfrac{3x + 1}{6x - 7} = \dfrac{2}{5}$.

7. Solve $(x - 3)^2 + 4 = (x + 1)^2$.

8. Graph $3x - 5y = 15$.

9. Graph $y = \sqrt{9 - x}$.

10. Find the equation of the line through $(2, 4)$ and $(3, -1)$.

11. Find the equation of the line with slope $m = 4/3$ and $y$-intercept $b = -3$.

12. Given $f(x) = x^2 - 2x + 1$, find $f(x - 3)$.

13. Given $f(x) = 4x - 11$, find $\dfrac{f(x) - f(3)}{x - 3}$.

14. Find the domain and range of $f(x) = \sqrt{36 - x^2}$.

15. Which equations determine $y$ as a function of $x$?

    (a) $6x - 5y + 4 = 0$

    (b) $x^2 + y^2 = 9$

    (c) $y^3 = x^2 + 6$

16. Sketch the graph of $f(x) = x^2 - 5$.

17. Sketch the graph of $f(x) = |x + 3|$.

18. Use the graph of $f(x) = |x|$ to graph the following:

    (a) $f(x + 2)$

    (b) $-f(x) + 2$

19. Given $f(x) = 3x + 7$ and $g(x) = 2x^2 - 5$, find the following:

    (a) $(g - f)(x)$

    (b) $(fg)(x)$

20. Given $f(x) = x^2 - 2x + 16$ and $g(x) = 2x + 3$, find $f(g(x))$.

**21.** Given $f(x) = x^3 + 7$, find $f^{-1}(x)$.

**22.** Which of the following functions have inverses?

    (a) $f(x) = |x - 6|$

    (b) $f(x) = ax + b, \ a \neq 0$

    (c) $f(x) = x^3 - 19$

**23.** Write the standard equation of the circle with center $(-3, 5)$ and radius 6.

**24.** If it costs a company \$32 to produce 5 units of a product and \$44 to produce 9 units, how much does it cost to produce 20 units? (Assume that the cost function is linear.)

**25.** The perimeter of a rectangle is 1100 feet. Find the dimension so that the enclosed area will be 60,000 square feet.

# CHAPTER 1
## Trigonometry

# CHAPTER 1
## Trigonometry

## Section 1.1   Radian and Degree Measure

---

You should know the following basic facts about angles, their measurement, and their applications.

- ■ Types of Angles:
  - (a) Acute: Measure between 0° and 90°.
  - (b) Right: Measure 90°.
  - (c) Obtuse: Measure between 90° and 180°.
  - (d) Straight: Measure 180°.
- ■ $\alpha$ and $\beta$ are complementary if $\alpha + \beta = 90°$. They are supplementary if $\alpha + \beta = 180°$.
- ■ Two angles in standard position that have the same terminal side are called coterminal angles.
- ■ To convert degrees to radians, use $1° = \pi/180$ radians.
- ■ To convert radians to degrees, use $1$ radian $= (180/\pi)°$.
- ■ $1' =$ one minute $= 1/60$ of $1°$.
- ■ $1'' =$ one second $= 1/60$ of $1' = 1/3600$ of $1°$.
- ■ The length of a circular arc is $s = r\theta$ where $\theta$ is measured in radians.
- ■ Speed = distance/time
- ■ Angular speed $= \theta/t = s/rt$

---

**Solutions to Odd-Numbered Exercises**

1.

   The angle shown is approximately 2 radians.

3.

   The angle shown is approximately $-3$ radians.

5. (a) Since $0 < \dfrac{\pi}{5} < \dfrac{\pi}{2}$; $\dfrac{\pi}{5}$ lies in Quadrant I.

   (b) Since $\pi < \dfrac{7\pi}{5} < \dfrac{3\pi}{2}$; $\dfrac{7\pi}{5}$ lies in Quadrant III.

7. (a) Since $-\dfrac{\pi}{2} < -\dfrac{\pi}{12} < 0$; $-\dfrac{\pi}{12}$ lies in Quadrant IV.

   (b) Since $-\dfrac{3\pi}{2} < -\dfrac{11\pi}{9} < -\pi$; $-\dfrac{11\pi}{9}$ lies in Quadrant II.

9. (a) Since $\pi < 3.5 < \dfrac{3\pi}{2}$; $3.5$ lies in Quadrant III.

   (b) Since $\dfrac{\pi}{2} < 2.25 < \pi$; $2.25$ lies in Quadrant II.

**11.** (a)

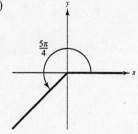

(b)

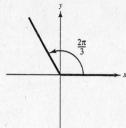

**13.** (a)

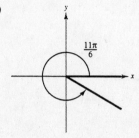

(b)

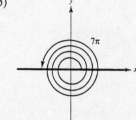

**15.** (a) Coterminal angles for $\dfrac{\pi}{12}$

$$\frac{\pi}{12} + 2\pi = \frac{25\pi}{12}$$

$$\frac{\pi}{12} - 2\pi = -\frac{23\pi}{12}$$

(b) Coterminal angles for $\dfrac{2\pi}{3}$

$$\frac{2\pi}{3} + 2\pi = \frac{8\pi}{3}$$

$$\frac{2\pi}{3} - 2\pi = -\frac{4\pi}{3}$$

**17.** (a) Coterminal angles for $-\dfrac{9\pi}{4}$

$$-\frac{9\pi}{4} + 4\pi = \frac{7\pi}{4}$$

$$\frac{7\pi}{4} - 2\pi = -\frac{\pi}{4}$$

(b) Coterminal angles for $-\dfrac{2\pi}{15}$

$$-\frac{2\pi}{15} + 2\pi = \frac{28\pi}{15}$$

$$-\frac{2\pi}{15} - 2\pi = -\frac{32\pi}{15}$$

**19.** (a) Complement: $\dfrac{\pi}{2} - \dfrac{\pi}{3} = \dfrac{\pi}{6}$

Supplement: $\pi - \dfrac{\pi}{3} = \dfrac{2\pi}{3}$

(b) Complement: Not possible; $\dfrac{3\pi}{4}$ is greater than $\dfrac{\pi}{2}$.

Supplement: $\pi - \dfrac{3\pi}{4} = \dfrac{\pi}{4}$

**21.**

The angle shown is approximately 210°.

**23.**

The angle shown is approximately −45°.

**25.** (a) Since 90° < 130° < 180°; 130° lies in Quadrant II.

(b) Since 270° < 285° < 360°; 285° lies in Quadrant IV.

**27.** (a) Since −180° < −132°50′ < −90°; −132° 50′ lies in Quadrant III.

(b) Since −360° < −336° < −270°; −336° lies in Quadrant I.

**29.** (a)
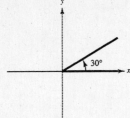
(b)
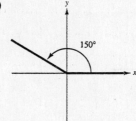

**31.** (a)
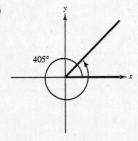
(b)
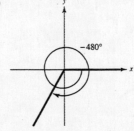

**33.** (a) Coterminal angles for $45°$
$45° + 360° = 405°$
$45° - 360° = -315°$

(b) Coterminal angles for $-36°$
$-36° + 360° = 324°$
$-36° - 360° = -396°$

**35.** (a) Coterminal angles for $300°$
$300° + 360° = 660°$
$300° - 360° = -60°$

(b) Coterminal angles for $740°$
$740° - 2(360°) = 20°$
$20° - 360° = -340°$

**37.** (a) Complement: $90° - 18° = 72°$
Supplement: $180° - 18° = 162°$

(b) Complement: Not possible; $115°$ is greater than $90°$.
Supplement: $180° - 115° = 65°$

**39.** (a) $30° = 30\left(\dfrac{\pi}{180}\right) = \dfrac{\pi}{6}$

(b) $150° = 150\left(\dfrac{\pi}{180}\right) = \dfrac{5\pi}{6}$

**41.** (a) $-20° = -20\left(\dfrac{\pi}{180}\right) = -\dfrac{\pi}{9}$

(b) $-240° = -240\left(\dfrac{\pi}{180}\right) = -\dfrac{4\pi}{3}$

**43.** (a) $\dfrac{3\pi}{2} = \dfrac{3\pi}{2}\left(\dfrac{180}{\pi}\right)° = 270°$

(b) $\dfrac{7\pi}{6} = \dfrac{7\pi}{6}\left(\dfrac{180}{\pi}\right)° = 210°$

**45.** (a) $\dfrac{7\pi}{3} = \dfrac{7\pi}{3}\left(\dfrac{180}{\pi}\right)° = 420°$

(b) $-\dfrac{11\pi}{30} = -\dfrac{11\pi}{30}\left(\dfrac{180}{\pi}\right)° = -66°$

**47.** $115° = 115\left(\dfrac{\pi}{180}\right) \approx 2.007 \text{ radians}$

**49.** $-216.35° = -216.35\left(\dfrac{\pi}{180}\right) \approx -3.776 \text{ radians}$

**51.** $532° = 532\left(\dfrac{\pi}{180}\right) \approx 9.285 \text{ radians}$

**53.** $-0.83° = -0.83\left(\dfrac{\pi}{180}\right) \approx -0.014 \text{ radian}$

**55.** $\dfrac{\pi}{7} = \dfrac{\pi}{7}\left(\dfrac{180}{\pi}\right) \approx 25.714°$

**57.** $\dfrac{15\pi}{8} = \dfrac{15\pi}{8}\left(\dfrac{180}{\pi}\right) = 337.5°$

**59.** $-4.2\pi = -4.2\pi\left(\dfrac{180}{\pi}\right) = -756°$

**61.** $-2 = -2\left(\dfrac{180}{\pi}\right) \approx -114.592°$

**63.** (a) $54° \, 45' = 54° + \left(\frac{45}{60}\right)° = 54.75°$

(b) $-128° \, 30' = -128° - \left(\frac{30}{60}\right)° = -128.5°$

**65.** (a) $85° \, 18' \, 30'' = \left(85 + \frac{18}{60} + \frac{30}{3600}\right)° \approx 85.308°$

(b) $330° \, 25'' = \left(330 + \frac{25}{3600}\right)° \approx 330.007°$

**67.** (a) $240.6° = 240° + 0.6(60)' = 240° \, 36'$

(b) $-145.8° = -[145° + 0.8(60')] = -145° \, 48'$

**69.** (a) $2.5 = 2.5\left(\frac{180}{\pi}\right)° \approx 143.23945° \approx 143° \, 14' \, 22''$

(b) $-3.58 = -3.58\left(\frac{180}{\pi}\right)° \approx -205.11889° \approx -205° \, 7' \, 8''$

**71.** $s = r\theta$

$6 = 5\theta$

$\theta = \frac{6}{5}$ radians

**73.** $s = r\theta$

$32 = 7\theta$

$\theta = \frac{32}{7} = 4\frac{4}{7}$ radians

**75.** $s = r\theta$

$4 = 15\theta$

$\theta = \frac{4}{15}$ radian

**77.** $s = r\theta$

$25 = 14.5\theta$

$\theta = \frac{25}{14.5} \approx 1.724$ radians

**79.** $s = r\theta$, $\theta$ in radians

$s = 15(180)\left(\frac{\pi}{180}\right) = 15\pi$ inches

$\approx 47.12$ inches

**81.** $s = r\theta$, $\theta$ in radians

$s = 6(2) = 12$ meters

**83.** $\theta = 41° \, 15' \, 42'' - 32° \, 47' \, 9'' = 8° \, 28' \, 33'' \approx 8.47583° \approx 0.14793$ radian

$s = r\theta = 4000(0.14793) \approx 591.72$ miles

**85.** $\theta = 42° \, 7' \, 15'' - 25° \, 46' \, 37'' = 16° \, 20' \, 38'' \approx 0.285255$ radian

$s = r\theta = 4000(0.285255) \approx 1141.02$ miles

**87.** $\theta = \frac{s}{r} = \frac{600}{6378} \approx 0.094$ radian $\approx 5.39°$

**89.** $\theta = \frac{s}{r} = \frac{2.5}{6} = \frac{25}{60} = \frac{5}{12}$ radian

**91.** (a) 50 miles per hour $= 50(5280)/60 = 4400$ feet per minute

The circumference of the tire is $C = 2.5\pi$ feet.

The number of revolutions per minute is $r = 4400/2.5\pi \approx 560.2$ rev/min.

(b) The angular speed is $\theta/t$.

$\theta = \frac{4400}{2.5\pi}(2\pi) = 3520$ radians

Angular speed $= \frac{3520 \text{ radians}}{1 \text{ minute}} = 3520$ rad/min

**93.** 1 Radian $= \left(\frac{180}{\pi}\right)° \approx 57.3°$, so one radian is much larger than one degree.

**95.** Circumference: $C = 2\pi(1.68) = 3.36\pi$ inches

360 rev/min $= 6$ rev/sec

Linear speed: $(3.36\pi)(6) = 20.16$ inches/sec

**97.** The area of a circle is $A = \pi r^2 \implies \pi = \dfrac{A}{r^2}$.

The circumference of a circle is $C = 2\pi r$.

$$C = 2\left(\dfrac{A}{r^2}\right)r$$

$$C = \dfrac{2A}{r}$$

$$\dfrac{Cr}{2} = A$$

For a sector, $C = s = r\theta$

Thus, $A = \dfrac{(r\theta)r}{2} = \dfrac{1}{2}\theta r^2$ for a sector.

**99.** $s = r\theta = (4000 \text{ miles})(0.031)\left(\dfrac{\pi}{180}\right) \approx 2.164 \text{ miles}$

# Section 1.2    Trigonometric Functions: The Unit Circle

- You should know the definition of the trigonometric functions in terms of the unit circle. Let $t$ be a real number and $(x, y)$ the point on the unit circle corresponding to $t$.

$$\sin t = y \qquad\qquad \csc t = \dfrac{1}{y}, \quad y \neq 0$$

$$\cos t = x \qquad\qquad \sec t = \dfrac{1}{x}, \quad x \neq 0$$

$$\tan t = \dfrac{y}{x}, \quad x \neq 0 \qquad\qquad \cot t = \dfrac{x}{y}, \quad y \neq 0$$

- The cosine and secant functions are even.

$$\cos(-t) = \cos t \qquad\qquad \sec(-t) = \sec t$$

- The other four trigonometric functions are odd.

$$\sin(-t) = -\sin t \qquad\qquad \csc(-t) = -\csc t$$

$$\tan(-t) = -\tan t \qquad\qquad \cot(-t) = -\cot t$$

- Be able to evaluate the trigonometric functions with a calculator.

## Solutions to Odd-Numbered Exercises

**1.** $x = -\dfrac{3}{5}, \quad y = \dfrac{4}{5}$

$$\sin t = y = \dfrac{4}{5} \qquad\qquad \csc t = \dfrac{1}{y} = \dfrac{5}{4}$$

$$\cos t = x = -\dfrac{3}{5} \qquad\qquad \sec t = \dfrac{1}{x} = -\dfrac{5}{3}$$

$$\tan t = \dfrac{y}{x} = -\dfrac{4}{3} \qquad\qquad \cot t = \dfrac{x}{y} = -\dfrac{3}{4}$$

**3.** $x = \dfrac{8}{17}, \quad y = -\dfrac{15}{17}$

$$\sin t = y = -\dfrac{15}{17} \qquad\qquad \csc t = \dfrac{1}{y} = -\dfrac{17}{15}$$

$$\cos t = x = \dfrac{8}{17} \qquad\qquad \sec t = \dfrac{1}{x} = \dfrac{17}{8}$$

$$\tan t = \dfrac{y}{x} = -\dfrac{15}{8} \qquad\qquad \cot t = \dfrac{x}{y} = -\dfrac{8}{15}$$

**5.** $t = \dfrac{\pi}{4}$ corresponds to $\left(\dfrac{\sqrt{2}}{2}, \dfrac{\sqrt{2}}{2}\right)$.

**7.** $t = \dfrac{5\pi}{6}$ corresponds to $\left(-\dfrac{\sqrt{3}}{2}, \dfrac{1}{2}\right)$.

**9.** $t = \dfrac{4\pi}{3}$ corresponds to $\left(-\dfrac{1}{2}, -\dfrac{\sqrt{3}}{2}\right)$.

**11.** $t = \dfrac{3\pi}{2}$ corresponds to $(0, -1)$.

**13.** $t = \dfrac{\pi}{4}$ corresponds to $\left(\dfrac{\sqrt{2}}{2}, \dfrac{\sqrt{2}}{2}\right)$.

$$\sin t = y = \frac{\sqrt{2}}{2}$$

$$\cos t = x = \frac{\sqrt{2}}{2}$$

$$\tan t = \frac{y}{x} = 1$$

**15.** $t = -\dfrac{\pi}{6}$ corresponds to $\left(\dfrac{\sqrt{3}}{2}, -\dfrac{1}{2}\right)$.

$$\sin t = y = -\frac{1}{2}$$

$$\cos t = x = \frac{\sqrt{3}}{2}$$

$$\tan t = \frac{y}{x} = -\frac{1}{\sqrt{3}}$$

**17.** $t = -\dfrac{5\pi}{4}$ corresponds to $\left(-\dfrac{\sqrt{2}}{2}, \dfrac{\sqrt{2}}{2}\right)$.

$$\sin t = y = \frac{\sqrt{2}}{2}$$

$$\cos t = x = -\frac{\sqrt{2}}{2}$$

$$\tan t = \frac{y}{x} = -1$$

**19.** $t = \dfrac{11\pi}{6}$ corresponds to $\left(\dfrac{\sqrt{3}}{2}, -\dfrac{1}{2}\right)$.

$$\sin t = y = -\frac{1}{2}$$

$$\cos t = x = \frac{\sqrt{3}}{2}$$

$$\tan t = \frac{y}{x} = -\frac{1}{\sqrt{3}}$$

**21.** $t = \dfrac{4\pi}{3}$ corresponds to $\left(-\dfrac{1}{2}, -\dfrac{\sqrt{3}}{2}\right)$.

$$\sin t = y = -\frac{\sqrt{3}}{2}$$

$$\cos t = x = -\frac{1}{2}$$

$$\tan t = \frac{y}{x} = \sqrt{3}$$

**23.** $t = -\dfrac{3\pi}{2}$ corresponds to $(0, 1)$.

$$\sin t = y = 1$$

$$\cos t = x = 0$$

$$\tan t = \frac{y}{x} \text{ is undefined.}$$

**25.** $t = \dfrac{3\pi}{4}$ corresponds to $\left(-\dfrac{\sqrt{2}}{2}, \dfrac{\sqrt{2}}{2}\right)$.

$$\sin t = y = \frac{\sqrt{2}}{2} \qquad \csc t = \frac{1}{y} = \sqrt{2}$$

$$\cos t = x = -\frac{\sqrt{2}}{2} \qquad \sec t = \frac{1}{x} = -\sqrt{2}$$

$$\tan t = \frac{y}{x} = -1 \qquad \cot t = \frac{x}{y} = -1$$

**27.** $t = \dfrac{\pi}{2}$ corresponds to $(0, 1)$.

$$\sin t = y = 1 \qquad \csc t = \frac{1}{y} = 1$$

$$\cos t = x = 0 \qquad \sec t = \frac{1}{x} \text{ is undefined.}$$

$$\tan t = \frac{y}{x} \text{ is undefined.} \qquad \cot t = \frac{x}{y} = 0$$

**29.** $t = -\dfrac{4\pi}{3}$ corresponds to $\left(-\dfrac{1}{2}, \dfrac{\sqrt{3}}{2}\right)$.

$$\sin t = y = \frac{\sqrt{3}}{2} \qquad \csc t = \frac{1}{y} = \frac{2\sqrt{3}}{3}$$

$$\cos t = x = -\frac{1}{2} \qquad \sec t = \frac{1}{x} = -2$$

$$\tan t = \frac{y}{x} = -\sqrt{3} \qquad \cot t = \frac{x}{y} = -\frac{\sqrt{3}}{3}$$

**31.** $\sin 3\pi = \sin \pi = 0$

**33.** $\cos\dfrac{8\pi}{3} = \cos\dfrac{2\pi}{3} = -\dfrac{1}{2}$

**35.** $\cos\dfrac{19\pi}{6} = \cos\dfrac{7\pi}{6} = -\dfrac{\sqrt{3}}{2}$

**37.** $\sin\left(-\dfrac{9\pi}{4}\right) = \sin\left(-\dfrac{\pi}{4}\right) = -\dfrac{\sqrt{2}}{2}$

**39.** $\sin t = \dfrac{1}{3}$

    (a) $\sin(-t) = -\sin t = -\dfrac{1}{3}$

    (b) $\csc(-t) = -\csc t = -3$

**41.** $\cos(-t) = -\dfrac{7}{8}$

    (a) $\cos t = \cos(-t) = -\dfrac{7}{8}$

    (b) $\sec(-t) = \dfrac{1}{\cos(-t)} = -\dfrac{8}{7}$

**43.** $\sin y = \dfrac{4}{5}$

    (a) $\sin(\pi - t) = \sin t = \dfrac{4}{5}$

    (b) $\sin(t + \pi) = -\sin t = -\dfrac{4}{5}$

**45.** $\sin\dfrac{\pi}{4} \approx 0.7071$

**47.** $\cos(-3) \approx -0.9900$

**49.** $\cos(-1.7) \approx -0.1288$

**51.** $\csc 0.8 = \dfrac{1}{\sin 0.8} \approx 1.3940$

**53.** $\sec 22.8 = \dfrac{1}{\cos 22.8} \approx -1.4486$

**55.** (a) $\sin 5 \approx -1$

    (b) $\cos 2 \approx -0.4$

**57.** (a) $\sin t = 0.25$

       $t \approx 0.25$ or $2.89$

    (b) $\cos t = -0.25$

       $t \approx 1.82$ or $4.46$

**59.** $\cos 1.5 \approx 0.0707$

    $2\cos 0.75 \approx 1.4634$

    $\cos 2t \neq 2\cos t$

**61.** (a) The points have $y$-axis symmetry.

    (b) $\sin t_1 = \sin(\pi - t_1)$ since they have the same $y$-value.

    (c) $-\cos t_1 = \cos(\pi - t_1)$ since the $x$-values have the opposite signs.

**63.** $y(t) = \frac{1}{4}\cos 6t$

    (a) $y(0) = \frac{1}{4}\cos 0 = 0.2500$ feet

    (b) $y\left(\frac{1}{4}\right) = \frac{1}{4}\cos\frac{3}{2} \approx 0.0177$ feet

    (c) $y\left(\frac{1}{2}\right) = \frac{1}{4}\cos 3 \approx -0.2475$ feet

**65.** $I = 5e^{-2(0.7)}\sin(0.7) \approx 0.794$

**67.** Let $h(t) = f(t)g(t)$

          $= \sin t \cos t.$

    Then, $h(-t) = \sin(-t)\cos(-t)$

               $= -\sin t \cos t$

               $= -h(t).$

    Thus, $h(t)$ is odd.

**69.** $\quad f(x) = \frac{1}{2}(3x - 2)$

          $y = \frac{1}{2}(3x - 2)$

          $x = \frac{1}{2}(3y - 2)$

        $2x = 3y - 2$

      $2x + 2 = 3y$

    $\frac{2}{3}(x + 1) = y$

      $f^{-1}(x) = \frac{2}{3}(x + 1)$

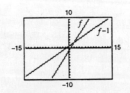

**71.**  
$$f(x) = \sqrt{x^2 - 4}, \quad x \geq 2, \quad y \geq 0$$
$$y = \sqrt{x^2 - 4}$$
$$x = \sqrt{y^2 - 4}$$
$$x^2 = y^2 - 4$$
$$x^2 + 4 = y^2$$
$$\sqrt{x^2 + 4} = y, \quad x \geq 0$$
$$f^{-1}(x) = \sqrt{x^2 + 4}, \quad x \geq 0$$

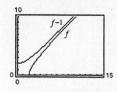

Note: The domain of $f^{-1}(x)$ equals the range of $f(x)$.

# Section 1.3    Right Triangle Trigonometry

- You should know the right triangle definition of trigonometric functions.

  (a) $\sin \theta = \dfrac{\text{opp}}{\text{hyp}}$    (b) $\cos \theta = \dfrac{\text{adj}}{\text{hyp}}$    (c) $\tan \theta = \dfrac{\text{opp}}{\text{adj}}$

  (d) $\csc \theta = \dfrac{\text{hyp}}{\text{opp}}$    (e) $\sec \theta = \dfrac{\text{hyp}}{\text{adj}}$    (f) $\cot \theta = \dfrac{\text{adj}}{\text{opp}}$

- You should know the following identities.

  (a) $\sin \theta = \dfrac{1}{\csc \theta}$    (b) $\csc \theta = \dfrac{1}{\sin \theta}$    (c) $\cos \theta = \dfrac{1}{\sec \theta}$

  (d) $\sec \theta = \dfrac{1}{\cos \theta}$    (e) $\tan \theta = \dfrac{1}{\cot \theta}$    (f) $\cot \theta = \dfrac{1}{\tan \theta}$

  (g) $\tan \theta = \dfrac{\sin \theta}{\cos \theta}$    (h) $\cot \theta = \dfrac{\cos \theta}{\sin \theta}$    (i) $\sin^2 \theta + \cos^2 \theta = 1$

  (j) $1 + \tan^2 \theta = \sec^2 \theta$    (k) $1 + \cot^2 \theta = \csc^2 \theta$

- You should know that two acute angles $\alpha$ and $\beta$ are complementary if $\alpha + \beta = 90°$, and that cofunctions of complementary angles are equal.

- You should know the trigonometric function values of 30°, 45°, and 60°, or be able to construct triangles from which you can determine them.

**Solutions to Odd-Numbered Exercises**

**1.**
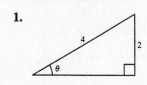

$$\text{adj} = \sqrt{4^2 - 2^2} = \sqrt{12} = 2\sqrt{3}$$

$$\sin \theta = \frac{\text{opp}}{\text{hyp}} = \frac{2}{4} = \frac{1}{2} \qquad \csc \theta = \frac{\text{hyp}}{\text{opp}} = \frac{4}{2} = 2$$

$$\cos \theta = \frac{\text{adj}}{\text{hyp}} = \frac{2\sqrt{3}}{4} = \frac{\sqrt{3}}{2} \qquad \sec \theta = \frac{\text{hyp}}{\text{adj}} = \frac{4}{2\sqrt{3}} = \frac{2\sqrt{3}}{3}$$

$$\tan \theta = \frac{\text{opp}}{\text{adj}} = \frac{2}{2\sqrt{3}} = \frac{\sqrt{3}}{3} \qquad \cot \theta = \frac{\text{adj}}{\text{opp}} = \frac{2\sqrt{3}}{2} = \sqrt{3}$$

**3.**
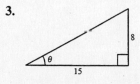

$$\text{hyp} = \sqrt{8^2 + 15^2} = 17$$

$$\sin \theta = \frac{\text{opp}}{\text{hyp}} = \frac{8}{17} \qquad \csc \theta = \frac{\text{hyp}}{\text{opp}} = \frac{17}{8}$$

$$\cos \theta = \frac{\text{adj}}{\text{hyp}} = \frac{15}{17} \qquad \sec \theta = \frac{\text{hyp}}{\text{adj}} = \frac{17}{15}$$

$$\tan \theta = \frac{\text{opp}}{\text{adj}} = \frac{8}{15} \qquad \cot \theta = \frac{\text{adj}}{\text{opp}} = \frac{15}{8}$$

**5.**

$$adj = \sqrt{3^2 - 1^2} = \sqrt{8} = 2\sqrt{2}$$

$$\sin \theta = \frac{opp}{hyp} = \frac{1}{3} \qquad\qquad \csc \theta = \frac{hyp}{opp} = 3$$

$$\cos \theta = \frac{adj}{hyp} = \frac{2\sqrt{2}}{3} \qquad\qquad \sec \theta = \frac{hyp}{adj} = \frac{3}{2\sqrt{2}} = \frac{3\sqrt{2}}{4}$$

$$\tan \theta = \frac{opp}{adj} = \frac{1}{2\sqrt{2}} = \frac{\sqrt{2}}{4} \qquad\qquad \cot \theta = \frac{adj}{opp} = 2\sqrt{2}$$

$$adj = \sqrt{6^2 - 2^2} = \sqrt{32} = 4\sqrt{2}$$

$$\sin \theta = \frac{opp}{hyp} = \frac{2}{6} = \frac{1}{3} \qquad\qquad \csc \theta = \frac{hyp}{opp} = \frac{6}{2} = 3$$

$$\cos \theta = \frac{adj}{hyp} = \frac{4\sqrt{2}}{6} = \frac{2\sqrt{2}}{3} \qquad\qquad \sec \theta = \frac{hyp}{adj} = \frac{6}{4\sqrt{2}} = \frac{3}{2\sqrt{2}} = \frac{3\sqrt{2}}{4}$$

$$\tan \theta = \frac{opp}{adj} = \frac{2}{4\sqrt{2}} = \frac{1}{2\sqrt{2}} = \frac{\sqrt{2}}{4} \qquad \cot \theta = \frac{adj}{opp} = \frac{4\sqrt{2}}{2} = 2\sqrt{2}$$

The function values are the same since the triangles are similar and the corresponding sides are proportional.

**7.**

$$opp = \sqrt{10^2 - 8^2} = 6$$

$$\sin \theta = \frac{opp}{hyp} = \frac{6}{10} = \frac{3}{5} \qquad\qquad \csc \theta = \frac{hyp}{opp} = \frac{10}{6} = \frac{5}{3}$$

$$\cos \theta = \frac{adj}{hyp} = \frac{8}{10} = \frac{4}{5} \qquad\qquad \sec \theta = \frac{hyp}{adj} = \frac{10}{8} = \frac{5}{4}$$

$$\tan \theta = \frac{opp}{adj} = \frac{6}{8} = \frac{3}{4} \qquad\qquad \cot \theta = \frac{adj}{opp} = \frac{8}{6} = \frac{4}{3}$$

$$opp = \sqrt{2.5^2 - 2^2} = 1.5$$

$$\sin \theta = \frac{opp}{hyp} = \frac{1.5}{2.5} = \frac{3}{5} \qquad\qquad \csc \theta = \frac{hyp}{opp} = \frac{2.5}{1.5} = \frac{5}{3}$$

$$\cos \theta = \frac{adj}{hyp} = \frac{2}{2.5} = \frac{4}{5} \qquad\qquad \sec \theta = \frac{hyp}{adj} = \frac{2.5}{2} = \frac{5}{4}$$

$$\tan \theta = \frac{opp}{adj} = \frac{1.5}{2} = \frac{3}{4} \qquad\qquad \cot \theta = \frac{adj}{opp} = \frac{2}{1.5} = \frac{4}{3}$$

The function values are the same since the triangles are similar and the corresponding sides are proportional.

**9.** Given: $\sin\theta = \dfrac{2}{3} = \dfrac{\text{opp}}{\text{hyp}}$

$$2^2 + (\text{adj})^2 = 3^2$$

$$\text{adj} = \sqrt{5}$$

$$\cos\theta = \frac{\sqrt{5}}{3}$$

$$\tan\theta = \frac{2\sqrt{5}}{5}$$

$$\cot\theta = \frac{\sqrt{5}}{2}$$

$$\sec\theta = \frac{3\sqrt{5}}{5}$$

$$\csc\theta = \frac{3}{2}$$

**11.** Given: $\sec\theta = 2 = \dfrac{2}{1} = \dfrac{\text{hyp}}{\text{adj}}$

$$(\text{opp})^2 + 1^2 = 2^2$$

$$\text{opp} = \sqrt{3}$$

$$\sin\theta = \frac{\sqrt{3}}{2}$$

$$\cos\theta = \frac{1}{2}$$

$$\tan\theta = \sqrt{3}$$

$$\cot\theta = \frac{\sqrt{3}}{3}$$

$$\csc\theta = \frac{2\sqrt{3}}{3}$$

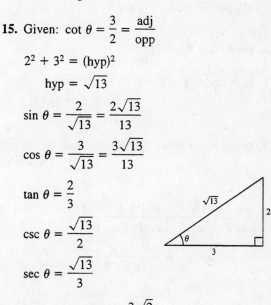

**13.** Given: $\tan\theta = 3 = \dfrac{3}{1} = \dfrac{\text{opp}}{\text{adj}}$

$$3^2 + 1^2 = (\text{hyp})^2$$

$$\text{hyp} = \sqrt{10}$$

$$\sin\theta = \frac{3\sqrt{10}}{10}$$

$$\cos\theta = \frac{\sqrt{10}}{10}$$

$$\cot\theta = \frac{1}{3}$$

$$\sec\theta = \sqrt{10}$$

$$\csc\theta = \frac{\sqrt{10}}{3}$$

**15.** Given: $\cot\theta = \dfrac{3}{2} = \dfrac{\text{adj}}{\text{opp}}$

$$2^2 + 3^2 = (\text{hyp})^2$$

$$\text{hyp} = \sqrt{13}$$

$$\sin\theta = \frac{2}{\sqrt{13}} = \frac{2\sqrt{13}}{13}$$

$$\cos\theta = \frac{3}{\sqrt{13}} = \frac{3\sqrt{13}}{13}$$

$$\tan\theta = \frac{2}{3}$$

$$\csc\theta = \frac{\sqrt{13}}{2}$$

$$\sec\theta = \frac{\sqrt{13}}{3}$$

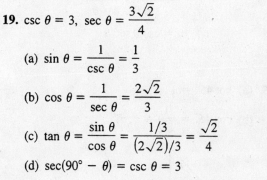

**17.** $\sin 60° = \dfrac{\sqrt{3}}{2}$, $\cos 60° = \dfrac{1}{2}$

(a) $\tan 60° = \dfrac{\sin 60°}{\cos 60°} = \sqrt{3}$

(b) $\sin 30° = \cos 60° = \dfrac{1}{2}$

(c) $\cos 30° = \sin 60° = \dfrac{\sqrt{3}}{2}$

(d) $\cot 60° = \dfrac{\cos 60°}{\sin 60°} = \dfrac{1}{\sqrt{3}} = \dfrac{\sqrt{3}}{3}$

**19.** $\csc\theta = 3$, $\sec\theta = \dfrac{3\sqrt{2}}{4}$

(a) $\sin\theta = \dfrac{1}{\csc\theta} = \dfrac{1}{3}$

(b) $\cos\theta = \dfrac{1}{\sec\theta} = \dfrac{2\sqrt{2}}{3}$

(c) $\tan\theta = \dfrac{\sin\theta}{\cos\theta} = \dfrac{1/3}{(2\sqrt{2})/3} = \dfrac{\sqrt{2}}{4}$

(d) $\sec(90° - \theta) = \csc\theta = 3$

**21.** $\cos \alpha = \dfrac{1}{4}$

(a) $\sec \alpha = \dfrac{1}{\cos \alpha} = 4$

(b) $\sin^2 \alpha + \cos^2 \alpha = 1$

$\sin^2 \alpha + \left(\dfrac{1}{4}\right)^2 = 1$

$\sin^2 \alpha = \dfrac{15}{16}$

$\sin \alpha = \dfrac{\sqrt{15}}{4}$

(c) $\cot \alpha = \dfrac{\cos \alpha}{\sin \alpha} = \dfrac{1/4}{\sqrt{15}/4} = \dfrac{1}{\sqrt{15}} = \dfrac{\sqrt{15}}{15}$

(d) $\sin(90° - \alpha) = \cos \alpha = \dfrac{1}{4}$

**23.** $\tan \theta \cot \theta = \tan \theta \left(\dfrac{1}{\tan \theta}\right) = 1$

**25.** $\tan \alpha \cos \alpha = \left(\dfrac{\sin \alpha}{\cos \alpha}\right)\cos \alpha = \sin \alpha$

**27.** $(1 + \cos \theta)(1 - \cos \theta) = 1 - \cos^2 \theta$
$= (\sin^2 \theta + \cos^2 \theta) - \cos^2 \theta$
$= \sin^2 \theta$

**29.** $(\sec \theta + \tan \theta)(\sec \theta - \tan \theta) = \sec^2 \theta - \tan^2 \theta$
$= (1 + \tan^2 \theta) - \tan^2 \theta$
$= 1$

**31.** $\dfrac{\sin \theta}{\cos \theta} + \dfrac{\cos \theta}{\sin \theta} = \dfrac{\sin^2 \theta + \cos^2 \theta}{\sin \theta \cos \theta}$

$= \dfrac{1}{\sin \theta \cos \theta}$

$= \dfrac{1}{\sin \theta} \cdot \dfrac{1}{\cos \theta}$

$= \csc \theta \sec \theta$

**33.** (a) $\cos 60° = \dfrac{1}{2}$

(b) $\tan \dfrac{\pi}{6} = \dfrac{1}{\sqrt{3}} = \dfrac{\sqrt{3}}{3}$

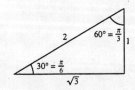

**35.** (a) $\cot 45° = 1$

(b) $\cos 45° = \dfrac{1}{\sqrt{2}} = \dfrac{\sqrt{2}}{2}$

**37.** (a) $\sin 10° \approx 0.1736$

(b) $\cos 80° \approx 0.1736$

Note: $\cos 80° = \sin(90° - 80°) = \sin 10°$

**39.** (a) $\sin 16.35° \approx 0.2815$

    (b) $\csc 16.35° = \dfrac{1}{\sin 16.35°} \approx 3.5523$

**41.** (a) $\sec 42° \, 12' = \sec 42.2° = \dfrac{1}{\cos 42.2°} \approx 1.3499$

    (b) $\csc 48° \, 7' = \dfrac{1}{\sin \left(48 + \frac{7}{60}\right)°} \approx 1.3432$

**43.** Make sure that your calculator is in radian mode.

    (a) $\cot \dfrac{\pi}{16} = \dfrac{1}{\tan (\pi/16)} \approx 5.0273$

    (b) $\tan \dfrac{\pi}{16} \approx 0.1989$

**45.** Make sure that your calculator is in radian mode.

    (a) $\csc 1 = \dfrac{1}{\sin 1} \approx 1.1884$

    (b) $\tan \dfrac{1}{2} \approx 0.5463$

**47.** (a) $\sin \theta = \dfrac{1}{2} \implies \theta = 30° = \dfrac{\pi}{6}$

    (b) $\csc \theta = 2 \implies \theta = 30° = \dfrac{\pi}{6}$

**49.** (a) $\sec \theta = 2 \implies \theta = 60° = \dfrac{\pi}{3}$

    (b) $\cot \theta = 1 \implies \theta = 45° = \dfrac{\pi}{4}$

**51.** (a) $\csc \theta = \dfrac{2\sqrt{3}}{3} \implies \theta = 60° = \dfrac{\pi}{3}$

    (b) $\sin \theta = \dfrac{\sqrt{2}}{2} \implies \theta = 45° = \dfrac{\pi}{4}$

**53.** (a) $\sin \theta = 0.8191 \implies \theta \approx 55° \approx 0.960$ radian

    (b) $\cos \theta = 0.0175 \implies \theta \approx 89° \approx 1.553$ radians

**55.** (a) $\tan \theta = 1.1920 \implies \theta \approx 50° \approx 0.873$ radian

    (b) $\tan \theta = 0.4663 \implies \theta \approx 25° \approx 0.436$ radian

**57.**   $\tan 30° = \dfrac{y}{75}$

$\dfrac{\sqrt{3}}{3} = \dfrac{y}{75}$

$75\left(\dfrac{\sqrt{3}}{3}\right) = y$

$25\sqrt{3} = y$

**59.** $\cot 60° = \dfrac{x}{32}$

$\dfrac{\sqrt{3}}{3} = \dfrac{x}{32}$

$\dfrac{32\sqrt{3}}{3} = x$

**61.** $\sin 40° = \dfrac{15}{r}$

$r = \dfrac{15}{\sin 40°} \approx 23.3$

**63.** $\sin 50° = \dfrac{y}{8}$

$y = 8 \sin 50° \approx 6.1$

**65.** $\dfrac{h}{23} = \dfrac{6}{8}$

$h = \dfrac{138}{8} = \dfrac{69}{4} = 17\tfrac{1}{4}$ ft

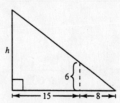

**67.** (a)

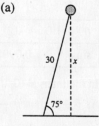

    (b) $\sin 75° = \dfrac{x}{30}$

    (c) $x = 30 \sin 75° \approx 29$ meters

**69.** Let $x =$ distance from the boat to the shoreline

$\tan 3° = \dfrac{60}{x}$

$x = \dfrac{60}{\tan 3°} \approx 1144.9$ feet

**71.**

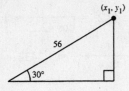

$$\sin 30° = \frac{y_1}{56}$$

$$y_1 = (\sin 30°)(56) = \left(\frac{1}{2}\right)(56) = 28$$

$$\cos 30° = \frac{x_1}{56}$$

$$x_1 = \cos 30°(56) = \frac{\sqrt{3}}{2}(56) = 28\sqrt{3}$$

$$(x_1, y_1) = (28\sqrt{3}, 28)$$

$$\sin 60° = \frac{y_2}{56}$$

$$y_2 = \sin 60°(56) = \left(\frac{\sqrt{3}}{2}\right)(56) = 28\sqrt{3}$$

$$\cos 60° = \frac{x_2}{56}$$

$$x_2 = (\cos 60°)(56) = \left(\frac{1}{2}\right)(56) = 28$$

$$(x_2, y_2) = (28, 28\sqrt{3})$$

**73.** $x \approx 9.397$, $y \approx 3.420$

$$\sin\theta = \frac{y}{10} \approx 0.34$$

$$\cos\theta = \frac{x}{10} \approx 0.94$$

$$\tan\theta = \frac{y}{x} \approx 0.36$$

$$\cot\theta = \frac{x}{y} \approx 2.75$$

$$\sec\theta = \frac{10}{x} \approx 1.06$$

$$\csc\theta = \frac{10}{y} \approx 2.92$$

**75.** (a)

| $\theta$ | 0 | 0.1 | 0.2 | 0.3 | 0.4 | 0.5 |
|---|---|---|---|---|---|---|
| $\sin\theta$ | 0 | 0.0998 | 0.1987 | 0.2955 | 0.3894 | 0.4794 |

(b) In the interval $(0, 0.5]$, $\theta > \sin\theta$

(c) As $\theta \to 0$, $\sin\theta \to 0$

**77.** True, $\csc x = \dfrac{1}{\sin x} \implies \sin 60° \csc 60° = \sin 60°\left(\dfrac{1}{\sin 60°}\right) = 1$

**79.** False, $\dfrac{\sqrt{2}}{2} + \dfrac{\sqrt{2}}{2} = \sqrt{2} \neq 1$

**81.** False, $\dfrac{\sin 60°}{\sin 30°} = \dfrac{\cos 30°}{\sin 30°} = \cot 30° \approx 1.7321$; $\sin 2° \approx 0.0349$

**83.** $\dfrac{x^2 - 6x}{x^2 + 4x - 12} \cdot \dfrac{x^2 + 12x + 36}{x^2 - 36} = \dfrac{x\cancel{(x-6)}}{\cancel{(x+6)}(x-2)} \cdot \dfrac{\cancel{(x+6)}(x+6)}{\cancel{(x+6)}\cancel{(x-6)}}$

$$= \dfrac{x}{x - 2}$$

**85.** $\dfrac{3}{x + 2} - \dfrac{2}{x - 2} + \dfrac{x}{x^2 + 4x + 4} = \dfrac{3(x + 2)(x - 2) - 2(x + 2)^2 + x(x - 2)}{(x - 2)(x + 2)^2}$

$$= \dfrac{3(x^2 - 4) - 2(x^2 + 4x + 4) + x^2 - 2x}{(x - 2)(x + 2)^2}$$

$$= \dfrac{2x^2 - 10x - 20}{(x - 2)(x + 2)^2} = \dfrac{2(x^2 - 5x - 10)}{(x - 2)(x + 2)^2}$$

# Section 1.4    Trigonometric Functions of Any Angle

---

■ Know the Definitions of Trigonometric Functions of Any Angle.

   If $\theta$ is in standard position, $(x, y)$ a point on the terminal side and $r = \sqrt{x^2 + y^2} \neq 0$, then

$$\sin \theta = \frac{y}{r} \qquad\qquad \csc \theta = \frac{r}{y},\ y \neq 0$$

$$\cos \theta = \frac{x}{r} \qquad\qquad \sec \theta = \frac{r}{x},\ x \neq 0$$

$$\tan \theta = \frac{y}{x},\ x \neq 0 \qquad \cot \theta = \frac{x}{y},\ y \neq 0$$

■ You should know the signs of the trigonometric functions in each quadrant.

■ You should know the trigonometric function values of the quadrant angles $0$, $\dfrac{\pi}{2}$, $\pi$, and $\dfrac{3\pi}{2}$.

■ You should be able to find reference angles.

■ You should be able to evaluate trigonometric functions of any angle. (Use reference angles.)

■ You should know that the period of sine and cosine is $2\pi$.

---

**Solutions to Odd-Numbered Exercises**

**1.** (a) $(x, y) = (4, 3)$

   $r = \sqrt{16 + 9} = 5$

   $\sin \theta = \dfrac{y}{r} = \dfrac{3}{5} \qquad \csc \theta = \dfrac{r}{y} = \dfrac{5}{3}$

   $\cos \theta = \dfrac{x}{r} = \dfrac{4}{5} \qquad \sec \theta = \dfrac{r}{x} = \dfrac{5}{4}$

   $\tan \theta = \dfrac{y}{x} = \dfrac{3}{4} \qquad \cot \theta = \dfrac{x}{y} = \dfrac{4}{3}$

   (b) $(x, y) = (-8, -15)$

   $r = \sqrt{64 + 225} = 17$

   $\sin \theta = \dfrac{y}{r} = -\dfrac{15}{17} \qquad \csc \theta = \dfrac{r}{y} = -\dfrac{17}{15}$

   $\cos \theta = \dfrac{x}{r} = -\dfrac{8}{17} \qquad \sec \theta = \dfrac{r}{x} = -\dfrac{17}{8}$

   $\tan \theta = \dfrac{y}{x} = \dfrac{15}{8} \qquad \cot \theta = \dfrac{x}{y} = \dfrac{8}{15}$

**3.** (a) $(x, y) = \left(-\sqrt{3}, -1\right)$

$r = \sqrt{3 + 1} = 2$

$\sin \theta = \dfrac{y}{r} = -\dfrac{1}{2}$ $\qquad$ $\csc \theta = \dfrac{r}{y} = -2$

$\cos \theta = \dfrac{x}{r} = -\dfrac{\sqrt{3}}{2}$ $\qquad$ $\sec \theta = \dfrac{r}{x} = -\dfrac{2\sqrt{3}}{3}$

$\tan \theta = \dfrac{y}{x} = \dfrac{\sqrt{3}}{3}$ $\qquad$ $\cot \theta = \dfrac{x}{y} = \sqrt{3}$

(b) $(x, y) = (-2, 2)$

$r = \sqrt{4 + 4} = 2\sqrt{2}$

$\sin \theta = \dfrac{y}{r} = \dfrac{\sqrt{2}}{2}$ $\qquad$ $\csc \theta = \dfrac{r}{y} = \sqrt{2}$

$\cos \theta = \dfrac{x}{r} = -\dfrac{\sqrt{2}}{2}$ $\qquad$ $\sec \theta = \dfrac{r}{x} = -\sqrt{2}$

$\tan \theta = \dfrac{y}{x} = -1$ $\qquad$ $\cot \theta = \dfrac{x}{y} = -1$

**5.** (a) $(x, y) = (7, 24)$

$r = \sqrt{49 + 576} = 25$

$\sin \theta = \dfrac{y}{r} = \dfrac{24}{25}$ $\qquad$ $\csc \theta = \dfrac{r}{y} = \dfrac{25}{24}$

$\cos \theta = \dfrac{x}{r} = \dfrac{7}{25}$ $\qquad$ $\sec \theta = \dfrac{r}{x} = \dfrac{25}{7}$

$\tan \theta = \dfrac{y}{x} = \dfrac{24}{7}$ $\qquad$ $\cot \theta = \dfrac{x}{y} = \dfrac{7}{24}$

(b) $(x, y) = (7, -24)$

$r = \sqrt{49 + 576} = 25$

$\sin \theta = \dfrac{y}{r} = -\dfrac{24}{25}$ $\qquad$ $\csc \theta = \dfrac{r}{y} = -\dfrac{25}{24}$

$\cos \theta = \dfrac{x}{r} = \dfrac{7}{25}$ $\qquad$ $\sec \theta = \dfrac{r}{x} = \dfrac{25}{7}$

$\tan \theta = \dfrac{y}{x} = -\dfrac{24}{7}$ $\qquad$ $\cot \theta = \dfrac{x}{y} = -\dfrac{7}{24}$

**7.** (a) $(x, y) = (-4, 10)$

$r = \sqrt{16 + 100} = 2\sqrt{29}$

$\sin \theta = \dfrac{y}{r} = \dfrac{5\sqrt{29}}{29}$ $\qquad$ $\csc \theta = \dfrac{r}{y} = \dfrac{\sqrt{29}}{5}$

$\cos \theta = \dfrac{x}{r} = -\dfrac{2\sqrt{29}}{29}$ $\qquad$ $\sec \theta = \dfrac{r}{x} = -\dfrac{\sqrt{29}}{2}$

$\tan \theta = \dfrac{y}{x} = -\dfrac{5}{2}$ $\qquad$ $\cot \theta = \dfrac{x}{y} = -\dfrac{2}{5}$

(b) $(x, y) = (3, -5)$

$r = \sqrt{9 + 25} = \sqrt{34}$

$\sin \theta = \dfrac{y}{r} = -\dfrac{5\sqrt{34}}{34}$ $\qquad$ $\csc \theta = \dfrac{r}{y} = -\dfrac{\sqrt{34}}{5}$

$\cos \theta = \dfrac{x}{r} = \dfrac{3\sqrt{34}}{34}$ $\qquad$ $\sec \theta = \dfrac{r}{x} = \dfrac{\sqrt{34}}{3}$

$\tan \theta = \dfrac{y}{x} = -\dfrac{5}{3}$ $\qquad$ $\cot \theta = \dfrac{x}{y} = -\dfrac{3}{5}$

**9.** (a) $\sin \theta < 0 \implies \theta$ lies in Quadrant III or in Quadrant IV.

$\cos \theta < 0 \implies \theta$ lies in Quadrant II or in Quadrant III.

$\sin \theta < 0$ *and* $\cos \theta < 0 \implies \theta$ lies in Quadrant III.

(b) $\sin \theta > 0 \implies \theta$ lies in Quadrant I or in Quadrant II.

$\cos \theta < \theta \implies \theta$ lies in Quadrant II or in Quadrant III.

$\sin \theta > 0$ *and* $\cos \theta < 0 \implies \theta$ lies in Quadrant II.

**11.** (a) $\sin \theta > 0 \implies \theta$ lies in Quadrant I or in Quadrant II.

$\tan \theta < 0 \implies \theta$ lies in Quadrant II or in Quadrant IV.

$\sin \theta > 0$ *and* $\tan \theta < 0 \implies \theta$ lies in Quadrant II.

(b) $\cos \theta > 0 \implies \theta$ lies in Quadrant I or in Quadrant IV.

$\tan \theta < 0 \implies \theta$ lies in Quadrant II or in Quadrant IV.

$\cos \theta > 0$ *and* $\tan \theta < 0 \implies \theta$ lies in Quadrant IV.

**13.** $\sin \theta = \dfrac{y}{r} = \dfrac{3}{5} \implies x^2 = 25 - 9 = 16$

$\theta$ in Quadrant II $\implies x = -4$

$\sin \theta = \dfrac{y}{r} = \dfrac{3}{5}$ $\qquad$ $\csc \theta = \dfrac{r}{y} = \dfrac{5}{3}$

$\cos \theta = \dfrac{x}{r} = -\dfrac{4}{5}$ $\qquad$ $\sec \theta = \dfrac{r}{x} = -\dfrac{5}{4}$

$\tan \theta = \dfrac{y}{x} = -\dfrac{3}{4}$ $\qquad$ $\cot \theta = \dfrac{x}{y} = -\dfrac{4}{3}$

**15.** $\sin \theta < 0 \implies y < 0$

$\tan \theta = \dfrac{y}{x} = \dfrac{-15}{8} \implies r = 17$

$\sin \theta = \dfrac{y}{r} = -\dfrac{15}{17}$ $\qquad$ $\csc \theta = \dfrac{r}{y} = -\dfrac{17}{15}$

$\cos \theta = \dfrac{x}{r} = \dfrac{8}{17}$ $\qquad$ $\sec \theta = \dfrac{r}{x} = \dfrac{17}{8}$

$\tan \theta = \dfrac{y}{x} = -\dfrac{15}{8}$ $\qquad$ $\cot \theta = \dfrac{x}{y} = -\dfrac{8}{15}$

**17.** $\cot \theta = \dfrac{x}{y} = -\dfrac{3}{1} = \dfrac{3}{-1}$

$\cos \theta > 0 \implies x$ is positive; $x = 3, y = -1, r = \sqrt{10}$

$\sin \theta = \dfrac{y}{r} = -\dfrac{\sqrt{10}}{10}$ $\qquad$ $\csc \theta = \dfrac{r}{y} = -\sqrt{10}$

$\cos \theta = \dfrac{x}{r} = \dfrac{3\sqrt{10}}{10}$ $\qquad$ $\sec \theta = \dfrac{r}{x} = \dfrac{\sqrt{10}}{3}$

$\tan \theta = \dfrac{y}{x} = -\dfrac{1}{3}$ $\qquad$ $\cot \theta = \dfrac{x}{y} = -3$

**19.** $\sec \theta = \dfrac{r}{x} = \dfrac{2}{-1} \implies y^2 = 4 - 1 = 3$

$\sin \theta > 0 \implies y = \sqrt{3}$

$\sin \theta = \dfrac{y}{r} = \dfrac{\sqrt{3}}{2}$ $\qquad$ $\csc \theta = \dfrac{r}{y} = \dfrac{2\sqrt{3}}{3}$

$\cos \theta = \dfrac{x}{r} = -\dfrac{1}{2}$ $\qquad$ $\sec \theta = \dfrac{r}{x} = -2$

$\tan \theta = \dfrac{y}{x} = -\sqrt{3}$ $\qquad$ $\cot \theta = \dfrac{x}{y} = -\dfrac{\sqrt{3}}{3}$

**21.** $\sin \theta = 0 \implies \theta = n\pi$

$\sec \theta = -1 \implies \theta = \pi$

$\sin \theta = \dfrac{y}{r} = \dfrac{0}{r} = 0$ $\qquad$ $\csc \theta = \dfrac{r}{y}$ is undefined.

$\cos \theta = \dfrac{x}{r} = \dfrac{-r}{r} = -1$ $\qquad$ $\sec \theta = \dfrac{r}{x} = -1$

$\tan \theta = \dfrac{y}{x} = \dfrac{0}{x} = 0$ $\qquad$ $\cot \theta = \dfrac{x}{y}$ is undefined.

**23.** To find a point on the terminal side of $\theta$, use any point on the line $y = -x$ that lies in Quadrant II. $(-1, 1)$ is one such point.

$x = -1, y = 1, r = \sqrt{2}$

$\sin \theta = \dfrac{1}{\sqrt{2}} = \dfrac{\sqrt{2}}{2}$ $\qquad$ $\csc \theta = \sqrt{2}$

$\cos \theta = -\dfrac{1}{\sqrt{2}} = -\dfrac{\sqrt{2}}{2}$ $\qquad$ $\sec \theta = -\sqrt{2}$

$\tan \theta = -1$ $\qquad$ $\cot \theta = -1$

**25.** To find a point on the terminal side of $\theta$, use any point on the line $y = 2x$ that lies in Quadrant III. $(-1, -2)$ is one such point.

$x = -1, y = -2, r = \sqrt{5}$

$\sin \theta = -\dfrac{2}{\sqrt{5}} = -\dfrac{2\sqrt{5}}{5}$ $\qquad$ $\csc \theta = \dfrac{\sqrt{5}}{-2} = -\dfrac{\sqrt{5}}{2}$

$\cos \theta = -\dfrac{1}{\sqrt{5}} = -\dfrac{\sqrt{5}}{5}$ $\qquad$ $\sec \theta = \dfrac{\sqrt{5}}{-1} = -\sqrt{5}$

$\tan \theta = \dfrac{-2}{-1} = 2$ $\qquad$ $\cot \theta = \dfrac{-1}{-2} = \dfrac{1}{2}$

**27.** $(x, y) = (-1, 0)$

$\cos \pi = \dfrac{x}{r} = \dfrac{-1}{1} = -1$

**29.** $(x, y) = (-1, 0)$

$\sec \pi = \dfrac{r}{x} = \dfrac{1}{-1} = -1$

**31.** $(x, y) = (0, 1)$

$\tan \dfrac{\pi}{2} = \dfrac{y}{x} = \dfrac{1}{0}$ undefined

**33.** $(x, y) = (0, 1)$

$\cot \dfrac{\pi}{2} = \dfrac{x}{y} = \dfrac{0}{1} = 0$

**35.** (a)   $\theta = 203°$

$\theta' = 203° - 180° = 23°$

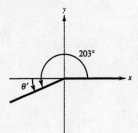

(b)   $\theta = 127°$

$\theta' = 180° - 127° = 53°$

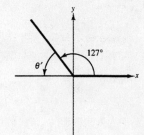

**37.** (a)                 $\theta = -245°$

$360° - 245° = 115°$  (coterminal angle)

$\theta' = 180° - 115° = 65°$

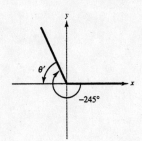

(b)   $\theta = -72°$

$\theta' = 72°$

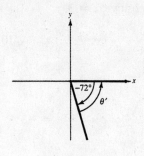

**39.** (a)   $\theta = \dfrac{2\pi}{3}$

$\theta' = \pi - \dfrac{2\pi}{3} = \dfrac{\pi}{3}$

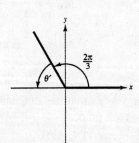

(b)   $\theta = \dfrac{7\pi}{6}$

$\theta' = \dfrac{7\pi}{6} - \pi = \dfrac{\pi}{6}$

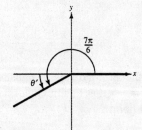

**41.** (a)   $\theta = 3.5$

$\theta' = 3.5 - \pi$

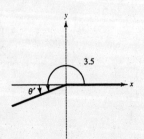

(b)   $\theta = 5.8$

$\theta' = 2\pi - 5.8$

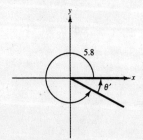

**43.** (a) $\theta' = 45°$, Quadrant III

$$\sin 225° = -\sin 45° = -\frac{\sqrt{2}}{2}$$

$$\cos 225° = -\cos 45° = -\frac{\sqrt{2}}{2}$$

$$\tan 225° = \tan 45° = 1$$

(b) $\theta' = 45°$, Quadrant II

$$\sin(-225°) = \sin 45° = \frac{\sqrt{2}}{2}$$

$$\cos(-225°) = -\cos 45° = -\frac{\sqrt{2}}{2}$$

$$\tan(-225°) = -\tan 45° = -1$$

**45.** (a) $\theta' = 30°$, Quadrant I

$$\sin 750° = \sin 30° = \frac{1}{2}$$

$$\cos 750° = \cos 30° = \frac{\sqrt{3}}{2}$$

$$\tan 750° = \tan 30° = \frac{\sqrt{3}}{3}$$

(b) $\theta' = 30°$, Quadrant II

$$\sin 510° = \sin 30° = \frac{1}{2}$$

$$\cos 510° = -\cos 30° = -\frac{\sqrt{3}}{2}$$

$$\tan 510° = -\tan 30° = -\frac{\sqrt{3}}{3}$$

**47.** (a) $\theta' = \frac{\pi}{3}$, Quadrant III

$$\sin \frac{4\pi}{3} = -\sin \frac{\pi}{3} = -\frac{\sqrt{3}}{2}$$

$$\cos \frac{4\pi}{3} = -\cos \frac{\pi}{3} = -\frac{1}{2}$$

$$\tan \frac{4\pi}{3} = \tan \frac{\pi}{3} = \sqrt{3}$$

(b) $\theta' = \frac{\pi}{3}$, Quadrant II

$$\sin \frac{2\pi}{3} = \sin \frac{\pi}{3} = \frac{\sqrt{3}}{2}$$

$$\cos \frac{2\pi}{3} = -\cos \frac{\pi}{3} = -\frac{1}{2}$$

$$\tan \frac{2\pi}{3} = -\tan \frac{\pi}{3} = -\sqrt{3}$$

**49.** (a) $\theta' = \frac{\pi}{6}$, Quadrant IV

$$\sin\left(-\frac{\pi}{6}\right) = -\sin \frac{\pi}{6} = -\frac{1}{2}$$

$$\cos\left(-\frac{\pi}{6}\right) = \cos \frac{\pi}{6} = \frac{\sqrt{3}}{2}$$

$$\tan\left(-\frac{\pi}{6}\right) = -\tan \frac{\pi}{6} = -\frac{\sqrt{3}}{3}$$

(b) $\theta' = \frac{\pi}{6}$, Quadrant II

$$\sin \frac{5\pi}{6} = \sin \frac{\pi}{6} = \frac{1}{2}$$

$$\cos \frac{5\pi}{6} = -\cos \frac{\pi}{6} = -\frac{\sqrt{3}}{2}$$

$$\tan \frac{5\pi}{6} = -\tan \frac{\pi}{6} = -\frac{\sqrt{3}}{3}$$

**51.** (a) $\theta' = \frac{\pi}{4}$, Quadrant II

$$\sin \frac{11\pi}{4} = \sin \frac{\pi}{4} = \frac{\sqrt{2}}{2}$$

$$\cos \frac{11\pi}{4} = -\cos \frac{\pi}{4} = -\frac{\sqrt{2}}{2}$$

$$\tan \frac{11\pi}{4} = -\tan \frac{\pi}{4} = -1$$

(b) $\theta' = \frac{\pi}{6}$, Quadrant IV

$$\sin\left(-\frac{13\pi}{6}\right) = -\sin \frac{\pi}{6} = -\frac{1}{2}$$

$$\cos\left(-\frac{13\pi}{6}\right) = \cos \frac{\pi}{6} = \frac{\sqrt{3}}{2}$$

$$\tan\left(-\frac{13\pi}{6}\right) = -\tan \frac{\pi}{6} = -\frac{\sqrt{3}}{3}$$

**53.** (a) $\sin 10° \approx 0.1736$

(b) $\csc 10° = \dfrac{1}{\sin 10°} \approx 5.7588$

**55.** (a) $\cos(-110°) \approx -0.3420$

(b) $\cos 250° \approx -0.3420$

**57.** (a) $\tan 240° \approx 1.7321$    (b) $\cot 210° = \dfrac{1}{\tan 210°} \approx 1.7321$

**59.** (a) $\tan \dfrac{\pi}{9} \approx 0.3640$    (b) $\tan \dfrac{10\pi}{9} \approx 0.3640$

**61.** (a) $\sin 0.65 \approx 0.6052$    (b) $\sin(-5.63) \approx 0.6077$

**63.** (a) $\sin \theta = \dfrac{1}{2} \implies$ reference angle is $30°$ or $\dfrac{\pi}{6}$ and $\theta$ is in Quadrant I or Quadrant II.

 Values in degrees:  $30°, 150°$

 Values in radian:  $\dfrac{\pi}{6}, \dfrac{5\pi}{6}$

 (b) $\sin \theta = -\dfrac{1}{2} \implies$ reference angle is $30°$ or $\dfrac{\pi}{6}$ and $\theta$ is in Quadrant III or Quadrant IV.

 Values in degrees:  $210°, 330°$

 Values in radians:  $\dfrac{7\pi}{6}, \dfrac{11\pi}{6}$

**65.** (a) $\csc \theta = \dfrac{2\sqrt{3}}{3} \implies$ reference angle is $60°$ or $\dfrac{\pi}{3}$ and $\theta$ is in Quadrant I or Quadrant II.

 Values in degrees:  $60°, 120°$

 Values in radians:  $\dfrac{\pi}{3}, \dfrac{2\pi}{3}$

 (b) $\cot \theta = -1 \implies$ reference angle is $45°$ or $\dfrac{\pi}{4}$ and $\theta$ is in Quadrant II or Quadrant IV.

 Values in degrees:  $135°, 315°$

 Values in radians:  $\dfrac{3\pi}{4}, \dfrac{7\pi}{4}$

**67.** (a) $\tan \theta = 1 \implies$ reference angle is $45°$ or $\dfrac{\pi}{4}$ and $\theta$ is in Quadrant I or Quadrant III.

 Values in degrees:  $45°, 225°$

 Values in radians:  $\dfrac{\pi}{4}, \dfrac{5\pi}{4}$

 (b) $\cot \theta = -\sqrt{3} \implies$ reference angle is $30°$ or $\dfrac{\pi}{6}$ and $\theta$ is in Quadrant II or Quadrant IV.

 Values in degrees:  $150°, 330°$

 Values in radians:  $\dfrac{5\pi}{6}, \dfrac{11\pi}{6}$

**69.** (a) $\sin \theta = 0.8191 \implies \theta' \approx 54.99°$

 Quadrant I: $\theta = \sin^{-1} 0.8191 \approx 54.99°$

 Quadrant II: $\theta = 180° - \sin^{-1} 0.8191 \approx 125.01°$

 (b) $\theta' = \sin^{-1} 0.2589 \approx 15.00°$

 Quadrant III: $\theta = 180° + 15° = 195°$

 Quadrant IV: $\theta = 360° - 15° = 345°$

**71.** (a) $\cos\theta = 0.9848 \implies \theta' \approx 0.175$

Quadrant I: $\theta = \cos^{-1}(0.9848) \approx 0.175$

Quadrant IV: $\theta = 2\pi - \theta' \approx 6.109$

(b) $\theta' = \cos^{-1} 0.5890 \approx 0.941$

Quadrant II: $\theta = \pi - 0.941 \approx 2.201$

Quadrant III: $\theta = \pi + 0.941 \approx 4.083$

**73.** (a) $\tan\theta = 1.192 \implies \theta' \approx 0.873$

Quadrant I: $\theta = \tan^{-1} 1.192 \approx 0.873$

Quadrant III: $\theta = \pi + \theta' \approx 4.014$

(b) $\theta' = \tan^{-1} 8.144 \approx 1.4486$

Quadrant II: $\theta = \pi - 1.4486 \approx 1.693$

Quadrant IV: $\theta = 2\pi - 1.4486 \approx 4.835$

**75.**    $\sin\theta = -\frac{3}{5}$

$\sin^2\theta + \cos^2\theta = 1$

$\cos^2\theta = 1 - \sin^2\theta$

$\cos^2\theta = 1 - \left(-\frac{3}{5}\right)^2$

$\cos^2\theta = 1 - \frac{9}{25}$

$\cos^2\theta = \frac{16}{25}$

$\cos\theta > 0$ in Quadrant IV.

$\cos\theta = \frac{4}{5}$

**77.**  $\tan\theta = \frac{3}{2}$

$\sec^2\theta = 1 + \tan^2\theta$

$\sec^2\theta = 1 + \left(\frac{3}{2}\right)^2$

$\sec^2\theta = 1 + \frac{9}{4}$

$\sec^2\theta = \frac{13}{4}$

$\sec\theta < 0$ in Quadrant III.

$\sec\theta = -\frac{\sqrt{13}}{2}$

**79.** $\cos\theta = \frac{5}{8}$

$\cos\theta = \frac{1}{\sec\theta} \implies \sec\theta = \frac{1}{\cos\theta}$

$\sec\theta = \frac{1}{5/8} = \frac{8}{5}$

**81.** (a) $t = 1$

$T = 45 - 23\cos\left[\frac{2\pi}{365}(1 - 32)\right] \approx 25.2°\,\text{F}$

(c) $t = 291$

$T = 45 - 23\cos\left[\frac{2\pi}{365}(291 - 32)\right] \approx 50.8°\,\text{F}$

(b) $t = 185$

$T = 45 - 23\cos\left[\frac{2\pi}{365}(185 - 32)\right] \approx 65.1°\,\text{F}$

**83.** $\sin\theta = \frac{6}{d} \implies d = \frac{6}{\sin\theta}$

(a) $d = \frac{6}{\sin 30°} = 12$ miles

(b) $d = \frac{6}{\sin 90°} = 6$ miles

(c) $d = \frac{6}{\sin 120°} \approx 6.9$ miles

**85.** $y = 2^{x-1}$

| $x$ | $-1$ | $0$ | $1$ | $2$ | $3$ |
|---|---|---|---|---|---|
| $y$ | $\frac{1}{4}$ | $\frac{1}{2}$ | $1$ | $2$ | $4$ |

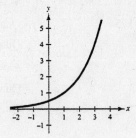

**87.** $y = \ln(x - 1)$

Domain: $x - 1 > 0 \Rightarrow x > 1$

| $x$ | $1.1$ | $1.5$ | $2$ | $3$ | $4$ |
|---|---|---|---|---|---|
| $y$ | $-2.30$ | $-0.69$ | $0$ | $0.69$ | $1.10$ |

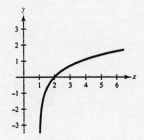

# Section 1.5   Graphs of Sine and Cosine Functions

- You should be able to graph $y = a\sin(bx - c)$ and $y = a\cos(bx - c)$.

- Amplitude: $|a|$

- Period: $\dfrac{2\pi}{|b|}$

- Shift: Solve $bx - c = 0$ and $bx - c = 2\pi$.

- Key Increments: $\dfrac{1}{4}$ (period)

**Solutions to Odd-Numbered Exercises**

**1.** $y = 3 \sin 2x$

Period: $\dfrac{2\pi}{2} = \pi$

Amplitude: $|3| = 3$

**3.** $y = \dfrac{5}{2} \cos \dfrac{x}{2}$

Period: $\dfrac{2\pi}{1/2} = 4\pi$

Amplitude: $\left|\dfrac{5}{2}\right| = \dfrac{5}{2}$

**5.** $y = \dfrac{2}{3} \sin \pi x$

Period: $\dfrac{2\pi}{\pi} = 2$

Amplitude: $\left|\dfrac{2}{3}\right| = \dfrac{2}{3}$

**7.** $y = -2 \sin x$

Period: $\dfrac{2\pi}{1} = 2\pi$

Amplitude: $|-2| = 2$

**9.** $y = 3 \sin 10x$

Period: $\dfrac{2\pi}{10} = \dfrac{\pi}{5}$

Amplitude: $|3| = 3$

**11.** $y = \dfrac{1}{2} \cos \dfrac{2\pi}{3}$

Period: $\dfrac{2\pi}{2/3} = 3\pi$

Amplitude: $\left|\dfrac{1}{2}\right| = \dfrac{1}{2}$

**13.** $y = 3 \sin 4\pi x$

Period: $\dfrac{2\pi}{4\pi} = \dfrac{1}{2}$

Amplitude: $|3| = 3$

**15.** $f(x) = \sin x$

$g(x) = \sin(x - \pi)$

The graph of $g$ is a horizontal shift to the right $\pi$ units of the graph of $f$ (a phase shift).

**17.** $f(x) = \cos 2x$

$g(x) = -\cos 2x$

The graph of $g$ is a reflection in the $x$-axis of the graph of $f$.

**19.** $f(x) = \cos x$

$g(x) = \cos 2x$

The period of $f$ is twice that of $g$.

**21.** $f(x) = \sin x$

$f(x) = 2 + \sin x$

The graph of $g$ is a vertical shift 2 units upward of the graph of $f$.

**23.** The graph of $g$ has twice the amplitude as the graph of $f$. The period is the same.

**25.** The graph of $g$ is a horizontal shift $\pi$ units to the right of the graph of $f$.

**27.** $y_1 = \dfrac{1}{2} \sin x$; $y_2 = \dfrac{3}{2} \sin x$; $y_3 = -3 \sin x$

Changing the value of $a$ changes the amplitude.

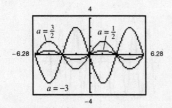

**29.** $y_1 = \sin\left(\dfrac{1}{2}x\right)$; $y_2 = \sin\left(\dfrac{3}{2}x\right)$; $y_3 = \sin(4x)$

Changing the value of $b$ changes the period.

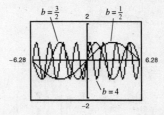

**31.** $f(x) = -2 \sin x$

Period: $2\pi$

Amplitude: 2

$g(x) = 4 \sin x$

Period: $2\pi$

Amplitude: 4

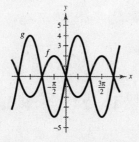

**33.** $f(x) = \cos x$

Period: $2\pi$

Amplitude: 1

$g(x) = 1 + \cos x$

is a vertical shift of the graph of $f(x)$ one unit upward.

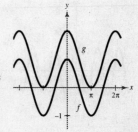

**35.** $f(x) = -\dfrac{1}{2}\sin\dfrac{x}{2}$

Period: $4\pi$

Amplitude: $\dfrac{1}{2}$

$g(x) = 3 - \dfrac{1}{2}\sin\dfrac{x}{2}$ is the graph of $f(x)$ shifted vertically three units upward.

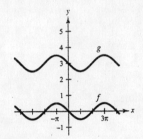

**37.** $f(x) = 2\cos x$

Period: $2\pi$

Amplitude: 2

$g(x) = 2\cos(x + \pi)$ is the graph of $f(x)$ shifted $\pi$ units to the left.

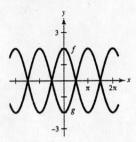

**39.** Since sine and cosine are cofunctions and $x$ and $x - (\pi/2)$ are complementary, we have

$$\sin x = \cos\left(x - \dfrac{\pi}{2}\right).$$

Period: $2\pi$

Amplitude: 1

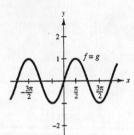

**41.** $f(x) = \cos x$

$g(x) = -\sin\left(x - \dfrac{\pi}{2}\right) = \sin\left(\dfrac{\pi}{2} - x\right) = \cos x$

Thus, $f(x) = g(x)$.

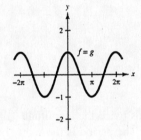

**43.** $y = -2\sin 6x$;   $a = -2$, $b = 6$, $c = 0$

Period: $\dfrac{2\pi}{6} = \dfrac{\pi}{3}$

Amplitude: $|-2| = 2$

Key points: $(0, 0)$, $\left(\dfrac{\pi}{12}, -2\right)$, $\left(\dfrac{\pi}{6}, 0\right)$, $\left(\dfrac{\pi}{4}, 2\right)$, $\left(\dfrac{\pi}{3}, 0\right)$

**45.** $y = \cos 2\pi x$

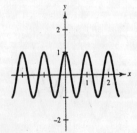

Period: $\dfrac{2\pi}{2\pi} = 1$

Amplitude: 1

Key points: $(1, 0)$, $\left(0, \dfrac{1}{4}\right)$, $\left(-1, \dfrac{1}{2}\right)$, $\left(0, \dfrac{3}{4}\right)$, $(1, 1)$

**47.** $y = -\sin\dfrac{2\pi x}{3}$; $a = -1$, $b = \dfrac{2\pi}{3}$, $c = 0$

Period: $\dfrac{2\pi}{2\pi/3} = 3$

Amplitude: 1

Key points: $(0, 0), \left(\dfrac{3}{4}, -1\right), \left(\dfrac{3}{2}, 0\right), \left(\dfrac{9}{4}, 1\right),$ (3, 0)

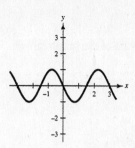

**49.** $y = \sin\left(x - \dfrac{\pi}{4}\right)$; $a = 1$, $b = 1$, $c = \dfrac{\pi}{4}$

Period: $2\pi$

Amplitude: 1

Shift: Set $x - \dfrac{\pi}{4} = 0$  and  $x - \dfrac{\pi}{4} = 2\pi$

$$x = \dfrac{\pi}{4} \qquad\qquad x = \dfrac{9\pi}{4}$$

Key points: $\left(\dfrac{\pi}{4}, 0\right), \left(\dfrac{3\pi}{4}, 1\right), \left(\dfrac{5\pi}{4}, 0\right), \left(\dfrac{7\pi}{4}, -1\right). \left(\dfrac{9\pi}{4}, 0\right)$

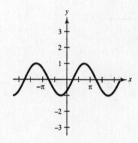

**51.** $y = 3\cos(x + \pi)$

Period: $2\pi$

Amplitude: 3

Shift: Set $x + \pi = \phantom{0}0$ and $x + \pi = 2\pi$

$$x = -\pi \qquad\qquad x = \pi$$

Key points: $(-\pi, 3), \left(-\dfrac{\pi}{2}, 0\right),$ (0, -3), $\left(\dfrac{\pi}{2}, 0\right),$ $(\pi, 3)$

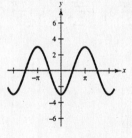

**53.** $y = \dfrac{1}{10}\cos(60\pi x)$; $a = \dfrac{1}{10}$, $b = 60\pi$, $c = 0$

Period: $\dfrac{2\pi}{60\pi} = \dfrac{1}{30}$

Amplitude: $\dfrac{1}{10}$

Key points: $\left(0, \dfrac{1}{10}\right), \left(\dfrac{1}{120}, 0\right), \left(\dfrac{1}{60}, -\dfrac{1}{10}\right), \left(\dfrac{1}{40}, 0\right), \left(\dfrac{1}{30}, \dfrac{1}{10}\right)$

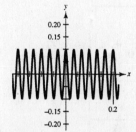

**55.** $y = 2 - \sin\dfrac{2\pi x}{3}$

Vertical shift 2 units upward of the graph in Exercise 47.

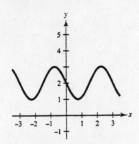

**57.** $y = 3\cos(x + \pi) - 3$

Vertical shift 3 units downward of the graph in Exercise 51.

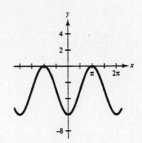

**59.** $y = \dfrac{2}{3}\cos\left(\dfrac{x}{2} - \dfrac{\pi}{4}\right);\ a = \dfrac{2}{3},\ b = \dfrac{1}{2},\ c = \dfrac{\pi}{4}$

Period: $4\pi$

Amplitude: $\dfrac{2}{3}$

$$\dfrac{x}{2} - \dfrac{\pi}{4} = 0 \quad\text{and}\quad \dfrac{x}{2} - \dfrac{\pi}{4} = 2\pi$$

$$x = \dfrac{\pi}{2} \qquad\qquad x = \dfrac{9\pi}{2}$$

Key points: $\left(\dfrac{\pi}{2}, \dfrac{2}{3}\right)$, $\left(\dfrac{3\pi}{2}, 0\right)$, $\left(\dfrac{5\pi}{2}, \dfrac{-2}{3}\right)$, $\left(\dfrac{7\pi}{2}, 0\right)$, $\left(\dfrac{9\pi}{2}, \dfrac{2}{3}\right)$

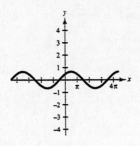

**61.** $y = -2\sin(4x + \pi)$

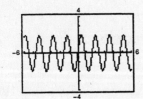

**63.** $y = \cos\left(2\pi x - \dfrac{\pi}{2}\right) + 1$

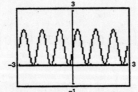

**65.** $y = -0.1\sin\left(\dfrac{\pi x}{10} + \pi\right)$

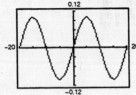

**67.** $y = 5\cos(\pi - 2x) + 2$

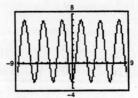

**69.** $f(x) = a\cos x + d$

Amplitude: $\frac{1}{2}[5 - (-1)] = 3 \implies a = 3$

Vertical shift 2 units upward of
$g(x) = 3\cos x \implies d = 2.$
Thus, $f(x) = 3\cos x + 2 = 2 + 3\cos x.$

**71.** $f(x) = a\cos x + d$

Amplitude: $\frac{1}{2}[8 - 0] = 4$

Since $f(x)$ is the graph of $g(x) = 4\cos x$ reflected about the $x$-axis and shifted vertically 4 units upward, we have $a = -4$ and $d = 4.$

Thus, $f(x) = -4\cos x + 4.$

**73.** $y = a \sin(bx - c)$

Amplitude: $|a| = |3|$ Since the graph is reflected about the $x$-axis, we have $a = -3$.

Period: $\dfrac{2\pi}{b} = \pi \implies b = 2$

Phase shift: $c = 0$

Thus, $y = -3 \sin 2x$.

**75.** $y = a \sin(bx + c)$

Amplitude: $a = 1$

Period: $2\pi \implies b = 1$

Phase shift:    $bx + c = 0$  when  $x = \dfrac{\pi}{4}$

$$(1)\left(\dfrac{\pi}{4}\right) + c = 0 \implies c = -\dfrac{\pi}{4}$$

Thus, $y = \sin\left(x - \dfrac{\pi}{4}\right)$.

**77.** $y_1 = \sin x$

$y_2 = -\dfrac{1}{2}$

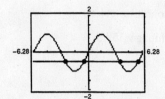

In the interval $[-2\pi, 2\pi]$, $\sin x = -\dfrac{1}{2}$ when

$x = -\dfrac{5\pi}{6},\ -\dfrac{\pi}{6},\ \dfrac{7\pi}{6},\ \dfrac{11\pi}{6}$.

**79.** $y_1 = \cos x$

$y_2 = \dfrac{\sqrt{2}}{2}$

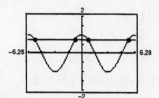

In the interval $[-2\pi, 2\pi]$, $\cos x = \dfrac{\sqrt{2}}{2}$ when

$x = \pm\dfrac{\pi}{4},\ \pm\dfrac{7\pi}{4}$.

**81.** (a) $h(x) = \cos^2 x$ is even.

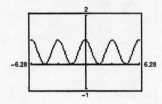

(b) $g(x) = \sin^2 x$ is even.

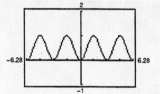

**83.** $y = 0.85 \sin \dfrac{\pi t}{3}$

(a) Time for one cycle = one period $= \dfrac{2\pi}{\pi/3} = 6$ sec

(b) Cycles per min $= \dfrac{60}{6} = 10$ cycles per min

(c) Amplitude: 0.85

Period: 6

Key points: $(0, 0),\ \left(\dfrac{3}{2}, 0.85\right),\ (3, 0),\ \left(\dfrac{9}{2}, -0.85\right),\ (6, 0)$

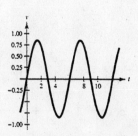

**85.** $y = 0.001 \sin 880\pi t$

(a) Period: $\dfrac{2\pi}{880\pi} = \dfrac{1}{440}$ seconds

(b) $f = \dfrac{1}{p} = 440$ cycles per second

**87.** $S = 22.3 - 3.4 \cos \dfrac{\pi t}{6},\ 1 \le t \le 12$

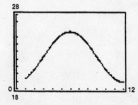

**89.** (a) $C(t) = 56.35 + 27.35 \sin\left(\dfrac{\pi t}{6} + 4.19\right)$

(b)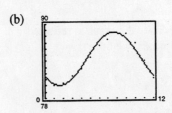

The model is a good fit for most months.

(c)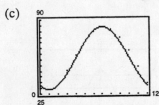

The model is a good fit.

(d) Use the constant term of each model to estimate the average annual temperature.

Honolulu: 84.40°

Chicago: 58.50°

(e) Each model has a period of 12. This corresponds to the 12 months in a year.

(f) Chicago has a greater variability in temperatures during the year. The amplitude of each model indicates this variability.

**91.** (a) $\sin x \approx x - \dfrac{x^3}{3!} + \dfrac{x^5}{5!}$

Near $x = 0$ the graphs are approximately the same. They appear to coincide from $-\pi/2$ to $\pi/2$.

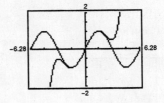

(b) $\cos x \approx 1 - \dfrac{x^2}{2!} + \dfrac{x^4}{4!}$

Near $x = 0$ the graphs are approximately the same. They appear to coincide from $-\pi/2$ to $\pi/2$.

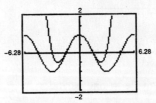

(c) $\sin x \approx x - \dfrac{x^3}{3!} + \dfrac{x^5}{5!} - \dfrac{x^7}{7!}$

$\cos x \approx 1 - \dfrac{x^2}{2!} + \dfrac{x^4}{4!} - \dfrac{x^6}{6!}$

The accuracy is increased.

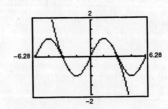

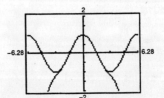

**93.** (a)

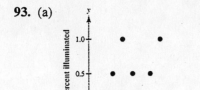

(b) $y = \dfrac{1}{2} + \dfrac{1}{2} \sin\left[\dfrac{\pi}{15}(t - 303)\right]$

(c)

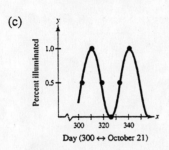

(d) $y(356) \approx 0.003 \approx 0$

The model is a good fit.

**95.** $\log_2 [x^2(x - 3)] = \log_2 x^2 + \log_2(x - 3)$

$$= 2 \log_2 x + \log_2(x - 3)$$

**97.** $\ln\sqrt{\dfrac{z}{z^2 + 1}} = \dfrac{1}{2} \ln\left(\dfrac{z}{z^2 + 1}\right) = \dfrac{1}{2}[\ln z - \ln(z^2 + 1)]$

$$= \dfrac{1}{2} \ln z - \dfrac{1}{2} \ln(z^2 + 1)$$

# Section 1.6    Graphs of Other Trigonometric Functions

■ You should be able to graph

$y = a \tan (bx - c)$            $y = a \cot (bx - c)$

$y = a \sec (bx - c)$            $y = a \csc (bx - c)$

■ When graphing $y = a \sec (bx - c)$ or $y = a \csc (bx - c)$ you should first graph $y = a \cos (bx - c)$ or $y = a \sin (bx - c)$ because

(a) The $x$-intercepts of sine and cosine are the vertical asymptotes of cosecant and secant.

(b) The maximums of sine and cosine are the local minimums of cosecant and secant.

(c) The minimums of sine and cosine are the local maximums of cosecant and secant.

■ You should be able to graph using a damping factor.

**Solutions to Odd-Numbered Exercises**

**1.** $y = \sec \dfrac{x}{2}$

Period: $\dfrac{2\pi}{1/2} = 4\pi$

Matches graph (g).

**3.** $y = \tan 2x$

Period: $\dfrac{\pi}{2}$

Matches graph (f).

**5.** $y = \cot \dfrac{\pi x}{2}$

Period: $\dfrac{\pi}{\pi/2} = 2$

Matches graph (b).

**7.** $y = -\csc x$

Period: $2\pi$

Matches graph (e).

**9.** $y = \dfrac{1}{3} \tan x$

Period: $\pi$

Two consecutive asymptotes:

$$x = -\dfrac{\pi}{2} \text{ and } x = \dfrac{\pi}{2}$$

| $x$ | $-\dfrac{\pi}{4}$ | $0$ | $\dfrac{\pi}{4}$ |
|---|---|---|---|
| $y$ | $-\dfrac{1}{3}$ | $0$ | $\dfrac{1}{3}$ |

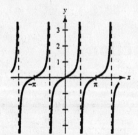

**11.** $y = \tan 2x$

Period: $\dfrac{\pi}{2}$

Two consecutive asymptotes: $2x = -\dfrac{\pi}{2} \Rightarrow x = -\dfrac{\pi}{4}$

$$2x = \dfrac{\pi}{2} \Rightarrow x = \dfrac{\pi}{4}$$

| $x$ | $-\dfrac{\pi}{8}$ | $0$ | $\dfrac{\pi}{8}$ |
|---|---|---|---|
| $y$ | $-1$ | $0$ | $1$ |

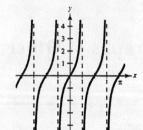

**13.** $y = -\dfrac{1}{2} \sec x$

Graph $y = -\dfrac{1}{2} \cos x$ first.

Period: $2\pi$

One cycle: $0$ to $2\pi$

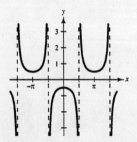

**15.** $y = \sec \pi x$

Graph $y = \cos \pi x$ first.

Period: $\dfrac{2\pi}{\pi} = 2$

One cycle: $0$ to $2$

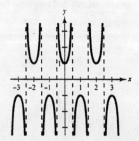

**17.** $y = \sec \pi x - 1$

Reflect the graph in Exercise 15 about the $x$-axis and then shift it vertically down one unit.

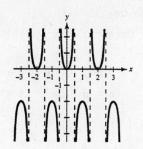

**19.** $y = \csc \dfrac{x}{2}$

Graph $y = \sin \dfrac{x}{2}$ first.

Period: $\dfrac{2\pi}{1/2} = 4\pi$

One cycle:  0 to $4\pi$

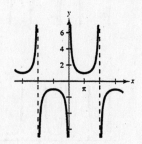

**21.** $y = \cot \dfrac{x}{2}$

Period: $\dfrac{\pi}{1/2} = 2\pi$

Two consecutive asymptotes: $\dfrac{x}{2} = 0 \implies x = 0$

$\dfrac{x}{2} = \pi \implies x = 2\pi$

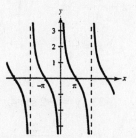

| $x$ | $\dfrac{\pi}{2}$ | $\pi$ | $\dfrac{3\pi}{2}$ |
|---|---|---|---|
| $y$ | 1 | 0 | $-1$ |

**23.** $y = \dfrac{1}{2} \sec 2x$

Graph $y = \dfrac{1}{2} \cos 2x$ first.

Period: $\dfrac{2\pi}{2} = \pi$

One cycle:  0 to $\pi$

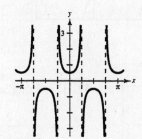

**25.** $y = \tan \dfrac{\pi x}{4}$

Period: $\dfrac{\pi}{\pi/4} = 4$

Two consecutive asymptotes: $\dfrac{\pi x}{4} = -\dfrac{\pi}{2} \implies x = -2$

$\dfrac{\pi x}{4} = \dfrac{\pi}{2} \implies x = 2$

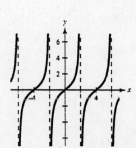

| $x$ | $-1$ | 0 | 1 |
|---|---|---|---|
| $y$ | $-1$ | 0 | 1 |

**27.** $y = \csc (\pi - x)$

Graph $y = \sin(\pi - x)$ first.

Period: $2\pi$

Shift: Set $\pi - x = 0$  and  $\pi - x = 2\pi$

$\qquad x = \pi \qquad\qquad x = -\pi$

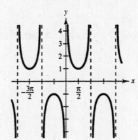

**29.** $y = \dfrac{1}{4} \csc\left(x + \dfrac{\pi}{4}\right)$

Graph $y = \dfrac{1}{4}\sin\left(x + \dfrac{\pi}{4}\right)$ first.

Period: $2\pi$

Shift: Set $x + \dfrac{\pi}{4} = 0$  and  $x + \dfrac{\pi}{4} = 2\pi$

$\qquad x = -\dfrac{\pi}{4}$  to  $\qquad x = \dfrac{7\pi}{4}$

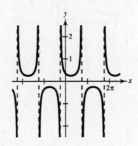

**31.** $y = \tan \dfrac{x}{3}$

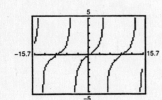

**33.** $y = -2 \sec 4x$

$= \dfrac{-2}{\cos 4x}$

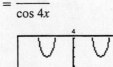
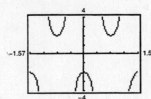

**35.** $y = \tan\left(x - \dfrac{\pi}{4}\right)$

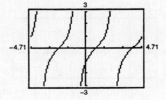

**37.** $y = \dfrac{1}{4} \cot\left(x - \dfrac{\pi}{2}\right)$

$= \dfrac{1}{4 \tan(x - \pi/2)}$

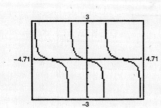

**39.** $y = 2 \sec(2x - \pi)$

$y = \dfrac{2}{\cos(2x - \pi)}$

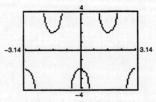

**41.** $\tan x = 1$

$$x = -\frac{7\pi}{4}, \ -\frac{3\pi}{4}, \ \frac{\pi}{4}, \ \frac{5\pi}{4}$$

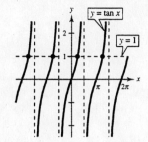

**43.** $\sec x = -2$

$$x = \pm\frac{2\pi}{3}, \ \pm\frac{4\pi}{3}$$

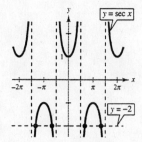

**45.** Thus graph of $f(x) = \sec x$ has $y$-axis symmetry. Thus, the function is even.

**47.** As $x \rightarrow \dfrac{\pi}{2}$ from the left, $f(x) = \tan x \rightarrow \infty$.

As $x \rightarrow \dfrac{\pi}{2}$ from the right, $f(x) = \tan x \rightarrow -\infty$.

**49.** $f(x) = 2\sin x$

$g(x) = \dfrac{1}{2}\csc x$

(a)

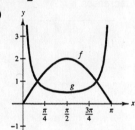

(b) $f > g$ on the interval, $\dfrac{\pi}{6} < x < \dfrac{5\pi}{6}$

(c) As $x \rightarrow \pi, f(x) = 2\sin x \rightarrow 0$ and $g(x) = \dfrac{1}{2}\csc x \rightarrow \pm\infty$ since $g(x)$ is the reciprocal of $f(x)$.

**51.** $y_1 = \sin x \csc x$ and $y_2 = 1$

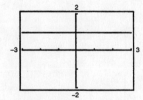

$$\sin x \csc x = \sin x \left(\frac{1}{\sin x}\right) = 1, \ \sin x \ne 0$$

**53.** $y_1 = \dfrac{\cos x}{\sin x}$ and $y_2 = \cot x = \dfrac{1}{\tan x}$

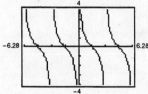

$$\cot x = \frac{\cos x}{\sin x}$$

**55.** $f(x) = x\cos x$

As $x \rightarrow 0, f(x) \rightarrow 0$.

Matches graph (d).

**57.** $g(x) = |x|\sin x$

As $x \rightarrow 0, g(x) \rightarrow 0$.

Matches graph (b).

**59.** $f(x) = \sin x + \cos\left(x + \dfrac{\pi}{2}\right)$, $g(x) = 0$

$f(x) = g(x)$   The graph is the line $y = 0$.

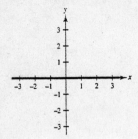

**61.** $f(x) = \sin^2 x$, $g(x) = \dfrac{1}{2}(1 - \cos 2x)$

$f(x) = g(x)$

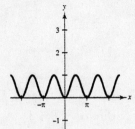

**63.** $f(x) = 2^{-x/4} \cos \pi x$

$-2^{-x/4} \le f(x) \le 2^{-2x/4}$

The damping factor is $y = 2^{-x/4}$.

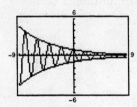

**65.** $g(x) = e^{-x^2/2} \sin x$

$-e^{-x^2/2} \le g(x) \le e^{-x^2/2}$

The damping factor is

$y = e^{-x^2/2}$.

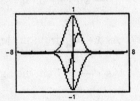

**67.** $\tan x = \dfrac{5}{d}$

$d = \dfrac{5}{\tan x} = 5 \cot x$

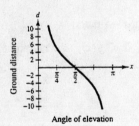

**69.** As the predator population increases, the number of prey decrease. When the number of prey is small, the number of predators decreases.

**71. (a)**

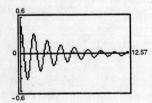

(b) The displacement function is a damped sine wave. It approaches 0 as $t$ increases.

**73.** $\tan x \approx x + \dfrac{2x^3}{3!} + \dfrac{16x^5}{5!}$

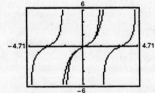

The graphs are approximately the same when $x$ is near zero. As $x$ gets larger, the graphs are further apart.

**75.** (a) $y_1 = \dfrac{4}{\pi}\left(\sin \pi x + \dfrac{1}{3}\sin 3\pi x\right)$

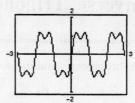
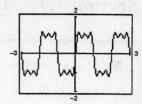

$y_2 = \dfrac{4}{\pi}\left(\sin \pi x + \dfrac{1}{3}\sin 3\pi x + \dfrac{1}{5}\sin 5\pi x\right)$

(b) $y_3 = \dfrac{4}{\pi}\left(\sin \pi x + \dfrac{1}{3}\sin 3\pi x + \dfrac{1}{5}\sin 5\pi x + \dfrac{1}{7}\sin 7\pi x\right)$

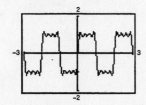

(c) $y_4 = \dfrac{4}{\pi}\left(\sin \pi x + \dfrac{1}{3}\sin 3\pi x + \dfrac{1}{5}\sin 5\pi x + \dfrac{1}{7}\sin 7\pi x + \dfrac{1}{9}\sin 9\pi x\right)$

**77.** $y = \dfrac{6}{x} + \cos x, \ x > 0$

As $x \to 0, \ y \to \infty$.

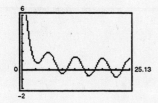

**79.** $g(x) = \dfrac{\sin x}{x}$

As $x \to 0, \ g(x) \to 1$.

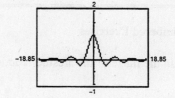

**81.** $f(x) = \sin \dfrac{1}{x}$

As $x \to 0, f(x)$ oscillates between $-1$ and $1$.

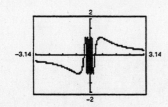

**83.** $e^{2x} = 54$

$2x = \ln 54$

$x = \dfrac{\ln 54}{2} \approx 1.994$

**85.** $\ln (x^2 + 1) = 3.2$

$x^2 + 1 = e^{3.2}$

$x^2 = e^{3.2} - 1$

$x = \pm\sqrt{e^{3.2} - 1} \approx \pm 4.851$

# Section 1.7    Inverse  Trigonometric Functions

■  You should know the definitions, domains, and ranges of $y = \arcsin x$, $y = \arccos x$, and $y = \arctan x$.

| Function | Domain | Range |
|---|---|---|
| $y = \arcsin x \implies x = \sin y$ | $-1 \le x \le 1$ | $-\dfrac{\pi}{2} \le y \le \dfrac{\pi}{2}$ |
| $y = \arccos x \implies x = \cos y$ | $-1 \le x \le 1$ | $0 \le y \le \pi$ |
| $y = \arctan x \implies x = \tan y$ | $-\infty < x < \infty$ | $-\dfrac{\pi}{2} < x < \dfrac{\pi}{2}$ |

■  You should know the inverse properties of the inverse trigonometric functions.

$\sin(\arcsin x) = x$  and  $\arcsin(\sin y) = y,\ -\dfrac{\pi}{2} \le y \le \dfrac{\pi}{2}$

$\cos(\arccos x) = x$  and  $\arccos(\cos y) = y,\ 0 \le y \le \pi$

$\tan(\arctan x) = x$  and  $\arctan(\tan y) = y,\ -\dfrac{\pi}{2} < y < \dfrac{\pi}{2}$

■  You should be able to use the triangle technique to convert trigonometric functions of inverse trigonometric functions into algebraic expressions.

## Solutions to Odd-Numbered Exercises

1.  False, $\dfrac{5\pi}{6}$ is not in the range of the arcsine function.

3.  $y = \arcsin \dfrac{1}{2} \implies \sin y = \dfrac{1}{2}$ for $-\dfrac{\pi}{2} \le y \le \dfrac{\pi}{2} \implies y = \dfrac{\pi}{6}$

5.  $y = \arccos \dfrac{1}{2} \implies \cos y = \dfrac{1}{2}$ for $0 \le y \le \pi \implies y = \dfrac{\pi}{3}$

7.  $y = \arctan \dfrac{\sqrt{3}}{3} \implies \tan y = \dfrac{\sqrt{3}}{3}$ for $-\dfrac{\pi}{2} < y < \dfrac{\pi}{2} \implies y = \dfrac{\pi}{6}$

9.  $y = \arccos\left(-\dfrac{\sqrt{3}}{2}\right) \implies \cos y = -\dfrac{\sqrt{3}}{2}$ for $0 \le y \le \pi \implies y = \dfrac{5\pi}{6}$

11.  $y = \arctan(-\sqrt{3}) \implies \tan y = -\sqrt{3}$ for $-\dfrac{\pi}{2} < y < \dfrac{\pi}{2} \implies y = -\dfrac{\pi}{3}$

13.  $y = \arccos\left(-\dfrac{1}{2}\right) \implies \cos y = -\dfrac{1}{2}$ for $0 \le y \le \pi \implies y = \dfrac{2\pi}{3}$

15.  $y = \arcsin \dfrac{\sqrt{3}}{2} \implies \sin y = \dfrac{\sqrt{3}}{2}$ for $-\dfrac{\pi}{2} \le y \le \dfrac{\pi}{2} \implies y = \dfrac{\pi}{3}$

17.  $y = \arctan 0 \implies \tan y = 0$ for $-\dfrac{\pi}{2} < y < \dfrac{\pi}{2} \implies y = 0$

19.  $\arccos 0.28 = \cos^{-1} 0.28 \approx 1.29$

21.  $\arcsin(-0.75) = \sin^{-1}(-0.75) \approx -0.85$

23.  $\arctan(-3) = \tan^{-1}(-3) \approx -1.25$

25.  $\arcsin 0.31 = \sin^{-1} 0.31 \approx 0.32$

**27.** $\arccos(-0.41) = \cos^{-1}(-0.41) \approx 1.99$

**29.** $\arctan 0.92 = \tan^{-1} 0.92 \approx 0.74$

**31.** This is the graph of $y = \arctan x$. The coordinates are $\left(-\sqrt{3}, -\dfrac{\pi}{3}\right)$, $\left(-\dfrac{1}{\sqrt{3}}, -\dfrac{\pi}{6}\right)$, and $\left(1, \dfrac{\pi}{4}\right)$.

**33.** $f(x) = \tan x$ and $g(x) = \arctan x$

Graph $\quad y_1 = \tan x$

$\qquad\quad y_2 = \tan^{-1} x$

$\qquad\quad y_3 = x$

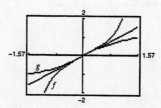

**35.** $\tan\theta = \dfrac{x}{4}$

$\quad \theta = \arctan \dfrac{x}{4}$

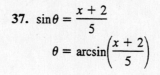

**37.** $\sin\theta = \dfrac{x+2}{5}$

$\quad \theta = \arcsin\left(\dfrac{x+2}{5}\right)$

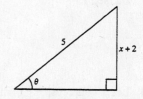

**39.** $\sin(\arcsin 0.3) = 0.3$

**41.** $\cos[\arccos(-0.1)] = -0.1$

**43.** $\arcsin(\sin 3\pi) = \arcsin(0) = 0$

Note: $3\pi$ is not in the range of the arcsine function.

**45.** Let $y = \arctan \dfrac{3}{4}$. Then,

$\quad \tan y = \dfrac{3}{4}, \; 0 < y < \dfrac{\pi}{2}$

and $\sin y = \dfrac{3}{5}$.

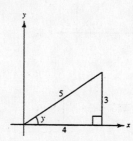

**47.** Let $y = \arctan 2$. Then,

$\quad \tan y = 2 = \dfrac{2}{1}, \; 0 < y < \dfrac{\pi}{2}$

and $\cos y = \dfrac{1}{\sqrt{5}} = \dfrac{\sqrt{5}}{5}$.

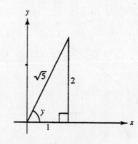

**49.** Let $y = \arcsin \dfrac{5}{13}$. Then,

$$\sin y = \frac{5}{13}, \ 0 < y < \frac{\pi}{2}$$

and $\cos y = \dfrac{12}{13}$.

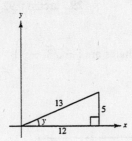

**51.** Let $y = \arctan\left(-\dfrac{3}{5}\right)$. Then,

$$\tan y = -\frac{3}{5}, \ -\frac{\pi}{2} < y < 0$$

and $\sec y = \dfrac{\sqrt{34}}{5}$.

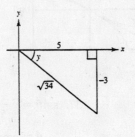

**53.** Let $y = \arccos\left(-\dfrac{2}{3}\right)$. Then,

$$\cos y = -\frac{2}{3}, \ \frac{\pi}{2} < y < \pi$$

and $\sin y = \dfrac{\sqrt{5}}{3}$.

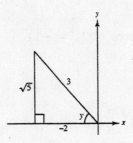

**55.** Let $y = \arctan x$. Then,

$$\tan y = x = \frac{x}{1}$$

and $\cot y = \dfrac{1}{x}$.

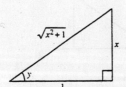

**57.** Let $y = \arcsin(2x)$. Then,

$$\sin y = 2x = \frac{2x}{1}$$

and $\cos y = \sqrt{1 - 4x^2}$.

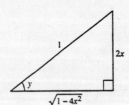

**59.** Let $y = \arccos x$. Then,

$$\cos y = x = \frac{x}{1}$$

and $\sin y = \sqrt{1 - x^2}$.

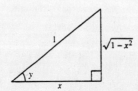

**61.** Let $y = \arccos\left(\dfrac{x}{3}\right)$. Then,

$$\cos y = \frac{x}{3}$$

and $\tan y = \dfrac{\sqrt{9 - x^2}}{x}$.

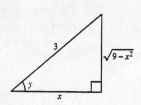

**63.** Let $y = \arctan\dfrac{x}{\sqrt{2}}$. Then,

$$\tan y = \frac{x}{\sqrt{2}}$$

and $\csc y = \dfrac{\sqrt{x^2 + 2}}{x}$.

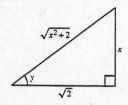

**65.** $f(x) = \sin(\arctan 2x)$, $g(x) = \dfrac{2x}{\sqrt{1 - 4x^2}}$

Let $y = \arctan 2x$. Then,

$$\tan y = 2x = \frac{2x}{1}$$

and $\sin y = \dfrac{2x}{\sqrt{1 + 4x^2}}$.

$$g(x) = \frac{2x}{\sqrt{1 + 4x^2}} = f(x)$$

The graph has horizontal asymptotes
at $y = \pm 1$.

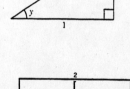

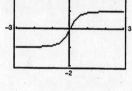

**67.** Let $y = \arctan\dfrac{9}{x}$. Then,

$$\tan y = \frac{9}{x} \text{ and } \sin y = \frac{9}{\sqrt{x^2 + 81}}.$$

Thus, $\arcsin y = \dfrac{9}{\sqrt{x^2 + 81}}$.

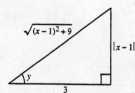

**69.** Let $y = \arccos\dfrac{3}{\sqrt{x^2 - 2x + 10}}$. Then,

$$\cos y = \frac{3}{\sqrt{x^2 - 2x + 10}} = \frac{3}{\sqrt{(x - 1)^2 + 9}}$$

and $\sin y = \dfrac{|x - 1|}{\sqrt{(x - 1)^2 + 9}}$.

Thus, $\arcsin y = \dfrac{|x - 1|}{\sqrt{(x - 1)^2 + 9}} = \arcsin\dfrac{|x - 1|}{\sqrt{x^2 - 2x + 10}}$.

**71.** $y = 2 \arccos x$

Domain: $-1 \leq x \leq 1$

Range: $0 \leq y \leq 2\pi$

Vertical stretch of $f(x) = \arccos x$

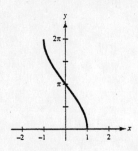

**73.** The graph of $f(x) = \arcsin(x - 1)$ is a horizontal translation of the graph of $y = \arcsin x$ by one unit.

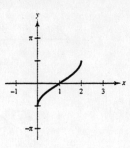

**75.** $f(x) = \arctan 2x$

Domain: all real numbers

Range: $-\dfrac{\pi}{2} < y < \dfrac{\pi}{2}$

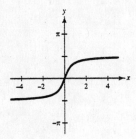

**77.** $h(v) = \tan(\arccos v) = \dfrac{\sqrt{1 - v^2}}{v}$

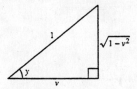

Domain: $-1 \leq v \leq 1, v \neq 0$

Range: all real numbers

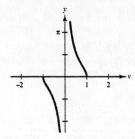

**79.** $f(t) = 3 \cos 2t + 3 \sin 2t = \sqrt{3^2 + 3^2} \sin\left(3t + \arctan \dfrac{3}{3}\right)$

$$= 3\sqrt{2} \sin(3t + \arctan 1)$$

$$= 3\sqrt{2} \sin\left(3t + \dfrac{\pi}{4}\right)$$

The graphs are the same.

**81.** $f(x) = \sin x$, $f^{-1}(x) = \arcsin x$

(a) $f \circ f^{-1} = f(f^{-1}(x)) = f(\arcsin x) = \sin(\arcsin x)$

$f^{-1} \circ f = f^{-1}(f(x)) = f^{-1}(\sin x) = \arcsin(\sin x)$

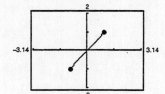    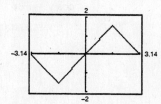

(b) Both the domain and range of $f \circ f^{-1} = \sin(\arcsin x)$ are the intervals of $[-1, 1]$.

The domain of $f^{-1} \circ f$ is all real numbers. The range is the interval $\left[-\dfrac{\pi}{2}, \dfrac{\pi}{2}\right]$.

Neither graph is the line $y = x$ because of these domain/range restrictions.

**83.** (a) $\sin \theta = \dfrac{10}{s}$

$\theta = \arcsin \dfrac{10}{s}$

(b) $s = 48$: $\theta = \arcsin \dfrac{10}{48} \approx 0.21$

$s = 24$: $\theta = \arcsin \dfrac{10}{24} \approx 0.43$

**85.** $\beta = \arctan \dfrac{3x}{x^2 + 4}$

(a)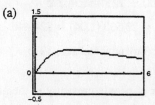

(b) $\beta$ is maximum when $x = 2$.

(c) The graph has a horizontal asymptote at $\beta = 0$. As $x$ increases, $\beta$ decreases.

**87.** (a) $\tan \theta = \dfrac{5}{x}$

$\theta = \arctan \dfrac{5}{x}$

(b) $x = 10$: $\theta = \arctan \dfrac{5}{10} \approx 26.6°$

$x = 3$: $\theta = \arctan \dfrac{5}{3} \approx 59.0°$

**89.** $y = \text{arccot } x$ if and only if $\cot y = x$.

Domain: $-\infty < x < \infty$

Range: $0 < x < \pi$

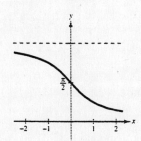

**91.** $y = \text{arccsc } x$ if and only if $\csc y = x$.

Domain: $(-\infty, -1] \cup [1, \infty)$

Range: $\left[-\dfrac{\pi}{2}, 0\right) \cup \left(0, \dfrac{\pi}{2}\right]$

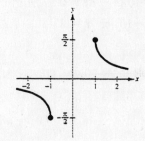

**93.** Let $y = \arcsin(-x)$. Then,

$$\sin y = -x$$
$$-\sin y = x$$
$$\sin(-y) = x$$
$$-y = \arcsin x$$
$$y = -\arcsin x.$$

Therefore, $\arcsin(-x) = -\arcsin x.$

**95.**
$$y = \pi - \arccos x$$
$$\cos y = \cos(\pi - \arccos x)$$
$$\cos y = \cos \pi \cos(\arccos x) + \sin \pi \sin(\arccos x)$$
$$\cos y = -x$$
$$y = \arccos(-x)$$

**97.** Let $\alpha = \arcsin x$ and $\beta = \arccos x$, then $\sin \alpha = x$ and $\cos \beta = x$. Thus, $\sin \alpha = \cos \beta$ which implies that $\alpha$ and $\beta$ are complementary angles and we have

$$\alpha + \beta = \frac{\pi}{2}$$
$$\arcsin x + \arccos x = \frac{\pi}{2}.$$

**99.** Now: Cost $= 23{,}500 + 725 = \$24{,}225$

Wait a month: Cost $= 23{,}500\,(1.04) = \$24{,}440$

The customer should buy now and save $215.

**101.** Let $x$ = the number of people presently in the group. Each person's share is now $\dfrac{250{,}000}{x}$.

If two more join the group, each person's share would then be $\dfrac{250{,}000}{x + 2}$.

$$\begin{array}{c} \text{Share per person with} \\ \text{two more people} \end{array} = \begin{array}{c} \text{Original share} \\ \text{per person} \end{array} - 6250$$

$$\frac{250{,}000}{x + 2} = \frac{250{,}000}{x} - 6250$$

$$250{,}000x = 250{,}000(x + 2) - 6250x(x + 2)$$

$$250{,}000x = 250{,}000x + 500{,}000 - 6250x^2 - 12500x$$

$$6250x^2 + 12500x - 500{,}000 = 0$$
$$6250(x^2 + 2x - 80) = 0$$
$$6250(x + 10)(x - 8) = 0$$
$$x = -10 \quad \text{or} \quad x = 8$$

Not possible

There were 8 people in the original group.

# Section 1.8   Applications and Models

**Solutions to Odd-Numbered Exercises**

- ■ You should be able to solve right triangles.
- ■ You should be able to solve right triangle applications.
- ■ You should be able to solve applications of simple harmonic motion.

---

**1.** Given:  $A = 20°$,  $b = 10$

$$\tan A = \frac{a}{b} \implies a = b \tan A = 10 \tan 20° \approx 3.64$$

$$\cos A = \frac{a}{c} \implies c = \frac{a}{\cos A} = \frac{10}{\cos 20°} \approx 10.64$$

$$B = 90° - 20° = 70°$$

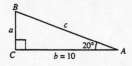

**3.** Given:  $B = 71°$,  $b = 24$

$$\tan B = \frac{b}{a} \implies a = \frac{b}{\tan B} = \frac{24}{\tan 71°} \approx 8.26$$

$$\sin B = \frac{b}{c} \implies c = \frac{b}{\sin B} = \frac{24}{\sin 71°} \approx 25.38$$

$$A = 90° - 71° = 19°$$

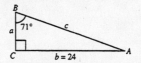

**5.** Given:  $a = 6$,  $b = 10$

$$c^2 = a^2 + b^2 \implies c = \sqrt{36 + 100}$$
$$= 2\sqrt{34} \approx 11.66$$

$$\tan A = \frac{a}{b} = \frac{6}{10} \implies A = \arctan \frac{3}{5} \approx 30.96°$$

$$B = 90° - 30.96° = 59.04°$$

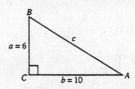

**7.**  $b = 16$,  $c = 52$

$$a = \sqrt{52^2 - 16^2}$$
$$= \sqrt{2448} = 12\sqrt{17} \approx 49.48$$

$$\cos A = \frac{16}{52}$$

$$A = \arccos \frac{16}{52} \approx 72.08°$$

$$B = 90° - 72.08° \approx 17.92°$$

**9.**  $A = 12°15'$,  $c = 430.5$

$$B = 90° - 12°15' = 77°45'$$

$$\sin 12°15' = \frac{a}{430.5}$$

$$a = 430.5 \sin 12°15' \approx 91.34$$

$$\cos 12°15' = \frac{b}{430.5}$$

$$b = 430.5 \cos 12°15' \approx 420.70$$

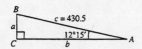

**11.**  $\tan \theta = \dfrac{h}{1/2\,b} \implies h = \dfrac{1}{2} b \tan \theta$

$$h = \frac{1}{2}(4) \tan 52° \approx 2.56 \text{ in.}$$

**13.** $\tan 30° = \dfrac{60}{x}$

$\dfrac{1}{\sqrt{3}} = \dfrac{60}{x}$

$x = 60\sqrt{3}$

$\approx 103.9$ feet

**15.**   $\sin 74° = \dfrac{h}{16}$

$16 \sin 74° = h$

$h \approx 15.4$ feet

**17. (a)**

**(b)** Let the height of the church $= x$ and the height of the church and steeple $= y$. Then,

$$\tan 35° = \dfrac{x}{50} \quad \text{and} \quad \tan 47°40' = \dfrac{y}{50}$$

$$x = 50 \tan 35° \text{ and } y = 50 \tan 47°40'$$

$$h = y - x = 50 \,(\tan 47°40' - \tan 35°).$$

**(c)** $h \approx 19.9$ feet

**19.** $\sin 34° = \dfrac{x}{4000}$

$x = 4000 \sin 34°$

$\approx 2236.8$ feet

**21.** $\tan \theta = \dfrac{75}{50}$

$\theta = \arctan \dfrac{3}{2} \approx 56.3°$

**23.** $\sin \theta = \dfrac{4000}{4150}$

$\theta = \arcsin\left(\dfrac{4000}{4150}\right)$

$\theta \approx 74.5°$

$\alpha = 90° - 74.5° = 15.5°$

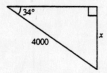

**25.** Since the airplane speed is

$$\left(275\dfrac{\text{ft}}{\text{sec}}\right)\left(60\dfrac{\text{sec}}{\text{min}}\right) = 16{,}500\dfrac{\text{ft}}{\text{min}},$$

after one minute its distance travelled in 16,500 feet.

$\sin 18° = \dfrac{a}{16{,}500}$

$a = 16{,}500 \sin 18°$

$\approx 5099$ ft

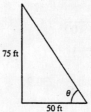

**27.** $\sin 10.5° = \dfrac{x}{4}$

$x = 4 \sin 10.5°$

$\approx 0.73$ mile

**29.** The plane has traveled 1.5 (550)= 825 miles.

$$\sin 38° = \frac{a}{825} \implies a \approx 508 \text{ miles north}$$

$$\cos 38° = \frac{b}{825} \implies b \approx 650 \text{ miles east}$$

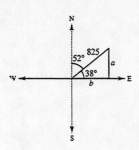

**31.** $\theta = 32°$, $\phi = 68°$

(a) $\alpha = 90° - 32° = 58°$

Bearing from $A$ to $C$:  N 58° E

(b)  $\beta = \theta = 32°$

$\gamma = 90° - \phi = 22°$

$C = \beta + \gamma = 54°$

$$\tan C = \frac{d}{50} \implies \tan 54° = \frac{d}{50} \implies d \approx 68.82 \text{ meters}$$

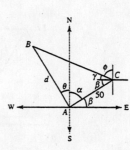

**33.** $\tan \theta = \frac{45}{30} \implies \theta \approx 56.3°$

Bearing:  N 56.3° W

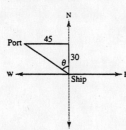

**35.** $\tan 6.5° = \dfrac{350}{d} \implies d \approx 3071.91$ ft

$\tan 4° = \dfrac{350}{D} \implies D \approx 5005.23$ ft

Distance between ships: $D - d \approx 1933.32$ ft

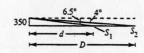

**37.**

$$\tan 57° = \frac{a}{x} \implies x = a \cot 57°$$

$$\tan 16° = \frac{a}{x + (55/6)}$$

$$\tan 16° = \frac{a}{a \cot 57° + (55/6)}$$

$$\cot 16° = \frac{a \cot 57° + (55/6)}{a}$$

$$a \cot 16° - a \cot 57° = \frac{55}{6} \implies a \approx 3.23 \text{ miles}$$

$$\approx 17{,}054 \text{ ft}$$

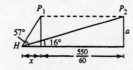

**39.** $L_1$: $3x - 2y = 5 \implies y = \frac{3}{2}x - \frac{5}{2} \implies m_1 = \frac{3}{2}$

$L_2$: $x - y = 1 \implies y = -x + 1 \implies m_2 = -1$

$\tan\alpha = \left| \dfrac{-1 - \frac{3}{2}}{1 + (-1)(\frac{3}{2})} \right| = \left| \dfrac{-\frac{5}{2}}{-\frac{1}{2}} \right| = 5$

$\alpha = \arctan 5 \approx 78.7°$

**41.** The diagonal of the base has a length of $\sqrt{a^2 + a^2} = \sqrt{2}a$.

Now, we have $\tan\theta = \dfrac{a}{\sqrt{2}a} = \dfrac{1}{\sqrt{2}}$

$\theta = \arctan\dfrac{1}{\sqrt{2}}$

$\theta \approx 35.3°$.

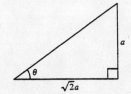

**43.** $\cos 30° = \dfrac{b}{r}$

$b = \cos 30° r$

$b = \dfrac{\sqrt{3}r}{2}$

$y = 2b = 2\left(\dfrac{\sqrt{3}r}{2}\right) = \sqrt{3}r$

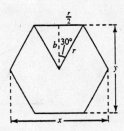

**45.** $\sin 36° = \dfrac{d}{25} \implies d \approx 14.69$

Length of side: $2d \approx 29.38$ inches

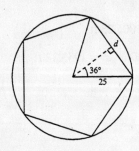

**47.** $\tan 35° = \dfrac{a}{10}$

$a = 10 \tan 35° \approx 7$

$\cos 33° = \dfrac{10}{c}$

$c = \dfrac{10}{\cos 35°} \approx 12.2$

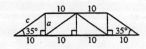

**49.** $d = 4 \cos 8\pi t$

(a) Maximum displacement = amplitude = 4

(b) Frequency $= \dfrac{\omega}{2\pi} = \dfrac{8\pi}{2\pi}$

$= 4$ cycles per unit of time

(c) $8\pi t = \dfrac{\pi}{2} \implies t = \dfrac{1}{16}$

**51.** $d = \dfrac{1}{16} \sin 120\pi t$

(a) Maximum displacement = amplitude = $\dfrac{1}{16}$

(b) Frequency $= \dfrac{\omega}{2\pi} = \dfrac{120\pi}{2\pi}$

$= 60$ cycles per unit of time

(c) $120\pi t = \pi \implies t = \dfrac{1}{120}$

**53.** $d = 0$ when $t = 0$, $a = 4$, Period $= 2$

Use $d = a \sin \omega t$ since $d = 0$ when $t = 0$.

$$\frac{2\pi}{\omega} = 2 \implies \omega = \pi$$

Thus, $d = 4 \sin \pi t$.

**55.** $d = 3$ when $t = 0$, $a = 3$, Period $= 1.5$

Use $d = a \cos \omega t$ since $d = 3$ when $t = 0$.

$$\frac{2\pi}{\omega} = 1.5 \implies \omega = \frac{4\pi}{3}$$

Thus, $d = 3 \cos\left(\frac{4\pi}{3}t\right) = 3 \cos\left(\frac{4\pi t}{3}\right)$.

**57.** $d = a \sin \omega t$

$$\text{Period} = \frac{2\pi}{\omega} = \frac{1}{\text{frequency}}$$

$$\frac{2\pi}{\omega} = \frac{1}{264}$$

$$\omega = 2\pi(264) = 528\pi$$

**59.** $y = \dfrac{1}{4} \cos 16t$, $t > 0$

(a)

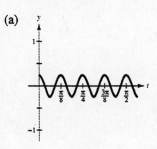

(b) Period: $\dfrac{2\pi}{16} = \dfrac{\pi}{8}$

(c) $\dfrac{1}{4} \cos 16t = 0$ when $16t = \dfrac{\pi}{2} \implies t = \dfrac{\pi}{32}$

**61.** (a) & (b)

| Base 1 | Base 2 | Altitude | Area |
|--------|--------|----------|------|
| 8 | $8 + 16 \cos 10°$ | $8 \sin 10°$ | 22.1 |
| 8 | $8 + 16 \cos 20°$ | $8 \sin 20°$ | 42.5 |
| 8 | $8 + 16 \cos 30°$ | $8 \sin 30°$ | 59.7 |
| 8 | $8 + 16 \cos 40°$ | $8 \sin 40°$ | 72.7 |
| 8 | $8 + 16 \cos 50°$ | $8 \sin 50°$ | 80.5 |
| 8 | $8 + 16 \cos 60°$ | $8 \sin 60°$ | 83.1 |
| 8 | $8 + 16 \cos 70°$ | $8 \sin 70°$ | 80.7 |

The maximum occurs when $\theta = 60°$ and is approximately 83.1 square feet.

(c) $A(\theta) = [8 + (8 + 16 \cos \theta)]\left[\dfrac{8 \sin \theta}{2}\right]$

$\quad = (16 + 16 \cos \theta)(4 \sin \theta)$

$\quad = 64 (1 + \cos \theta)(\sin \theta)$

(d)

The maximum occurs when $\theta = \dfrac{\pi}{3} = 60°$.

**63. (a)**

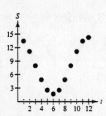

**(b)** $a = \dfrac{1}{2}(14.3 - 1.7) = 6.3$

$\dfrac{2\pi}{b} = 12 \implies b = \dfrac{\pi}{6}$

Shift: $d = 14.3 - 6.3 = 8$

$S = d + a \cos bt$

$S = 8 + 6.3 \cos\left(\dfrac{\pi t}{6}\right)$

The model is a good fit.

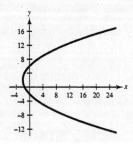

**(c)** Period: $\dfrac{2\pi}{\pi/6} = 12$

This corresponds to the 12 months in a year. Since the sales of outerwear is seasonal this is reasonable.

**(d)** The amplitude represents the maximum displacement from average sales of 8 million dollars. Sales are greatest in December (cold weather + Christmas) and least in June.

**65.** $(y - 2)^2 = 8(x + 2)$

Parabola

Vertex: $(-2, 2)$

$p = 2 > 0,$ opens to the right

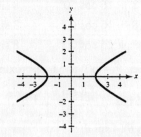

**67.** $\dfrac{x^2}{4} - y^2 = 1$

Hyperbola

Horizontal major axis

Center: $(0, 0)$

Vertices: $(\pm 2, 0)$

# ❑ Review Exercises for Chapter 1

**Solutions to Odd-Numbered Exercises**

**1.** $\theta = \dfrac{11\pi}{4}$

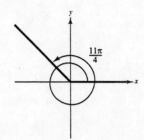

Coterminal angles: $\dfrac{11\pi}{4} - 2\pi = \dfrac{3\pi}{4}$

$$\dfrac{3\pi}{4} - 2\pi = -\dfrac{5\pi}{4}$$

**3.** $\theta = -110°$

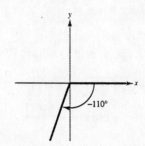

Coterminal angles: $-110° + 360° = 250°$

$$-110° - 360° = -470°$$

**5.** $135°\ 16'\ 45'' = \left(135 + \frac{16}{60} + \frac{45}{3600}\right)° \approx 135.28°$

**7.** $5°\ 22'\ 53'' = \left(5 + \frac{22}{60} + \frac{53}{3600}\right)° \approx 5.38°$

**9.** $135.27° = 135° + (0.27)(60)'$

$\qquad = 135° + 16' + 0.2(60)''$

$\qquad = 135°\ 16'\ 12''$

**11.** $-85.15° = -[85° + (0.15)(60)'] = -85°\ 9'$

**13.** $\dfrac{5\pi\ \text{rad}}{7} = \dfrac{5\pi\ \text{rad}}{7} \cdot \dfrac{180°}{\pi\ \text{rad}} \approx 128.57°$

**15.** $-3.5\ \text{rad} = -3.5\ \text{rad} \cdot \dfrac{180°}{\pi\ \text{rad}} \approx -200.54°$

**17.** $480° = 480° \cdot \dfrac{\pi\ \text{rad}}{180°} = \dfrac{8\pi}{3}\ \text{rad} \approx 8.3776\ \text{rad}$

**19.** $-33°\ 45' = -33.75° = -33.75° \cdot \dfrac{\pi\ \text{rad}}{180°} = -\dfrac{3\pi}{16}\ \text{rad} \approx -0.5890\ \text{rad}$

**21.** $252°$ is in Quadrant III.

Reference angle $= 252° - 180° = 72°$

**23.** $-\dfrac{6\pi}{5}$ is in Quadrant II and is coterminal to $\dfrac{4\pi}{5}$.

Reference angle $= \pi - \dfrac{4\pi}{5} = \dfrac{\pi}{5}$

**25.** $t = \dfrac{7\pi}{6}$ corresponds to the point $\left(-\dfrac{\sqrt{3}}{2}, -\dfrac{1}{2}\right)$.

$\sin \dfrac{7\pi}{6} = -\dfrac{1}{2}$

$\cos \dfrac{7\pi}{6} = -\dfrac{\sqrt{3}}{2}$

$\tan \dfrac{7\pi}{6} = \dfrac{1}{\sqrt{3}}$

**27.** $t = -\dfrac{\pi}{3}$ corresponds to the point $\left(\dfrac{1}{2}, -\dfrac{\sqrt{3}}{2}\right)$.

$\sin\left(-\dfrac{\pi}{3}\right) = -\dfrac{\sqrt{3}}{2}$

$\cos\left(-\dfrac{\pi}{3}\right) = \dfrac{1}{2}$

$\tan\left(-\dfrac{\pi}{3}\right) = -\sqrt{3}$

**29.** $x = 12, y = 16, r = \sqrt{144 + 256} = \sqrt{400} = 20$

$\sin \theta = \dfrac{y}{r} = \dfrac{4}{5}$     $\csc \theta = \dfrac{r}{y} = \dfrac{5}{4}$

$\cos \theta = \dfrac{x}{r} = \dfrac{3}{5}$     $\sec \theta = \dfrac{r}{x} = \dfrac{5}{3}$

$\tan \theta = \dfrac{y}{x} = \dfrac{4}{3}$     $\cot \theta = \dfrac{x}{y} = \dfrac{3}{4}$

**31.** $\sec \theta = \dfrac{6}{5}, \tan \theta < 0 \implies \theta$ is in Quadrant IV.

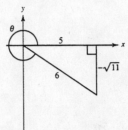

$r = 6, x = 5, y = -\sqrt{36 - 25} = -\sqrt{11}$

$\sin \theta = \dfrac{y}{r} = -\dfrac{\sqrt{11}}{6}$     $\csc \theta = -\dfrac{6\sqrt{11}}{11}$

$\cos \theta = \dfrac{x}{r} = \dfrac{5}{6}$     $\sec \theta = \dfrac{6}{5}$

$\tan \theta = \dfrac{y}{x} = -\dfrac{\sqrt{11}}{5}$     $\cot \theta = -\dfrac{5\sqrt{11}}{11}$

**33.** $\tan \dfrac{\pi}{3} = \sqrt{3}$

**35.** $\cos 495° = -\cos 45° = -\dfrac{\sqrt{2}}{2}$

**37.** $\tan 33° \approx 0.65$

**39.** $\sec \dfrac{12\pi}{5} = \dfrac{1}{\cos\left(\dfrac{12\pi}{5}\right)} \approx 3.24$

**41.** $\cos \theta = -\dfrac{\sqrt{2}}{2} \implies \theta$ is in Quadrant II or III.

Reference angle: $\dfrac{\pi}{4}$

$\theta = \dfrac{3\pi}{4}, \dfrac{5\pi}{4}$ or $\theta = 135°, 225°$

**43.** $\sin \theta = 0.8387 \implies \theta$ is in Quadrant I or II.

Reference angle: $\arcsin 0.8387 \approx 0.9949$

$\theta \approx 0.9949$ rad or $0.9949 \cdot \dfrac{180}{\pi} \approx 57.0°$

$\theta \approx \pi - 0.9949 \approx 2.1467$ rad or

$2.1467 \cdot \dfrac{180}{\pi} \approx 123.0°$

**45.** $\theta = 120°$

$m = \tan 120° = -\sqrt{3}$

**47.** $x + y - 10 = 0$

$y = -x + 1 \implies m = -1$

$\tan \theta = -1$

$\theta = 135°$

**49.** $y = 3 \cos 2\pi x$

Amplitude: 3

Period: $\dfrac{2\pi}{2\pi} = 1$

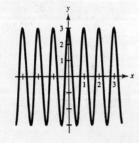

**51.** $f(x) = 5 \sin \dfrac{2x}{5}$.

Amplitude: 5

Period: $\dfrac{2\pi}{2/5} = 5\pi$

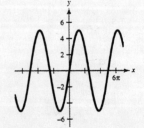

**53.** $f(x) = -\dfrac{1}{4}\cos\dfrac{\pi x}{4}$

Amplitude: $\left|-\dfrac{1}{4}\right| = \dfrac{1}{4}$

Period: $\dfrac{2\pi}{\pi/4} = 8$

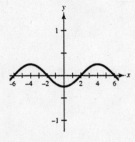

**55.** $g(t) = \dfrac{5}{2}\sin(t - \pi)$

Amplitude: $\dfrac{5}{2}$

Period: $2\pi$

Shift: $t - \pi = 0$ to $t - \pi = 2\pi$

$\qquad\qquad t = \pi \qquad\qquad t = 3\pi$

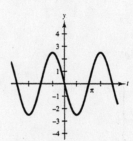

**57.** $h(t) = \tan\left(t - \dfrac{\pi}{4}\right)$

Period: $\pi$

Two consecutive asymptotes: $t - \dfrac{\pi}{4} = -\dfrac{\pi}{2}$ and $t - \dfrac{\pi}{4} = \dfrac{\pi}{2}$

$\qquad\qquad\qquad\qquad t = -\dfrac{\pi}{4} \qquad\qquad t = \dfrac{3\pi}{4}$

| $t$ | $0$ | $\dfrac{\pi}{4}$ | $\dfrac{\pi}{2}$ |
|---|---|---|---|
| $h(t)$ | $-1$ | $0$ | $1$ |

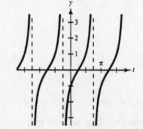

**59.** $y = \arcsin\dfrac{x}{2}$

Domain: $-2 \le x \le 2$

Range: $-\dfrac{\pi}{2} \le y \le \dfrac{\pi}{2}$

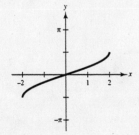

**61.** $f(x) = \dfrac{x}{4} - \sin x$

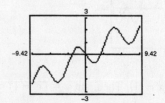

Not periodic.

**63.** $f(x) = \dfrac{\pi}{2} + \arctan x$

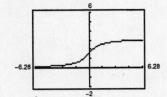

Not periodic.

**67.** $f(t) = 2.5e^{-t/4}\sin 2\pi t$

$-2.5e^{-t/4} \le f(t) \le 2.5e^{-t/4}$

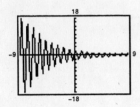

Not periodic.

**65.** $h(\theta) = \theta \sin \pi\theta$

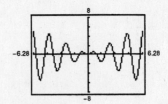

Not periodic.

**69.** $f(x) = e^{\sin x}$

The graph is periodic. The period is $2\pi$.

Maximum: $\left(\dfrac{\pi}{2}, e\right)$

Minimum: $\left(\dfrac{3\pi}{2}, e^{-1}\right)$

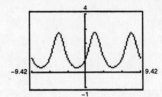

**71.** $g(x) = 2\sin x \cos^2 x$

The graph is periodic. The period is $2\pi$.

Relative Minimum: $\left(\dfrac{\pi}{2}, 0\right)$, $(3.76, -0.77)$, $(5.67, -0.77)$

Relative Maximum: $(0.61, 0.77)$, $(2.53, 0.77)$, $\left(\dfrac{3\pi}{2}, 0\right)$

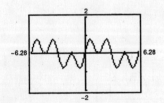

**73.** Let $y = \arcsin(x-1)$. Then, $\sin y = (x-1) = \dfrac{x-1}{1}$, and

$$\sec y = \frac{1}{\sqrt{1-(x-1)^2}} = \frac{1}{\sqrt{-x^2+2x}} = \frac{\sqrt{-x^2+2x}}{-x^2+2x}.$$

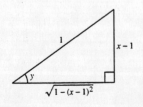

**75.** Let $y = \arccos \dfrac{x^2}{4 - x^2}$. Then $\cos y = \dfrac{x}{4 - x^2}$, and

$$\sin y = \frac{\sqrt{(4 - x^2)^2 - (x^2)^2}}{4 - x^2}$$

$$= \frac{\sqrt{16 - 8x^2}}{4 - x^2}$$

$$= \frac{2\sqrt{4 - 2x^2}}{4 - x^2}.$$

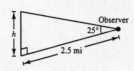

**77.** $\sin 50° = \dfrac{h}{12}$

$h = 12 \sin 50°$

$\approx 9.2$ m

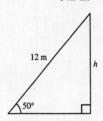

**79.** $\tan 25° = \dfrac{h}{2.5}$

$h = 2.5 \tan 25°$

$\approx 1.2$ miles

**81.** $\tan 1° \, 10' = \dfrac{a}{3.5}$

$a = 3.5 \tan 1° \, 10'$

$\approx 0.07$ km

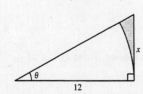

**83.** (a) $\tan \theta = \dfrac{x}{12}$

$x = 12 \tan \theta$

Area = Area of triangle $-$ Area of sector

$= \left(\tfrac{1}{2}bh\right) - \left(\tfrac{1}{2}r^2\theta\right)$

$= \tfrac{1}{2}(12)(12 \tan \theta) - \tfrac{1}{2}(12^2)(\theta)$

$= 72 \tan \theta - 72\theta$

$= 72(\tan \theta - \theta)$

(b)

As $\theta \to \dfrac{\pi}{2}$, $A \to \infty$. The area increases without bound as $\theta$ approaches $\dfrac{\pi}{2}$.

# ❏ Practice Test for Chapter 1

**1.** Express 350° in radian measure.

**2.** Express $(5\pi)/9$ in degree measure.

**3.** Convert 135° 14′ 12″ to decimal form.

**4.** Convert −22.569° to D° M′ S″ form.

**5.** If $\cos \theta = \frac{2}{3}$, use the trigonometric identities to find $\tan \theta$.

**6.** Find $\theta$ given $\sin \theta = 0.9063, 0 \le \theta < 2\pi$.

**7.** Solve for $x$ in the figure below.

**8.** Find the magnitude of the reference angle for $\theta = (6\pi)/5$.

**9.** Evaluate csc 3.92.

**10.** Find sec $\theta$ given that $\theta$ lies in Quadrant III and tan $\theta = 6$.

**11.** Graph $y = 3 \sin \dfrac{x}{2}$.

**12.** Graph $y = -2 \cos(x - \pi)$.

**13.** Graph $y = \tan 2x$.

**14.** Graph $y = -\csc\left(x + \dfrac{\pi}{4}\right)$.

**15.** Graph $y = 2x + \sin x$, using a graphing calculator.

**16.** Graph $y = 3x \cos x$, using a graphing calculator.

**17.** Evaluate arcsin 1.

**18.** Evaluate arctan(−3).

**19.** Evaluate $\sin\left(\arccos \dfrac{4}{\sqrt{35}}\right)$.

**20.** Write an algebraic expression for $\cos\left(\arcsin \dfrac{x}{4}\right)$.

**For Exercises 21–23, solve the right triangle.**

**21.** $A = 40°, c = 12$

**22.** $B = 6.84°, a = 21.3$

**23.** $a = 5, b = 9$

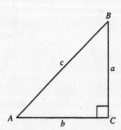

**24.** A 20-foot ladder leans against the side of a barn. Find the height of the top of the ladder if the angle of elevation of the ladder is 67°.

**25.** An observer in a lighthouse 250 feet above sea level spots a ship off the shore. If the angle of depression to the ship is 5°, how far out is the ship?

# CHAPTER 2
## Analytic Trigonometry

# CHAPTER 2
## Analytic Trigonometry

## Section 2.1    Using Fundamental Identities

■ You should know the fundamental trigonometric identities.

(a) Reciprocal Identities

$$\sin u = \frac{1}{\csc u} \qquad\qquad \csc u = \frac{1}{\sin u}$$

$$\cos u = \frac{1}{\sec u} \qquad\qquad \sec u = \frac{1}{\cos u}$$

$$\tan u = \frac{1}{\cot u} = \frac{\sin u}{\cos u} \qquad \cot u = \frac{1}{\tan u} = \frac{\cos u}{\sin u}$$

(b) Pythagorean Identities

$$\sin^2 u + \cos^2 u = 1$$
$$1 + \tan^2 u = \sec^2 u$$
$$1 + \cot^2 u = \csc^2 u$$

(c) Cofunction Identities

$$\sin\left(\frac{\pi}{2} - u\right) = \cos u \qquad\qquad \cos\left(\frac{\pi}{2} - u\right) = \sin u$$

$$\tan\left(\frac{\pi}{2} - u\right) = \cot u \qquad\qquad \cot\left(\frac{\pi}{2} - u\right) = \tan u$$

$$\sec\left(\frac{\pi}{2} - u\right) = \csc u \qquad\qquad \csc\left(\frac{\pi}{2} - u\right) = \sec u$$

(d) Negative Angle Identities

$$\sin(-x) = -\sin x \qquad\qquad \csc(-x) = -\csc x$$
$$\cos(-x) = \cos x \qquad\qquad \sec(-x) = \sec x$$
$$\tan(-x) = -\tan x \qquad\qquad \cot(-x) = -\cot x$$

■ You should be able to use these fundamental identities to find function values.

■ You should be able to convert trigonometric expressions to equivalent forms by using the fundamental identities.

**Solutions to Odd-Numbered Exercises**

**1.** $\sin x = \dfrac{1}{2}$, $\cos x = \dfrac{\sqrt{3}}{2} \implies x$ is in

Quadrant I.

$\tan x = \dfrac{\sin x}{\cos x} = \dfrac{1/2}{\sqrt{3}/2} = \dfrac{1}{\sqrt{3}} = \dfrac{\sqrt{3}}{3}$

$\cot x = \dfrac{1}{\tan x} = \sqrt{3}$

$\sec x = \dfrac{1}{\cos x} = \dfrac{2}{\sqrt{3}} = \dfrac{2\sqrt{3}}{3}$

$\csc x = \dfrac{1}{\sin x} = 2$

**3.** $\sec \theta = \sqrt{2}$, $\sin \theta = -\dfrac{\sqrt{2}}{2} \implies \theta$ is in

Quadrant IV.

$\cos \theta = \dfrac{1}{\sec \theta} = \dfrac{1}{\sqrt{2}} = \dfrac{\sqrt{2}}{2}$

$\tan \theta = \dfrac{\sin \theta}{\cos \theta} = \dfrac{-\sqrt{2}/2}{\sqrt{2}/2} = -1$

$\cot \theta = \dfrac{1}{\tan \theta} = -1$

$\csc \theta = \dfrac{1}{\sin \theta} = -\sqrt{2}$

**5.** $\tan x = \dfrac{5}{12}$, $\sec x = -\dfrac{13}{12} \implies x$ is in

Quadrant III.

$\cos x = \dfrac{1}{\sec x} = -\dfrac{12}{13}$

$\sin x = -\sqrt{1 - \cos^2 x} = -\sqrt{1 - \dfrac{144}{169}} = -\dfrac{5}{13}$

$\cot x = \dfrac{1}{\tan x} = \dfrac{12}{5}$

$\csc x = \dfrac{1}{\sin x} = -\dfrac{13}{5}$

**7.** $\sec \phi = -1$, $\sin \phi = 0 \implies \phi = \pi$

$\cos \phi = -1$

$\tan \phi = 0$

$\cot \phi$ is undefined.

$\csc \phi$ is undefined.

**9.** $\sin(-x) = -\sin x = -\dfrac{2}{3} \implies \sin x = \dfrac{2}{3}$

$\sin x = \dfrac{2}{3}$, $\tan x = -\dfrac{2\sqrt{5}}{5} \implies x$ is in

Quadrant II.

$\cos x = -\sqrt{1 - \sin^2 x} = -\sqrt{1 - \dfrac{4}{9}} = -\dfrac{\sqrt{5}}{3}$

$\cot x = \dfrac{1}{\tan x} = -\dfrac{\sqrt{5}}{2}$

$\sec x = \dfrac{1}{\cos x} = -\dfrac{3\sqrt{5}}{5}$

$\csc x = \dfrac{1}{\sin x} = \dfrac{3}{2}$

**11.** $\tan \theta = 2$, $\sin \theta < 0 \implies \theta$ is in Quadrant III.

$\sec \theta = -\sqrt{\tan^2 \theta + 1} = -\sqrt{4 + 1} = -\sqrt{5}$

$\cos \theta = \dfrac{1}{\sec \theta} = -\dfrac{1}{\sqrt{5}} = -\dfrac{\sqrt{5}}{5}$

$\sin \theta = -\sqrt{1 - \cos^2 \theta}$

$\quad = -\sqrt{1 - \dfrac{1}{5}} = -\dfrac{2}{\sqrt{5}} = -\dfrac{2\sqrt{5}}{5}$

$\csc \theta = \dfrac{1}{\sin \theta} = -\dfrac{\sqrt{5}}{2}$

$\cot \theta = \dfrac{1}{\tan \theta} = \dfrac{1}{2}$

**13.** $\sin \theta = -1$, $\cot \theta = 0 \implies \theta = \dfrac{3\pi}{2}$

$\cos \theta = \sqrt{1 - \sin^2 \theta} = 0$

$\sec \theta$ is undefined.

$\tan \theta$ is undefined.

$\csc \theta = -1$

**15.** By looking at the basic graphs of $\sin x$ and $\csc x$, we see that as $x \to \dfrac{\pi -}{2}$, $\sin x \to 1$ and $\csc x \to 1$.

**17.** By looking at the basic graphs of $\tan x$ and $\cot x$, we see that as

$x \to \dfrac{\pi -}{2}$, $\tan x \to \infty$ and $\cot x \to 0$.

**19.** $\sec x \cos x = \sec x \cdot \dfrac{1}{\sec x} = 1$

The expression is matched with (d).

**21.** $\tan^2 x - \sec^2 x = \tan^2 x - (\tan^2 x + 1) = -1$

The expression is matched with (a).

**23.** $\dfrac{\sin(-x)}{\cos(-x)} = \dfrac{-\sin x}{\cos x} = -\tan x$

The expression is matched with (e).

**25.** $\sin x \sec x = \sin x \cdot \dfrac{1}{\cos x} = \tan x$

The expression is matched with (b).

**27.** $\sec^4 x - \tan^4 x = (\sec^2 x + \tan^2 x)(\sec^2 x - \tan^2 x)$

$\qquad = (\sec^2 x + \tan^2 x)(1) = \sec^2 x + \tan^2 x$

The expression is matched with (f).

**29.** $\dfrac{\sec^2 x - 1}{\sin^2 x} = \dfrac{\tan^2 x}{\sin^2 x} = \dfrac{\sin^2 x}{\cos^2 x} \cdot \dfrac{1}{\sin^2 x} = \sec^2 x$

The expression is matched with (e).

**31.** $\tan \phi \csc \phi = \dfrac{\sin \phi}{\cos \phi} \cdot \dfrac{1}{\sin \phi} = \dfrac{1}{\cos \phi} = \sec \phi$

**33.** $\cos \beta \tan \beta = \cos \beta \left( \dfrac{\sin \beta}{\cos \beta} \right)$

$\qquad = \sin \beta$

**35.** $\dfrac{\cot x}{\csc x} = \dfrac{\cos x / \sin x}{1 / \sin x}$

$\qquad = \dfrac{\cos x}{\sin x} \cdot \dfrac{\sin x}{1} = \cos x$

**37.** $\sec \alpha \dfrac{\sin \alpha}{\tan \alpha} = \dfrac{1}{\cos \alpha} (\sin \alpha) \cot \alpha$

$\qquad = \dfrac{1}{\cos \alpha} (\sin \alpha) \left( \dfrac{\cos \alpha}{\sin \alpha} \right) = 1$

**39.** $\dfrac{\sin(-x)}{\cos x} = -\dfrac{\sin x}{\cos x} = -\tan x$

**41.** $\cos \left( \dfrac{\pi}{2} - x \right) \sec x = (\sin x)(\sec x)$

$\qquad = (\sin x) \left( \dfrac{1}{\cos x} \right)$

$\qquad = \dfrac{\sin x}{\cos x}$

$\qquad = \tan x$

**43.** $\dfrac{\cos^2 y}{1 - \sin y} = \dfrac{1 - \sin^2 y}{1 - \sin y}$

$\qquad = \dfrac{(1 + \sin y)(1 - \sin y)}{1 - \sin y}$

$\qquad = 1 + \sin y$

**45.** $\tan^2 x - \tan^2 x \sin^2 x = \tan^2 x (1 - \sin^2 x)$

$\qquad = \tan^2 x \cos^2 x$

$\qquad = \dfrac{\sin^2 x}{\cos^2 x} \cdot \cos^2 x$

$\qquad = \sin^2 x$

**47.** $\sin^2 x \sec^2 x - \sin^2 x = \sin^2 x(\sec^2 x - 1)$
$$= \sin^2 x \tan^2 x$$

**49.** $\tan^4 x + 2\tan^2 x + 1 = (\tan^2 x + 1)^2$
$$= (\sec^2 x)^2$$
$$= \sec^4 x$$

**51.** $\sin^4 x - \cos^4 x = (\sin^2 x + \cos^2 x)(\sin^2 x - \cos^2 x)$
$$= (1)(\sin^2 x - \cos^2 x)$$
$$= \sin^2 x - \cos^2 x$$

**53.** $(\sin x + \cos x)^2 = \sin^2 x + 2\sin x \cos x + \cos^2 x$
$$= (\sin^2 x + \cos^2 x) + 2 \sin x \cos x$$
$$= 1 + 2 \sin x \cos x$$

**55.** $(\sec x + 1)(\sec x - 1) = \sec^2 x - 1 = \tan^2 x$

**57.** $\dfrac{1}{1 + \cos x} + \dfrac{1}{1 - \cos x} = \dfrac{1 - \cos x + 1 + \cos x}{(1 + \cos x)(1 - \cos x)}$
$$= \dfrac{2}{1 - \cos^2 x}$$
$$= \dfrac{2}{\sin^2 x}$$
$$= 2 \csc^2 x$$

**59.** $\dfrac{\cos x}{1 + \sin x} + \dfrac{1 + \sin x}{\cos x} = \dfrac{\cos^2 x + (1 + \sin x)^2}{\cos x(1 + \sin x)} = \dfrac{\cos^2 x + 1 + 2 \sin x + \sin^2 x}{\cos x(1 + \sin x)}$
$$= \dfrac{2 + 2 \sin x}{\cos x(1 + \sin x)}$$
$$= \dfrac{2(1 + \sin x)}{\cos x(1 + \sin x)}$$
$$= \dfrac{2}{\cos x}$$
$$= 2 \sec x$$

**61.** $\dfrac{\sin^2 y}{1 - \cos y} = \dfrac{1 - \cos^2 y}{1 - \cos y}$
$$= \dfrac{(1 + \cos y)(1 - \cos y)}{1 - \cos y}$$
$$= 1 + \cos y$$

**63.** $\dfrac{3}{\sec x - \tan x} \cdot \dfrac{\sec x + \tan x}{\sec x + \tan x} = \dfrac{3(\sec x + \tan x)}{\sec^2 x - \tan^2 x}$
$$= \dfrac{3(\sec x + \tan x)}{1}$$
$$= 3(\sec x + \tan x)$$

**65.** $y_1 = \cos\left(\dfrac{\pi}{2} - x\right)$, $y_2 = \sin x$

| $x$ | 0.2 | 0.4 | 0.6 | 0.8 | 1.0 | 1.2 | 1.4 |
|-----|-----|-----|-----|-----|-----|-----|-----|
| $y_1$ | 0.1987 | 0.3894 | 0.5646 | 0.7174 | 0.8415 | 0.9320 | 0.9854 |
| $y_2$ | 0.1987 | 0.3894 | 0.5646 | 0.7174 | 0.8415 | 0.9320 | 0.9854 |

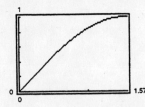

Conclusion: $y_1 = y_2$

**67.** $y_1 = \dfrac{\cos x}{1 - \sin x}$, $y_2 = \dfrac{1 + \sin x}{\cos x}$

| $x$ | 0.2 | 0.4 | 0.6 | 0.8 | 1.0 | 1.2 | 1.4 |
|-----|-----|-----|-----|-----|-----|-----|-----|
| $y_1$ | 1.2230 | 1.5085 | 1.8958 | 2.4650 | 3.4082 | 5.3319 | 11.6814 |
| $y_2$ | 1.2230 | 1.5085 | 1.8958 | 2.4650 | 3.4082 | 5.3319 | 11.6814 |

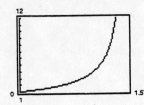

Conclusion: $y_1 = y_2$

**69.** $y_1 = \cos x \cot x + \sin x = \csc x$

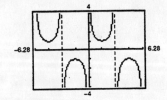

**71.** $\sqrt{25 - x^2} = \sqrt{25 - (5\sin\theta)^2}$, $x = 5\sin\theta$

$\qquad\qquad = \sqrt{25 - 25\sin^2\theta}$

$\qquad\qquad = \sqrt{25(1 - \sin^2\theta)}$

$\qquad\qquad = \sqrt{25\cos^2\theta}$

$\qquad\qquad = 5\cos\theta$

**73.** $\sqrt{x^2 - 9} = \sqrt{(3\sec\theta)^2 - 9}$, $x = 3\sec\theta$

$\qquad\quad\;\, = \sqrt{9\sec^2\theta - 9}$

$\qquad\quad\;\, = \sqrt{9(\sec^2\theta - 1)}$

$\qquad\quad\;\, = \sqrt{9\tan^2\theta}$

$\qquad\quad\;\, = 3\tan\theta$

**75.** $\sqrt{x^2 + 25} = \sqrt{(5\tan\theta)^2 + 25}$, $x = 5\tan\theta$

$\qquad\qquad = \sqrt{25\tan^2\theta + 25}$

$\qquad\qquad = \sqrt{25(\tan^2\theta + 1)}$

$\qquad\qquad = \sqrt{25\sec^2\theta}$

$\qquad\qquad = 5\sec\theta$

**77.** $\sin\theta = \sqrt{1 - \cos^2\theta}$

Let $y_1 = \sin x$ and $y_2 = \sqrt{1 - \cos^2 x}$, $0 \le x \le 2\pi$.

$y_1 = y_2$ for $0 \le x \le \pi$, so we have

$\sin\theta = \sqrt{1 - \cos^2\theta}$ for $0 \le \theta \le \pi$.

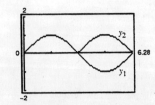

**79.** $\sec\theta = \sqrt{1 + \tan^2\theta}$

Let $y_1 = \dfrac{1}{\cos x}$ and $y_2 = \sqrt{1 + \tan^2 x}$, $0 \le x \le 2\pi$.

$y_1 = y_2$ for $0 \le x < \dfrac{\pi}{2}$ and $\dfrac{3\pi}{2} < x \le 2\pi$, so we have

$\sec\theta = \sqrt{1 + \tan^2\theta}$ for $0 \le \theta < \dfrac{\pi}{2}$ and $\dfrac{3\pi}{2} < \theta \le 2\pi$.

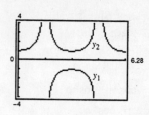

**81.** $\ln|\cos\theta| - \ln|\sin\theta| = \ln\dfrac{|\cos\theta|}{|\sin\theta|} = \ln|\cot\theta|$

**83.** False; $\dfrac{\sin k\theta}{\cos k\theta} = \tan k\theta$

**85.** True; $\sin\theta\csc\theta = \sin\theta\left(\dfrac{1}{\sin\theta}\right) = 1$, provided $\sin\theta \ne 0$.

**87.** (a) $\csc^2 132° - \cot^2 132° \approx 1.8107 - 0.8107 = 1$

(b) $\csc^2 \dfrac{2\pi}{7} - \cot^2 \dfrac{2\pi}{7} \approx 1.6360 - 0.6360 = 1$

**89.** $\cos\left(\dfrac{\pi}{2} - \theta\right) = \sin\theta$

(a) $\theta = 80°$

$\cos(90° - 80°) = \sin 80°$

$0.9848 = 0.9848$

(b) $\theta = 0.8$

$\cos\left(\dfrac{\pi}{2} - 0.8\right) = \sin 0.8$

$0.7174 = 0.7174$

**91.** Since $\sin^2\theta + \cos^2\theta = 1$ and $\cos^2\theta = 1 - \sin^2\theta$:

$\cos\theta = \pm\sqrt{1 - \sin^2\theta}$

$\tan\theta = \dfrac{\sin\theta}{\cos\theta} = \pm\dfrac{\sin\theta}{\sqrt{1 - \sin^2\theta}}$

$\cot\theta = \dfrac{1}{\tan\theta} = \pm\dfrac{\sqrt{1 - \sin^2\theta}}{\sin\theta}$

$\sec\theta = \dfrac{1}{\cos\theta} = \pm\dfrac{1}{\sqrt{1 - \sin^2\theta}}$

$\csc\theta = \dfrac{1}{\sin\theta}$

**93.** $(\sqrt{x} + 5)(\sqrt{x} - 5) = (\sqrt{x})^2 - (5)^2 = x - 25$

**95.** $(2\sqrt{z} + 3)^2 = (2\sqrt{z})^2 + 2(2\sqrt{z})(3) + (3)^2$

$= 4z + 12\sqrt{z} + 9$

# Section 2.2    Verifying Trigonometric Identities

- You should know the difference between an expression, a conditional equation, and an identity.
- You should be able to solve trigonometric identities, using the following techniques.
  (a) Work with *one* side at a time. Do not "cross" the equal sign.
  (b) Use algebraic techniques such as combining fractions, factoring expressions, rationalizing denominators, and squaring binomials.
  (c) Use the fundamental identities.
  (d) Convert all the terms into sines and cosines.

**Solutions to Odd-Numbered Exercises**

**1.** $\sin t \csc t = \sin t\left(\dfrac{1}{\sin t}\right) = 1$

**3.** $(1 + \sin\alpha)(1 - \sin\alpha) = (1 - \sin^2\alpha) = \cos^2\alpha$

**5.** $\cos^2\beta - \sin^2\beta = (1 - \sin^2\beta) - \sin^2\beta$

$\qquad\qquad\qquad = 1 - 2\sin^2\beta$

**7.** $\quad\tan^2\theta + 4 = (\sec^2\theta - 1) + 4$

$\qquad\qquad\qquad = \sec^2\theta + 3$

**9.** $\sin^2\alpha - \sin^4\alpha = \sin^2\alpha(1 - \sin^2\alpha)$

$\qquad\qquad\qquad = (1 - \cos^2\alpha)(\cos^2\alpha)$

$\qquad\qquad\qquad = \cos^2\alpha - \cos^4\alpha$

**11.** $\dfrac{\sec^2 x}{\tan x} = \sec^2 x \cdot \dfrac{1}{\tan x} = \dfrac{1}{\cos^2 x} \cdot \dfrac{\cos x}{\sin x}$

$\qquad\quad = \dfrac{1}{\cos x} \cdot \dfrac{1}{\sin x}$

$\qquad\quad = \sec x \csc x$

**13.** $\dfrac{\cot^2 t}{\csc t} = \dfrac{\cos^2 t}{\sin^2 t} \cdot \sin t$

$\qquad\quad = \dfrac{\cos^2 t}{\sin t}$

$\qquad\quad = \dfrac{1 - \sin^2 t}{\sin t} = \dfrac{1}{\sin t} - \dfrac{\sin^2 t}{\sin t}$

$\qquad\quad = \csc t - \sin t$

**15.** $\sin^{1/2} x \cos x - \sin^{5/2} x \cos x = \sin^{1/2} x \cos x(1 - \sin^2 x) = \sin^{1/2} x \cos x \cdot \cos^2 x = \cos^3 x \sqrt{\sin x}$

**17.** $\dfrac{1}{\sec x \tan x} = \cos x \cot x = \cos x \cdot \dfrac{\cos x}{\sin x}$

$\qquad\qquad\qquad = \dfrac{\cos^2 x}{\sin x}$

$\qquad\qquad\qquad = \dfrac{1 - \sin^2 x}{\sin x}$

$\qquad\qquad\qquad = \dfrac{1}{\sin x} - \sin x$

$\qquad\qquad\qquad = \csc x - \sin x$

**19.** $\csc x - \sin x = \dfrac{1}{\sin x} - \sin x$

$\qquad\qquad\qquad = \dfrac{1 - \sin^2 x}{\sin x}$

$\qquad\qquad\qquad = \dfrac{\cos^2 x}{\sin x}$

$\qquad\qquad\qquad = \cos x \cdot \dfrac{\cos x}{\sin x}$

$\qquad\qquad\qquad = \cos x \cot x$

**21.** $\cos x + \sin x \tan x = \cos x + \sin x \cdot \dfrac{\sin x}{\cos x}$

$\qquad\qquad\qquad = \dfrac{\cos^2 x + \sin^2 x}{\cos x}$

$\qquad\qquad\qquad = \dfrac{1}{\cos x}$

$\qquad\qquad\qquad = \sec x$

**23.** $\dfrac{1}{\tan x} + \dfrac{1}{\cot x} = \dfrac{\cot x + \tan x}{\tan x \cot x}$

$\qquad\qquad\qquad = \dfrac{\cot x + \tan x}{1}$

$\qquad\qquad\qquad = \tan x + \cot x$

**25.** $\dfrac{\cos\theta \cot\theta}{1 - \sin\theta} - 1 = \dfrac{\cos\theta \cot\theta - (1 - \sin\theta)}{1 - \sin\theta}$

$\qquad\qquad = \dfrac{\cos\theta(\cos\theta/\sin\theta) - 1 + \sin\theta}{1 - \sin\theta} \cdot \dfrac{\sin\theta}{\sin\theta}$

$\qquad\qquad = \dfrac{\cos^2\theta - \sin\theta + \sin^2\theta}{\sin\theta(1 - \sin\theta)}$

$\qquad\qquad = \dfrac{1 - \sin\theta}{\sin\theta(1 - \sin\theta)}$

$\qquad\qquad = \dfrac{1}{\sin\theta}$

$\qquad\qquad = \csc\theta$

**27.** $\dfrac{1}{\cot x + 1} + \dfrac{1}{\tan x + 1} = \dfrac{\tan x + 1 + \cot x + 1}{(\cot x + 1)(\tan x + 1)}$

$\qquad\qquad = \dfrac{\tan x + \cot x + 2}{\cot x \tan x + \cot x + \tan x + 1}$

$\qquad\qquad = \dfrac{\tan x + \cot x + 2}{\tan x + \cot x + 2}$

$\qquad\qquad = 1$

**29.** $\cos\left(\dfrac{\pi}{2} - x\right)\csc x = \sin x \left(\dfrac{1}{\sin x}\right) = 1$

**31.** $\dfrac{\csc(-x)}{\sec(-x)} = \dfrac{1/\sin(-x)}{1/\cos(-x)}$

$\qquad = \dfrac{\cos(-x)}{\sin(-x)}$

$\qquad = \dfrac{\cos x}{-\sin x}$

$\qquad = -\cot x$

**33.** $\dfrac{\cos(-\theta)}{1 + \sin(-\theta)} = \dfrac{\cos\theta}{1 - \sin\theta} \cdot \dfrac{1 + \sin\theta}{1 + \sin\theta}$

$\qquad = \dfrac{\cos\theta(1 + \sin\theta)}{1 - \sin^2\theta}$

$\qquad = \dfrac{\cos\theta(1 + \sin\theta)}{\cos^2\theta}$

$\qquad = \dfrac{1 + \sin\theta}{\cos\theta}$

$\qquad = \dfrac{1}{\cos\theta} + \dfrac{\sin\theta}{\cos\theta}$

$\qquad = \sec\theta + \tan\theta$

**35.** $\dfrac{\sin x \cos y + \cos x \sin y}{\cos x \cos y - \sin x \sin y} = \dfrac{\dfrac{\sin x \cos y}{\cos x \cos y} + \dfrac{\cos x \sin y}{\cos x \cos y}}{\dfrac{\cos x \cos y}{\cos x \cos y} - \dfrac{\sin x \sin y}{\cos x \cos y}} = \dfrac{\tan x + \tan y}{1 - \tan x \tan y}$

**37.** $\dfrac{\tan x + \cot y}{\tan x \cot y} = \dfrac{\dfrac{1}{\cot x} + \dfrac{1}{\tan y}}{\dfrac{1}{\cot x} \cdot \dfrac{1}{\tan y}} \cdot \dfrac{\cot x \tan y}{\cot x \tan y} = \tan y + \cot x$

**39.** $\sqrt{\dfrac{1 + \sin\theta}{1 - \sin\theta}} = \sqrt{\dfrac{1 + \sin\theta}{1 - \sin\theta} \cdot \dfrac{1 + \sin\theta}{1 + \sin\theta}}$

$\qquad = \sqrt{\dfrac{(1 + \sin\theta)^2}{1 - \sin^2\theta}}$

$\qquad = \sqrt{\dfrac{(1 + \sin\theta)^2}{\cos^2\theta}}$

$\qquad = \dfrac{1 + \sin\theta}{|\cos\theta|}$

**41.** $\sin^2 x + \sin^2\left(\dfrac{\pi}{2} - x\right) = \sin^2 x + \cos^2 x = 1$

**43.** $\csc x \cos\left(\dfrac{\pi}{2} - x\right) = \dfrac{1}{\sin x} \cdot \sin x = 1$

**45.** $2\sec^2 x - 2\sec^2 x\sin^2 x - \sin^2 x - \cos^2 x = 2\sec^2 x(1 - \sin^2 x) - (\sin^2 x + \cos^2 x)$

$$= 2\sec^2 x(\cos^2 x) - 1$$

$$= 2 \cdot \frac{1}{\cos^2 x} \cdot \cos^2 x - 1$$

$$= 2 - 1$$

$$= 1$$

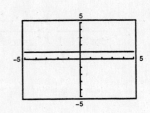

**47.** $2 + \cos^2 x - 3\cos^4 x = (1 - \cos^2 x)(2 + 3\cos^2 x)$

$$= \sin^2 x(2 + 3\cos^2 x)$$

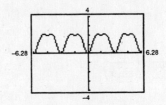

**49.** $\csc^4 x - 2\csc^2 x + 1 = (\csc^2 x - 1)^2$

$$= (\cot^2 x)^2 = \cot^4 x$$

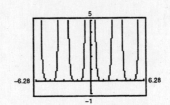

**51.** $\sec^4 \theta - \tan^4 \theta = (\sec^2 \theta + \tan^2 \theta)(\sec^2 \theta - \tan^2 \theta)$

$$= (1 + \tan^2 \theta + \tan^2 \theta)(1)$$

$$= 1 + 2\tan^2 \theta$$

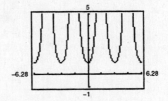

**53.** $\dfrac{\sin \beta}{1 - \cos \beta} \cdot \dfrac{1 + \cos \beta}{1 + \cos \beta} = \dfrac{\sin \beta(1 + \cos \beta)}{1 - \cos^2 \beta}$

$$= \frac{1 + \cos \beta}{\sin \beta}$$

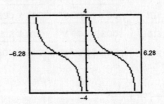

**55.** $\dfrac{\tan^3 \alpha - 1}{\tan \alpha - 1} = \dfrac{(\tan \alpha - 1)(\tan^2 \alpha + \tan \alpha + 1)}{\tan \alpha - 1} = \tan^2 \alpha + \tan \alpha + 1$

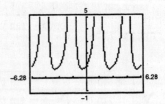

**57.** $\ln|\tan \theta| = \ln\left|\dfrac{\sin \theta}{\cos \theta}\right|$

$$= \ln\frac{|\sin \theta|}{|\cos \theta|}$$

$$= \ln|\sin \theta| - \ln|\cos \theta|$$

**59.** $-\ln(1 + \cos \theta) = \ln(1 + \cos \theta)^{-1}$

$$= \ln\frac{1}{1 + \cos \theta} \cdot \frac{1 - \cos \theta}{1 - \cos \theta}$$

$$= \ln\frac{1 - \cos \theta}{1 - \cos^2 \theta}$$

$$= \ln\frac{1 - \cos \theta}{\sin^2 \theta}$$

$$= \ln(1 - \cos \theta) - \ln \sin^2 \theta$$

$$= \ln(1 - \cos \theta) - 2\ln|\sin \theta|$$

**61.** Since $\sin^2 \theta = 1 - \cos^2 \theta$, then
$\sin \theta = \pm\sqrt{1 - \cos^2 \theta}$; $\sin \theta \neq \sqrt{1 - \cos \theta}$
if $\theta$ lies in Quadrant III or IV.

One such angle is $\theta = \dfrac{7\pi}{4}$.

**63.** $\sqrt{\tan^2 \theta} = |\tan \theta|$

$\sqrt{\tan^2 \theta} \neq \tan \theta$ if $\theta$ lies in Quadrant II or IV.

One such angle is $\theta = \dfrac{3\pi}{4}$.

**65.** $\sin^2 25° + \cos^2 25° = 1$

**67.** $\cos^2 20° + \cos^2 52° + \cos^2 38° + \cos^2 70° = \cos^2 20° + \cos^2 52^2 + \sin^2(90° - 38°) + \sin^2(90° - 70°)$
$$= \cos^2 20° + \cos^2 52^2 + \sin^2 52° + \sin^2 20°$$
$$= (\cos^2 20° + \sin^2 20°) + (\cos^2 52° + \sin^2 52°)$$
$$= 1 + 1$$
$$= 2$$

**69.** When $n$ is even,
$$\cos\left[\frac{(2n + 1)\pi}{2}\right] = \cos\frac{\pi}{2} = 0.$$
When $n$ is odd,
$$\cos\left[\frac{(2n + 1)\pi}{2}\right] = \cos\frac{3\pi}{2} = 0.$$

Thus, $\cos\left[\dfrac{(2n + 1)\pi}{2}\right] = 0$ for all integers $n$.

**71.** $\cos x - \csc x \cot x = \cos x - \dfrac{1}{\sin x}\dfrac{\cos x}{\sin x}$

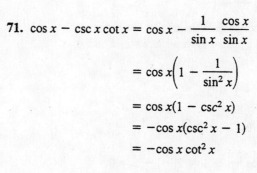

$$= \cos x\left(1 - \frac{1}{\sin^2 x}\right)$$
$$= \cos x(1 - \csc^2 x)$$
$$= -\cos x(\csc^2 x - 1)$$
$$= -\cos x \cot^2 x$$

**73.**

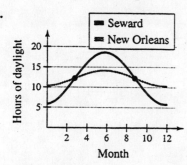

From the graph, you can see that Seward has the greater variation in the number of daylight hours.

Constant in Seward model: 6.4

Constant in New Orleans model: 1.9

**75.** $(2 + 3i) - \sqrt{-26} = 2 + 3i - \sqrt{26}i = 2 + (3 - \sqrt{26})i$

**77.** $\sqrt{-16}(1 + \sqrt{-4}) = 4i(1 + 2i) = 4i + 8i^2 = 4i - 8$

# Section 2.3    Solving Trigonometric Equations

> - You should be able to identify and solve trigonometric equations.
> - A trigonometric equation is a conditional equation. It is true for a specific set of values.
> - To solve trigonometric equations, use algebraic techniques such as collecting like terms, taking square roots, factoring, squaring, converting to quadratic form, using formulas, and using inverse functions. Study the examples in this section.

**Solutions to Odd-Numbered Exercises**

**1.** $y = \sin \dfrac{\pi x}{2} + 1$

From the graph in the textbook we see that the curve has $x$-intercepts at $x = -1$ and at $x = 3$.

**3.** $y = \tan^2\left(\dfrac{\pi x}{6}\right) - 3$

From the graph in the textbook we see that the curve has $x$-intercepts at $x = \pm 2$.

**5.** $2 \cos x - 1 = 0$

(a) $2 \cos \dfrac{\pi}{3} - 1 = 2\left(\dfrac{1}{2}\right) - 1 = 0$

(b) $2 \cos \dfrac{5\pi}{3} - 1 = 2\left(\dfrac{1}{2}\right) - 1 = 0$

**7.** $3 \tan^2 2x - 1 = 0$

(a) $3\left[\tan 2\left(\dfrac{\pi}{12}\right)\right]^2 - 1 = 3 \tan^2 \dfrac{\pi}{6} - 1$

$$= 3\left(\dfrac{1}{\sqrt{3}}\right)^2 - 1$$

$$= 0$$

(b) $3\left[\tan 2\left(\dfrac{5\pi}{12}\right)\right]^2 - 1 = 3 \tan^2 \dfrac{5\pi}{6} - 1$

$$= 3\left(-\dfrac{1}{\sqrt{3}}\right)^2 - 1$$

$$= 0$$

**9.** $2 \sin^2 x - \sin x - 1 = 0$

(a) $2 \sin^2 \dfrac{\pi}{2} - \sin \dfrac{\pi}{2} - 1 = 2(1)^2 - 1 - 1$

$$= 0$$

(b) $2 \sin^2 \dfrac{7\pi}{6} - \sin \dfrac{7\pi}{6} - 1 = 2\left(-\dfrac{1}{2}\right)^2 - \left(-\dfrac{1}{2}\right) - 1$

$$= \dfrac{1}{2} + \dfrac{1}{2} - 1$$

$$= 0$$

**11.** $2 \cos x + 1 = 0$

$$2 \cos x = -1$$

$$\cos x = -\frac{1}{2}$$

$$x = \frac{2\pi}{3} + 2n\pi$$

$$\text{or } x = \frac{4\pi}{3} + 2n\pi$$

**13.** $\sqrt{3} \csc x - 2 = 0$

$$\sqrt{3} \csc x = 2$$

$$\csc x = \frac{2}{\sqrt{3}}$$

$$x = \frac{\pi}{3} + 2n\pi$$

$$\text{or } x = \frac{2\pi}{3} + 2n\pi$$

**15.** $3 \sec^2 x - 4 = 0$

$$\sec x = \pm\frac{2}{\sqrt{3}}$$

$$x = \frac{\pi}{6} + n\pi$$

$$\text{or } x = \frac{5\pi}{6} + n\pi$$

**17.** $2 \sin^2 2x = 1$

$$\sin 2x = \pm\frac{1}{\sqrt{2}} = \pm\frac{\sqrt{2}}{2}$$

$$2x = \frac{\pi}{4} + 2n\pi, \ 2x = \frac{3\pi}{4} + 2n\pi, \ 2x = \frac{5\pi}{4} + 2n\pi, \ 2x = \frac{7\pi}{4} + 2n\pi,$$

$$2x = \frac{9\pi}{4} + 2n\pi, \ 2x = \frac{11\pi}{4} + 2n\pi, \ 2x = \frac{13\pi}{4} + 2n\pi, \ 2x = \frac{15\pi}{4} + 2n\pi,$$

$$\text{Thus, } x = \frac{\pi}{8} + n\pi, \ \frac{3\pi}{8} + n\pi, \ \frac{5\pi}{8} + n\pi, \ \frac{7\pi}{8} + n\pi, \ \frac{9\pi}{8} + n\pi, \ \frac{11\pi}{8} + n\pi,$$

$$\frac{13\pi}{8} + n\pi, \ \frac{15\pi}{8} + n\pi.$$

**19.** $4 \sin^2 x - 3 = 0$

$$\sin x = \pm\frac{\sqrt{3}}{2}$$

$$x = \frac{\pi}{3} + n\pi$$

$$\text{or } x = \frac{2\pi}{3} + n\pi$$

**21.**

$$\sin^2 x = 3 \cos^2 x$$

$$\sin^2 x - 3(1 - \sin^2 x) = 0$$

$$4 \sin^2 x = 3$$

$$\sin x = \pm\frac{\sqrt{3}}{2}$$

$$x = \frac{\pi}{3} + n\pi$$

$$\text{or } x = \frac{2\pi}{3} + n\pi$$

**23.** $(3 \tan^2 x - 1)(\tan^2 x - 3) = 0$

$3 \tan^2 x - 1 = 0$ or $\tan^2 x - 3 = 0$

$$\tan x = \pm\frac{1}{\sqrt{3}} \qquad\qquad \tan x = \pm\sqrt{3}$$

$$x = \frac{\pi}{6} + n\pi \qquad\qquad x = \frac{\pi}{3} + n\pi$$

$$\text{or } x = \frac{5\pi}{6} + n\pi \qquad\qquad \text{or } x = \frac{2\pi}{3} + n\pi$$

**25.**
$$\cos^3 x = \cos x$$
$$\cos^3 x - \cos x = 0$$
$$\cos x(\cos^2 x - 1) = 0$$
$$\cos x = 0 \qquad \text{or} \quad \cos^2 x - 1 = 0$$
$$x = \frac{\pi}{2}, \frac{3\pi}{2} \qquad\qquad \cos x = \pm 1$$
$$x = 0, \pi$$

**27.**
$$3\tan^3 x - \tan x = 0$$
$$\tan x(3\tan^2 x - 1) = 0$$
$$\tan x = 0 \qquad \text{or} \quad 3\tan^2 x - 1 = 0$$
$$x = 0, \pi \qquad\qquad \tan x = \pm\frac{\sqrt{3}}{3}$$
$$x = \frac{\pi}{6}, \frac{5\pi}{6}, \frac{7\pi}{6}, \frac{11\pi}{6}$$

**29.**
$$\sec^2 x - \sec x - 2 = 0$$
$$(\sec x - 2)(\sec x + 1) = 0$$
$$\sec x - 2 = 0 \qquad \text{or} \quad \sec x + 1 = 0$$
$$\sec x = 2 \qquad\qquad\qquad \sec x = -1$$
$$x = \frac{\pi}{3}, \frac{5\pi}{3} \qquad\qquad\qquad x = \pi$$

**31.**
$$2\sin x + \csc x = 0$$
$$2\sin x + \frac{1}{\sin x} = 0$$
$$2\sin^2 x + 1 = 0$$
$$\sin^2 x = -\frac{1}{2} \Longrightarrow \text{No solution}$$

**33.**
$$\csc x + \cot x = 1$$
$$\frac{1}{\sin x} + \frac{\cos x}{\sin x} = 1$$
$$1 + \cos x = \sin x$$
$$(1 + \cos x)^2 = \sin^2 x$$
$$1 + 2\cos x + \cos^2 x = 1 - \cos^2 x$$
$$2\cos^2 x + 2\cos x = 0$$
$$2\cos x(\cos x + 1) = 0$$
$$\cos x = 0 \qquad \text{or} \qquad \cos x = -1$$
$$x = \frac{\pi}{2}, \frac{3\pi}{2} \qquad\qquad x = \pi$$
$$(3\pi/2 \text{ is extraneous.}) \qquad (\pi \text{ is extraneous.})$$
$$x = \pi/2 \text{ is the only solution.}$$

**35.** $\cos\left(\dfrac{x}{2}\right) = \dfrac{\sqrt{2}}{2}$
$$\frac{x}{2} = \frac{\pi}{4} \qquad \text{or} \qquad \frac{x}{2} = \frac{7\pi}{4}$$
$$x = \frac{\pi}{2} \qquad\qquad\qquad x = \frac{7\pi}{2}$$
$$\text{(Disregard since it is}$$
$$\text{outside the interval } [0, 2\pi).)$$

**37.** $\dfrac{1 + \cos x}{1 - \cos x} = 0$
$$1 + \cos x = 0$$
$$\cos x = -1$$
$$x = \pi$$

**39.**
$$2\sec^2 x + \tan^2 x - 3 = 0$$
$$2(\tan^2 x + 1) + \tan^2 x - 3 = 0$$
$$3\tan^2 x - 1 = 0$$
$$\tan x = \pm\frac{\sqrt{3}}{3}$$
$$x = \frac{\pi}{6}, \frac{5\pi}{6}, \frac{7\pi}{6}, \frac{11\pi}{6}$$

**41.**   $6y^2 - 13y + 6 = 0$

$(3y - 2)(2y - 3) = 0$

$3y - 2 = 0$   or   $2y - 3 = 0$

$y = \frac{2}{3}$          $y = \frac{3}{2}$

$6\cos^2 x - 13\cos x + 6 = 0$

$(3\cos x - 2)(2\cos x - 3) = 0$

$3\cos x - 2 = 0$          or   $2\cos x - 3 = 0$

$\cos x = \frac{2}{3}$                    $\cos x = \frac{3}{2}$

$x \approx 0.8411, 5.4421$          (No solution)

**43.**   $2\cos x - \sin x = 0$

$$2 = \frac{\sin x}{\cos x}$$

$$2 = \tan x$$

$$x = \arctan 2 \approx 1.1071$$

$$\text{or } x = \arctan 2 + \pi \approx 4.2487$$

Graph $y_1 = 2\cos x - \sin x$.

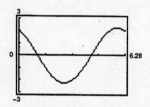

The $x$-intercepts occur at $x \approx 1.1071$ and $x \approx 4.2387$.

**45.**   $$\frac{1 + \sin x}{\cos x} + \frac{\cos x}{1 + \sin x} = 4$$

$$\frac{(1 + \sin x)^2 + \cos^2 x}{\cos x(1 + \sin x)} = 4$$

$$\frac{1 + 2\sin x + \sin^2 x + \cos^2 x}{\cos x(1 + \sin x)} = 4$$

$$\frac{2 + 2\sin x}{\cos x(1 + \sin x)} = 4$$

$$\frac{2}{\cos x} = 4$$

$$\cos x = \frac{1}{2}$$

$$x = \frac{\pi}{3}, \frac{5\pi}{3}$$

Graph $y_1 = \dfrac{1 + \sin x}{\cos x} + \dfrac{\cos x}{1 + \sin x} - 4$.

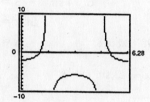

The $x$-intercepts occur at $x = \dfrac{\pi}{3} \approx 1.0472$ and $x = \dfrac{5\pi}{3} \approx 5.2360$.

**47.**   $2\sin x - x = 0$

Graph $y_1 = 2\sin x - x$.

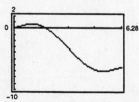

The $x$-intercepts occur at $x = 0$ and $x \approx 1.8955$.

**49.**   $\sec^2 x + 0.5\tan x - 1 = 0$

Graph $y_1 = \dfrac{1}{(\cos x)^2} + 0.5\tan x - 1$.

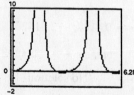

The $x$-intercepts occur at $x = 0$, $x \approx 2.6779$, $x = \pi \approx 3.1416$, and $x \approx 5.8195$.

**51.**  $2 \tan^2 x + 7 \tan x - 15 = 0$

$(2 \tan x - 3)(\tan x + 5) = 0$

$2 \tan x - 3 = 0$      or    $\tan x + 5 = 0$

    $\tan x = 1.5$          $\tan x = -5$

       $x \approx 0.9828, 4.1244$       $x \approx 1.7682, \ 4.9098$

Graph $y_1 = 2 \tan^2 x + 7 \tan x - 15$.

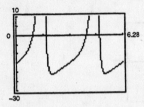

The $x$-intercepts occur at $x \approx 0.9828$, $x \approx 1.7682$, $x \approx 4.1244$, and $x \approx 4.9098$.

**53.**  $12 \sin^2 x - 13 \sin x + 3 = 0$

$(3 \sin x - 1)(4 \sin x - 3) = 0$

$3 \sin x - 1 = 0$      or    $4 \sin x - 3 = 0$

    $\sin x = \frac{1}{3}$          $\sin x = \frac{3}{4}$

       $x \approx 0.3398, 2.8018$       $x \approx 0.8481, \ 2.2935$

Graph $y_1 = 12 \sin^2 x - 13 \sin x + 3$.

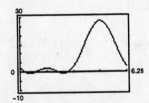

The $x$-intercepts occur at $x \approx 0.3398$, $x \approx 0.8481$, $x \approx 2.2935$, and $x \approx 2.8018$.

**55.**  $\sin^2 x + 2 \sin x - 1 = 0$

$\sin x = \dfrac{-2 \pm \sqrt{4+4}}{2} \approx 0.4142, \ -2.4142$

$\sin x \approx 0.4142$      or    $\sin x \approx -2.4142$

   $x \approx 0.4271, \ 2.7145$      (No solution)

Graph $y_1 = \sin^2 x + 2 \sin x - 1$.

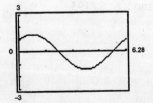

The $x$-intercepts occur at $x \approx 0.4271$ and $x \approx 2.7145$.

**57.** (a) $f(x) = \sin x + \cos x$

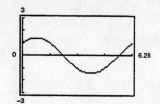

Maximum: $\left(\dfrac{\pi}{4},\ \sqrt{2}\right)$

Minimum: $\left(\dfrac{5\pi}{4},\ -\sqrt{2}\right)$

(b) $\cos x - \sin x = 0$

$$\cos x = \sin x$$

$$1 = \frac{\sin x}{\cos x}$$

$$\tan x = 1$$

$$x = \frac{\pi}{4},\ \frac{5\pi}{4}$$

$$f\left(\frac{\pi}{4}\right) = \sin\frac{\pi}{4} + \cos\frac{\pi}{4} = \frac{\sqrt{2}}{2} + \frac{\sqrt{2}}{2} = \sqrt{2}$$

$$f\left(\frac{5\pi}{4}\right) = \sin\frac{5\pi}{4} + \cos\frac{5\pi}{4} = -\sin\frac{\pi}{4} + \left(-\cos\frac{\pi}{4}\right) = -\frac{\sqrt{2}}{2} - \frac{\sqrt{2}}{2} = -\sqrt{2}$$

Therefore, the maximum point in the interval $[0, 2\pi)$ is $\left(\pi/4,\ \sqrt{2}\right)$ and the minimum point is $\left(5\pi/4,\ -\sqrt{2}\right)$.

**59.** $f(x) = \tan\dfrac{\pi x}{4}$

Since $\tan 0 = 0$, $x = 1$ is the smallest nonnegative fixed point.

**61.** $f(x) = \cos\dfrac{1}{x}$

(a) The domain of $f(x)$ is all real numbers except 0.

(b) The graph has $y$-axis symmetry and a horizontal asymptote at $y = 1$.

(c) As $x \to 0$, $f(x)$ oscillates between $-1$ and 1.

(d) There are infinitely many solutions in the interval $[-1, 1]$.

(e) The greatest solution appears to occur at $x \approx 0.6366$.

**63.**
$$y = \frac{1}{12}(\cos 8t - 3 \sin 8t)$$

$$\frac{1}{12}(\cos 8t - 3 \sin 8t) = 0$$

$$\cos 8t = 3 \sin 8t$$

$$\frac{1}{3} = \tan 8t$$

$$8t \approx 0.32175 + n\pi$$

$$t \approx 0.04 + \frac{n\pi}{8}$$

In the interval $0 \leq t \leq 1$, $t \approx 0.04, 0.43,$ and $0.83$.

**65.** $r = \frac{1}{32}v_0^2 \sin 2\theta$, $r = 300$, $v_0 = 100$

$$300 = \frac{1}{32}(100)^2 \sin 2\theta$$

$$\sin 2\theta = 0.96$$

$$2\theta \approx 1.2870 \qquad \text{or} \quad 2\theta \approx \pi - 1.287 \approx 1.855$$

$$\theta \approx 0.6435 \approx 37° \qquad \theta \approx 0.9275 \approx 53°$$

**67.** $A = 2x \cos x$, $0 < x < \dfrac{\pi}{2}$

(a)

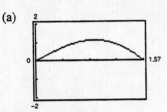

The maximum area of $A \approx 1.12$ occurs when $x \approx 0.86$.

(b) $A \geq 1$ for $0.6 < x < 1.1$

**69.** (a)

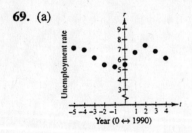

Year $(0 \leftrightarrow 1990)$

(b) By checking the graphs we see that
$(iii)\ r = 1.05 \sin (0.95)(t + 6.32) + 6.20$
best fits the data.

(c) The constant term gives the rate of 6.20%.

(d) Period: $\dfrac{2\pi}{0.95} \approx 7$ years.

(e) $r \approx 6.00$ when $t \approx 10$ which corresponds to 2000.

# Section 2.4    Sum and Difference Formulas

■  You should memorize the sum and difference formulas.

$$\sin(u \pm v) = \sin u \cos v \pm \cos u \sin v$$

$$\cos(u \pm v) = \cos u \cos v \mp \sin u \sin v$$

$$\tan(u \pm v) = \frac{\tan u \pm \tan v}{1 \mp \tan u \tan v}$$

■  You should be able to use these formulas to find the values of the trigonometric functions of angles whose sums or differences are special angles.

■  You should be able to use these formulas to solve trigonometric equations.

## Solutions to Odd-Numbered Exercises

**1.** (a) $\cos\left(\dfrac{\pi}{4} + \dfrac{\pi}{3}\right) = \cos\dfrac{\pi}{4}\cos\dfrac{\pi}{3} - \sin\dfrac{\pi}{4}\sin\dfrac{\pi}{3}$

$$= \frac{\sqrt{2}}{2} \cdot \frac{1}{2} - \frac{\sqrt{2}}{2} \cdot \frac{\sqrt{3}}{2}$$

$$= \frac{\sqrt{2} - \sqrt{6}}{4}$$

(b) $\cos\dfrac{\pi}{4} + \cos\dfrac{\pi}{3} = \dfrac{\sqrt{2}}{2} + \dfrac{1}{2} = \dfrac{\sqrt{2} + 1}{2}$

**3.** (a) $\sin\left(\dfrac{7\pi}{6} - \dfrac{\pi}{3}\right) = \sin\dfrac{5\pi}{6} = \sin\dfrac{\pi}{6} = \dfrac{1}{2}$

(b) $\sin\dfrac{7\pi}{6} - \sin\dfrac{\pi}{3} = -\dfrac{1}{2} - \dfrac{\sqrt{3}}{2} = \dfrac{-1 - \sqrt{3}}{2}$

**5.** Both statements are false. Parts (a) and (b) are unequal in Exercises 1–4.

**7.** $\sin 75° = \sin(30° + 45°)$

$= \sin 30° \cos 45° + \sin 45° \cos 30°$

$= \dfrac{1}{2} \cdot \dfrac{\sqrt{2}}{2} + \dfrac{\sqrt{2}}{2} \cdot \dfrac{\sqrt{3}}{2}$

$= \dfrac{\sqrt{2}}{4}\left(1 + \sqrt{3}\right)$

$\cos 75° = \cos(30° + 45°)$

$= \cos 30° \cos 45° - \sin 30° \sin 45°$

$= \dfrac{\sqrt{3}}{2} \cdot \dfrac{\sqrt{2}}{2} - \dfrac{1}{2} \cdot \dfrac{\sqrt{2}}{2}$

$= \dfrac{\sqrt{2}}{4}\left(\sqrt{3} - 1\right)$

$\tan 75° = \tan(30° + 45°)$

$= \dfrac{\tan 30° + \tan 45°}{1 - \tan 30° \tan 45°}$

$= \dfrac{\left(\sqrt{3}/3\right) + 1}{1 - \left(\sqrt{3}/3\right)} = \dfrac{\sqrt{3} + 3}{3 - \sqrt{3}} \cdot \dfrac{3 + \sqrt{3}}{3 + \sqrt{3}}$

$= \dfrac{6\sqrt{3} + 12}{6} = \sqrt{3} + 2$

**9.** $\sin 105° = \sin(60° + 45°)$

$= \sin 60° \cos 45° + \sin 45° \cos 60°$

$= \dfrac{\sqrt{3}}{2} \cdot \dfrac{\sqrt{2}}{2} + \dfrac{\sqrt{2}}{2} \cdot \dfrac{1}{2}$

$= \dfrac{\sqrt{2}}{4}\left(\sqrt{3} + 1\right)$

$\cos 105° = \cos(60° + 45°)$

$= \cos 60° \cos 45° - \sin 60° \sin 45°$

$= \dfrac{1}{2} \cdot \dfrac{\sqrt{2}}{2} - \dfrac{\sqrt{3}}{2} \cdot \dfrac{\sqrt{2}}{2}$

$= \dfrac{\sqrt{2}}{4}\left(1 - \sqrt{3}\right)$

$\tan 105° = \tan(60° + 45°)$

$= \dfrac{\tan 60° + \tan 45°}{1 - \tan 60° \tan 45°}$

$= \dfrac{\sqrt{3} + 1}{1 - \sqrt{3}} = \dfrac{\sqrt{3} + 1}{1 - \sqrt{3}} \cdot \dfrac{1 + \sqrt{3}}{1 + \sqrt{3}}$

$= \dfrac{4 + 2\sqrt{3}}{-2} = -2 - \sqrt{3}$

**11.** $\sin 195° = \sin(225° - 30°)$

$= \sin 225° \cos 30° - \sin 30° \cos 225°$

$= -\sin 45° \cos 30° + \sin 30° \cos 45°$

$= -\dfrac{\sqrt{2}}{2} \cdot \dfrac{\sqrt{3}}{2} + \dfrac{1}{2} \cdot \dfrac{\sqrt{2}}{2}$

$= \dfrac{\sqrt{2}}{4}\left(\sqrt{3} - 1\right)$

$\cos 195° = \cos(225° - 30°)$

$= \cos 225° \cos 30° + \sin 225° \sin 30°$

$= -\cos 45° \cos 30° - \sin 45° \sin 30°$

$= -\dfrac{\sqrt{2}}{2} \cdot \dfrac{\sqrt{3}}{2} - \dfrac{\sqrt{2}}{2} \cdot \dfrac{1}{2}$

$= -\dfrac{\sqrt{2}}{4}\left(\sqrt{3} + 1\right)$

$\tan 195° = \tan(225° - 30°)$

$= \dfrac{\tan 225° - \tan 30°}{1 + \tan 225° \tan 30°}$

$= \dfrac{\tan 45° - \tan 30°}{1 + \tan 45° \tan 30°}$

$= \dfrac{1 - \left(\sqrt{3}/3\right)}{1 + \left(\sqrt{3}/3\right)} = \dfrac{3 - \sqrt{3}}{3 + \sqrt{3}} \cdot \dfrac{3 - \sqrt{3}}{3 - \sqrt{3}}$

$= \dfrac{12 - 6\sqrt{3}}{6} = 2 - \sqrt{3}$

**13.**  $\sin \dfrac{11\pi}{12} = \sin\left(\dfrac{3\pi}{4} + \dfrac{\pi}{6}\right)$

$\qquad = \sin \dfrac{3\pi}{4} \cos \dfrac{\pi}{6} + \sin \dfrac{\pi}{6} \cos \dfrac{3\pi}{4}$

$\qquad = \dfrac{\sqrt{2}}{2} \cdot \dfrac{\sqrt{3}}{2} + \dfrac{1}{2}\left(-\dfrac{\sqrt{2}}{2}\right)$

$\qquad = \dfrac{\sqrt{2}}{4}\left(\sqrt{3} - 1\right)$

$\cos \dfrac{11\pi}{12} = \cos\left(\dfrac{3\pi}{4} + \dfrac{\pi}{6}\right)$

$\qquad = \cos \dfrac{3\pi}{4} \cos \dfrac{\pi}{6} - \sin \dfrac{3\pi}{4} \sin \dfrac{\pi}{6}$

$\qquad = -\dfrac{\sqrt{2}}{2} \cdot \dfrac{\sqrt{3}}{2} - \dfrac{\sqrt{2}}{2} \cdot \dfrac{1}{2}$

$\qquad = -\dfrac{\sqrt{2}}{4}\left(\sqrt{3} + 1\right)$

$\tan \dfrac{11\pi}{4} = \tan\left(\dfrac{3\pi}{4} + \dfrac{\pi}{6}\right)$

$\qquad = \dfrac{\tan(3\pi/4) + \tan(\pi/6)}{1 - \tan(3\pi/4)\tan(\pi/6)}$

$\qquad = \dfrac{-1 + \left(\sqrt{3}/3\right)}{1 - (-1)\left(\sqrt{3}/3\right)}$

$\qquad = \dfrac{-3 + \sqrt{3}}{3 + \sqrt{3}} \cdot \dfrac{3 - \sqrt{3}}{3 - \sqrt{3}}$

$\qquad = \dfrac{-12 + 6\sqrt{3}}{6} = -2 + \sqrt{3}$

**15.**  $\sin \dfrac{17\pi}{12} = \sin\left(\dfrac{9\pi}{4} - \dfrac{5\pi}{6}\right)$

$\qquad = \sin \dfrac{9\pi}{4} \cos \dfrac{5\pi}{6} - \sin \dfrac{5\pi}{6} \cos \dfrac{9\pi}{4}$

$\qquad = \dfrac{\sqrt{2}}{2}\left(-\dfrac{\sqrt{3}}{2}\right) - \left(\dfrac{1}{2}\right)\left(\dfrac{\sqrt{2}}{2}\right)$

$\qquad = -\dfrac{\sqrt{2}}{4}\left(\sqrt{3} + 1\right)$

$\cos \dfrac{17\pi}{12} = \cos\left(\dfrac{9\pi}{4} - \dfrac{5\pi}{6}\right)$

$\qquad = \cos \dfrac{9\pi}{4} \cos \dfrac{5\pi}{6} + \sin \dfrac{9\pi}{4} \sin \dfrac{5\pi}{6}$

$\qquad = \dfrac{\sqrt{2}}{2}\left(-\dfrac{\sqrt{3}}{2}\right) + \dfrac{\sqrt{2}}{2}\left(\dfrac{1}{2}\right)$

$\qquad = \dfrac{\sqrt{2}}{4}\left(1 - \sqrt{3}\right)$

$\tan \dfrac{17\pi}{12} = \tan\left(\dfrac{9\pi}{4} - \dfrac{5\pi}{6}\right)$

$\qquad = \dfrac{\tan(9\pi/4) - \tan(5\pi/6)}{1 + \tan(9\pi/4)\tan(5\pi/6)}$

$\qquad = \dfrac{1 - \left(-\sqrt{3}/3\right)}{1 + \left(-\sqrt{3}/3\right)}$

$\qquad = \dfrac{3 + \sqrt{3}}{3 - \sqrt{3}} \cdot \dfrac{3 + \sqrt{3}}{3 + \sqrt{3}}$

$\qquad = \dfrac{12 + 6\sqrt{3}}{6} = 2 + \sqrt{3}$

**17.**  $\quad 285° = 225° + 60°$

$\sin 285° = \sin(225° + 60°)$

$\qquad = \sin 225° \cos 60° + \cos 225° \sin 60°$

$\qquad = -\dfrac{\sqrt{2}}{2}\left(\dfrac{1}{2}\right) - \dfrac{\sqrt{2}}{2}\left(\dfrac{\sqrt{3}}{2}\right) = -\dfrac{\sqrt{2}}{4}\left(1 + \sqrt{3}\right)$

$\cos 285° = \cos(225° + 60°)$

$\qquad = \cos 225° \cos 60° - \sin 225° \sin 60°$

$\qquad = -\dfrac{\sqrt{2}}{2}\left(\dfrac{1}{2}\right) - \left(-\dfrac{\sqrt{2}}{2}\right)\left(\dfrac{\sqrt{3}}{2}\right) = \dfrac{\sqrt{2}}{4}\left(-1 + \sqrt{3}\right)$

$\qquad = \dfrac{\sqrt{2}}{4}\left(\sqrt{3} - 1\right)$

$\tan 285° = \tan(225° + 60°)$

$\qquad = \dfrac{\tan 225° + \tan 60°}{1 - \tan 225° \tan 60°} = \dfrac{1 + \sqrt{3}}{1 - \sqrt{3}} \cdot \dfrac{1 + \sqrt{3}}{1 + \sqrt{3}}$

$\qquad = \dfrac{4 + 2\sqrt{3}}{-2} = -2 - \sqrt{3} = -\left(2 + \sqrt{3}\right)$

**19.**
$$-\frac{13\pi}{12} = -\left(\frac{3\pi}{4} + \frac{\pi}{3}\right)$$

$$\sin\left[-\left(\frac{3\pi}{4} + \frac{\pi}{3}\right)\right] = -\sin\left(\frac{3\pi}{4} + \frac{\pi}{3}\right)$$

$$= -\left[\sin\frac{3\pi}{4}\cos\frac{\pi}{3} + \cos\frac{3\pi}{4}\sin\frac{\pi}{3}\right]$$

$$= -\left[\frac{\sqrt{2}}{2}\left(\frac{1}{2}\right) + \left(-\frac{\sqrt{2}}{2}\right)\left(\frac{\sqrt{3}}{2}\right)\right]$$

$$= -\frac{\sqrt{2}}{4}\left(1 - \sqrt{3}\right) = \frac{\sqrt{2}}{4}\left(\sqrt{3} - 1\right)$$

$$\cos\left[-\left(\frac{3\pi}{4} + \frac{\pi}{3}\right)\right] = \cos\left(\frac{3\pi}{4} + \frac{\pi}{3}\right)$$

$$= \cos\frac{3\pi}{4}\cos\frac{\pi}{3} - \sin\frac{3\pi}{4}\sin\frac{\pi}{3}$$

$$= -\frac{\sqrt{2}}{2}\left(\frac{1}{2}\right) - \frac{\sqrt{2}}{2}\left(\frac{\sqrt{3}}{2}\right) = -\frac{\sqrt{2}}{4}\left(1 + \sqrt{3}\right)$$

$$\tan\left[-\left(\frac{3\pi}{4} + \frac{\pi}{3}\right)\right] = -\tan\left(\frac{3\pi}{4} + \frac{\pi}{3}\right)$$

$$= -\frac{\tan(3\pi/4) + \tan(\pi/3)}{1 - \tan(3\pi/4)\tan(\pi/3)} = -\frac{-1 + \sqrt{3}}{1 - (-\sqrt{3})}$$

$$= \frac{1 - \sqrt{3}}{1 + \sqrt{3}} \cdot \frac{1 - \sqrt{3}}{1 - \sqrt{3}} = \frac{4 - 2\sqrt{3}}{-2} = -2 + \sqrt{3}$$

**21.** $\cos 25° \cos 15° - \sin 25° \sin 15° = \cos(25° + 15°) = \cos 40°$

**23.** $\sin 230° \cos 30° - \cos 230° \sin 30° = \sin(230° - 30°) = \sin 200°$

**25.** $\dfrac{\tan 325° - \tan 86°}{1 + \tan 325° \tan 86°} = \tan(325° - 86°) = \tan 239°$

**27.** $\sin 3 \cos 1.2 - \cos 3 \sin 1.2 = \sin(3 - 1.2) = \sin 1.8$

**29.** $\dfrac{\tan 2x + \tan x}{1 - \tan 2x \tan x} = \tan(2x + x) = \tan 3x$

**For Exercises 31 – 37, we have:**

$\sin u = \frac{5}{13}$, *u* **in Quadrant II** $\implies$ $\cos u = -\frac{12}{13}$

$\cos v = -\frac{3}{5}$, *v* **in Quadrant II** $\implies$ $\sin v = \frac{4}{5}$

**31.** $\sin(u + v) = \sin u \cos v + \cos u \sin v$
$$= \left(\tfrac{5}{13}\right)\left(-\tfrac{3}{5}\right) + \left(-\tfrac{12}{13}\right)\left(\tfrac{4}{5}\right)$$
$$= -\tfrac{63}{65}$$

**33.** $\cos(u + v) = \cos u \cos v - \sin u \sin v$
$$= \left(-\tfrac{12}{13}\right)\left(-\tfrac{3}{5}\right) - \left(\tfrac{5}{13}\right)\left(\tfrac{4}{5}\right)$$
$$= \tfrac{16}{65}$$

**35.** $\sec(u + v) = \dfrac{1}{\cos(u + v)} = \dfrac{1}{16/65} = \dfrac{65}{16}$

Use Exercise 33 for $\cos(u + v)$.

**37.** $\tan(u - v) = \dfrac{\tan u - \tan v}{1 + \tan u \tan v} = \dfrac{-5/12 - (-4/3)}{1 + (-5/12)(-4/3)}$

$= \dfrac{11/12}{14/9} = \dfrac{33}{56}$

**For Exercises 39–43, we have:**

$\sin u = -\frac{7}{25}$, $u$ **in Quadrant III** $\Longrightarrow \cos u = -\frac{24}{25}$

$\cos v = -\frac{4}{5}$, $v$ **in Quadrant III** $\Longrightarrow \sin v = -\frac{3}{5}$

**39.** $\cos(u + v) = \cos u \cos v - \sin u \sin v$

$= \left(-\frac{24}{25}\right)\left(-\frac{4}{5}\right) - \left(-\frac{7}{25}\right)\left(-\frac{3}{5}\right)$

$= \frac{3}{5}$

**41.** $\sin(v - u) = \sin v \cos u - \cos v \sin u$

$= \left(-\frac{3}{5}\right)\left(-\frac{24}{25}\right) - \left(-\frac{4}{5}\right)\left(-\frac{7}{25}\right)$

$= \frac{44}{125}$

**43.** $\csc(u + v) = \dfrac{1}{\sin(u + v)} = \dfrac{1}{\sin u \cos v + \cos u \sin v}$

$= \dfrac{1}{\left(-\frac{7}{25}\right)\left(-\frac{4}{5}\right) + \left(-\frac{24}{25}\right)\left(-\frac{3}{5}\right)} = \dfrac{1}{\frac{4}{5}} = \dfrac{5}{4}$

**45.** $\sin(3\pi - x) = \sin 3\pi \cos x - \sin x \cos 3\pi = (0)(\cos x) - (\sin x)(-1) = \sin x$

**47.** $\sin\left(\dfrac{\pi}{6} + x\right) = \sin \dfrac{\pi}{6} \cos x + \sin x \cos \dfrac{\pi}{6} = \dfrac{1}{2}\left(\cos x + \sqrt{3} \sin x\right)$

**49.** $\cos(\pi - \theta) + \sin\left(\dfrac{\pi}{2} + \theta\right) = \cos \pi \cos \theta + \sin \pi \sin \theta + \sin \dfrac{\pi}{2} \cos \theta + \sin \theta \cos \dfrac{\pi}{2}$

$= (-1)(\cos \theta) + (0)(\sin \theta) + (1)(\cos \theta) + (\sin \theta)(0)$

$= -\cos \theta + \cos \theta$

$= 0$

**51.** $\cos(x + y) \cos(x - y) = (\cos x \cos y - \sin x \sin y)(\cos x \cos y + \sin x \sin y)$

$= \cos^2 x \cos^2 y - \sin^2 x \sin^2 y$

$= \cos^2 x(1 - \sin^2 y) - \sin^2 x \sin^2 y$

$= \cos^2 x - \cos^2 x \sin^2 y - \sin^2 x \sin^2 y$

$= \cos^2 x - \sin^2 y(\cos^2 x + \sin^2 x)$

$= \cos^2 x - \sin^2 y$

**53.** $\sin(x + y) + \sin(x - y) = \sin x \cos y + \sin y \cos x + \sin x \cos y - \sin y \cos x$

$$= 2 \sin x \cos y$$

**55.** $\cos(n\pi + \theta) = \cos n\pi \cos \theta - \sin n\pi \sin \theta$

$$= (-1)^n (\cos \theta) - (0)(\sin \theta)$$

$$= (-1)^n (\cos \theta), \text{ where } n \text{ is an integer.}$$

**57.** $C = \arctan \dfrac{b}{a} \implies \sin C = \dfrac{b}{\sqrt{a^2 + b^2}}, \cos C = \dfrac{a}{\sqrt{a^2 + b^2}}$

$$\sqrt{a^2 + b^2} \sin(B\theta + C) = \sqrt{a^2 + b^2}\left(\sin B\theta \cdot \frac{a}{\sqrt{a^2 + b^2}} + \frac{b}{\sqrt{a^2 + b^2}} \cdot \cos \beta\theta\right) = a \sin B\theta + b \cos B\theta$$

**59.** $\cos\left(\dfrac{3\pi}{2} - x\right) = \cos \dfrac{3\pi}{2} \cos x + \sin \dfrac{3\pi}{2} \sin x$

$$= (0)(\cos x) + (-1)(\sin x)$$

$$= -\sin x$$

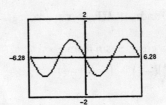

**61.** $\sin\left(\dfrac{3\pi}{2} + \theta\right) + \sin(\pi - \theta)$

$$= \sin \frac{3\pi}{2} \cos \theta + \cos \frac{3\pi}{2} \sin \theta + \sin \pi \cos \theta - \cos \pi \sin \theta$$

$$= (-1)(\cos \theta) + (0)(\sin \theta) + (0)(\cos \theta) - (-1)(\sin \theta)$$

$$= -\cos \theta + \sin \theta$$

$$= \sin \theta - \cos \theta$$

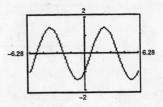

**63.** $\sin \theta + \cos \theta$

$a = 1, b = 1, B = 1$

(a) $C = \arctan \dfrac{b}{a} = \arctan 1 = \dfrac{\pi}{4}$

$\sin \theta + \cos \theta = \sqrt{a^2 + b^2} \sin(B\theta + C)$

$$= \sqrt{2} \sin\left(\theta + \frac{\pi}{4}\right)$$

(b) $C = \arctan \dfrac{a}{b} = \arctan 1 = \dfrac{\pi}{4}$

$\sin \theta + \cos \theta = \sqrt{a^2 + b^2} \cos(B\theta - C)$

$$= \sqrt{2} \cos\left(\theta - \frac{\pi}{4}\right)$$

**65.** $12 \sin 3\theta + 5 \cos 3\theta$

$a = 12, \ b = 5, \ B = 3$

(a) $C = \arctan \dfrac{b}{a} = \arctan \dfrac{5}{12} \approx 0.3948$

$12 \sin 3\theta + 5 \cos 3\theta = \sqrt{a^2 + b^2} \sin(B\theta + C)$

$\approx 13 \sin(3\theta + 0.3948)$

(b) $C = \arctan \dfrac{a}{b} = \arctan \dfrac{12}{5} \approx 1.1760$

$12 \sin 3\theta + 5 \cos 3\theta = \sqrt{a^2 + b^2} \cos(B\theta - C)$

$\approx 13 \cos(3\theta - 1.1760)$

**67.** $C = \arctan \dfrac{b}{a} = \dfrac{\pi}{2} \ \Longrightarrow \ a = 0$

$\sqrt{a^2 + b^2} = 2 \ \Longrightarrow \ b = 2$

$B = 1$

$2 \sin\left(\theta + \dfrac{\pi}{2}\right) = (0)(\sin\theta) + (2)(\cos\theta) = 2 \cos \theta$

**69.** $\sin(\arcsin x + \arccos x) = \sin(\arcsin x)\cos(\arccos x) + \sin(\arccos x)\cos(\arcsin x)$

$= x \cdot x + \sqrt{1 - x^2} \cdot \sqrt{1 - x^2}$

$= x^2 + 1 - x^2$

$= 1$

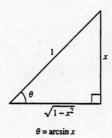

$\theta = \arcsin x$

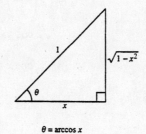

$\theta = \arccos x$

**71.**
$$\sin\left(x + \dfrac{\pi}{3}\right) + \sin\left(x - \dfrac{\pi}{3}\right) = 1$$

$$\sin x \cos \dfrac{\pi}{3} + \cos x \sin \dfrac{\pi}{3} + \sin x \cos \dfrac{\pi}{3} - \cos x \sin \dfrac{\pi}{3} = 1$$

$$2 \sin x(0.5) = 1$$

$$\sin x = 1$$

$$x = \dfrac{\pi}{2}$$

**73.** $\cos\left(x + \dfrac{\pi}{4}\right) - \cos\left(x - \dfrac{\pi}{4}\right) = 1$

$$\cos x \cos\frac{\pi}{4} - \sin x \sin\frac{\pi}{4} - \left(\cos x \cos\frac{\pi}{4} + \sin x \sin\frac{\pi}{4}\right) = 1$$

$$-2\sin x\left(\frac{\sqrt{2}}{2}\right) = 1$$

$$-\sqrt{2}\sin x = 1$$

$$\sin x = -\frac{1}{\sqrt{2}}$$

$$\sin x = -\frac{\sqrt{2}}{2}$$

$$x = \frac{5\pi}{4}, \frac{7\pi}{4}$$

**75.**          Analytically: $\cos\left(x + \dfrac{\pi}{4}\right) + \cos\left(x - \dfrac{\pi}{4}\right) = 1$

$$\cos x \cos\frac{\pi}{4} - \sin x \sin\frac{\pi}{4} + \cos x \cos\frac{\pi}{4} + \sin x \sin\frac{\pi}{4} = 1$$

$$2\cos x\left(\frac{\sqrt{2}}{2}\right) = 1$$

$$\sqrt{2}\cos x = 1$$

$$\cos x = \frac{1}{\sqrt{2}}$$

$$\cos x = \frac{\sqrt{2}}{2}$$

$$x = \frac{\pi}{4}, \frac{7\pi}{4}$$

Graphically: Graph $y_1 = \cos\left(x + \dfrac{\pi}{4}\right) + \cos\left(x - \dfrac{\pi}{4}\right)$ and $y_2 = 1$.

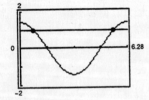

The points of intersection occur at $x = \dfrac{\pi}{4}$ and $x = \dfrac{7\pi}{4}$.

**77.** $\sin^2\left(\theta + \dfrac{\pi}{4}\right) + \sin^2\left(\theta - \dfrac{\pi}{4}\right) = \left[\sin\theta\cos\dfrac{\pi}{4} + \cos\theta\sin\dfrac{\pi}{4}\right]^2 + \left[\sin\theta\cos\dfrac{\pi}{4} - \cos\theta\sin\dfrac{\pi}{4}\right]^2$

$= \left[\dfrac{\sin\theta}{\sqrt{2}} + \dfrac{\cos\theta}{\sqrt{2}}\right]^2 + \left[\dfrac{\sin\theta}{\sqrt{2}} - \dfrac{\cos\theta}{\sqrt{2}}\right]^2$

$= \dfrac{\sin^2\theta}{2} + \sin\theta\cos\theta + \dfrac{\cos^2\theta}{2} + \dfrac{\sin^2\theta}{2} - \sin\theta\cos\theta + \dfrac{\cos^2\theta}{2}$

$= \sin^2\theta + \cos^2\theta$

$= 1$

**79.** $y = \dfrac{1}{3}\sin 2t + \dfrac{1}{4}\cos 2t$

(a) $a = \dfrac{1}{3},\ b = \dfrac{1}{4},\ B = 2$

$C = \arctan\dfrac{b}{a} = \arctan\dfrac{3}{4} \approx 0.6435$

$y \approx \sqrt{\left(\tfrac{1}{3}\right)^2 + \left(\tfrac{1}{4}\right)^2}\ \sin(2t + 0.6435)$

$= \dfrac{5}{12}\sin(2t + 0.6435)$

(b) Amplitude: $\dfrac{5}{12}$ feet

(c) Frequency: $\dfrac{1}{\text{period}} = \dfrac{b}{2\pi} = \dfrac{2}{2\pi} = \dfrac{1}{\pi}$ cycles per second

# Section 2.5    Multiple-Angle and Product-to-Sum Formulas

■ You should know the following double-angle formulas.

(a) $\sin 2u = 2 \sin u \cos u$

(b) $\cos 2u = \cos^2 u - \sin^2 u$

$\qquad = 2 \cos^2 u - 1$

$\qquad = 1 - 2 \sin^2 u$

(c) $\tan 2u = \dfrac{2 \tan u}{1 - \tan^2 u}$

■ You should be able to reduce the power of a trigonometric function.

(a) $\sin^2 u = \dfrac{1 - \cos 2u}{2}$

(b) $\cos^2 u = \dfrac{1 + \cos 2u}{2}$

(c) $\tan^2 u = \dfrac{1 - \cos 2u}{1 + \cos 2u}$

■ You should be able to use the half-angle formulas.

(a) $\sin \dfrac{u}{2} = \pm \sqrt{\dfrac{1 - \cos u}{2}}$

(b) $\cos \dfrac{u}{2} = \pm \sqrt{\dfrac{1 + \cos u}{2}}$

(c) $\tan \dfrac{u}{2} = \dfrac{1 - \cos u}{\sin u} = \dfrac{\sin u}{1 + \cos u}$

■ You should be able to use the product-sum formulas.

(a) $\sin u \sin v = \dfrac{1}{2} [\cos(u - v) - \cos(u + v)]$  (b) $\cos u \cos v = \dfrac{1}{2} [\cos(u - v) + \cos(u + v)]$

(c) $\sin u \cos v = \dfrac{1}{2} [\sin(u + v) + \sin(u - v)]$  (d) $\cos u \sin v = \dfrac{1}{2} [\sin(u + v) - \sin(u - v)]$

■ You should be able to use the sum-product formulas.

(a) $\sin x + \sin y = 2 \sin\left(\dfrac{x + y}{2}\right) \cos\left(\dfrac{x - y}{2}\right)$  (b) $\sin x - \sin y = 2 \cos\left(\dfrac{x + y}{2}\right) \sin\left(\dfrac{x - y}{2}\right)$

(c) $\cos x + \cos y = 2 \cos\left(\dfrac{x + y}{2}\right) \cos\left(\dfrac{x - y}{2}\right)$  (d) $\cos x - \cos y = -2 \sin\left(\dfrac{x + y}{2}\right) \sin\left(\dfrac{x - y}{2}\right)$

**Solutions to Odd-Numbered Exercises**

**Figure for Exercises 1–7**

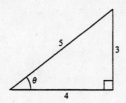

$\sin \theta = \frac{3}{5}$

$\cos \theta = \frac{4}{5}$

$\tan \theta = \frac{3}{4}$

**1.** $\sin \theta = \frac{3}{5}$

**3.** $\cos 2\theta = 2 \cos^2 \theta - 1$

$\qquad = 2\left(\frac{4}{5}\right)^2 - 1$

$\qquad = \frac{32}{25} - \frac{25}{25}$

$\qquad = \frac{7}{25}$

**5.** $\tan 2\theta = \dfrac{2 \tan \theta}{1 - \tan^2 \theta}$

$\qquad = \dfrac{2(3/4)}{1 - (3/4)^2}$

$\qquad = \dfrac{3/2}{1 - 9/16}$

$\qquad = \dfrac{3}{2} \cdot \dfrac{16}{7}$

$\qquad = \dfrac{24}{7}$

**7.** $\csc 2\theta = \dfrac{1}{\sin 2\theta}$

$\qquad = \dfrac{1}{2 \sin \theta \cos \theta}$

$\qquad = \dfrac{1}{2(3/5)(4/5)}$

$\qquad = \dfrac{25}{24}$

**9.** $\qquad \sin 2x - \sin x = 0$

$\quad 2 \sin x \cos x - \sin x = 0$

$\quad \sin x (2 \cos x - 1) = 0$

$\sin x = 0 \quad$ or $\quad 2 \cos x - 1 = 0$

$x = 0, \ \pi \qquad\qquad \cos x = \dfrac{1}{2}$

$\qquad\qquad\qquad\qquad x = \dfrac{\pi}{3}, \dfrac{5\pi}{3}$

$x = 0, \ \dfrac{\pi}{3}, \ \pi, \ \dfrac{5\pi}{3}$

**11.** $4 \sin x \cos x = 1$

$\quad 2 \sin 2x = 1$

$\qquad \sin 2x = \dfrac{1}{2}$

$2x = \dfrac{\pi}{6} + 2n\pi \quad$ or $\quad 2x = \dfrac{5\pi}{6} + 2n\pi$

$x = \dfrac{\pi}{12} + n\pi \qquad\qquad x = \dfrac{5\pi}{12} + n\pi$

$x = \dfrac{\pi}{12}, \dfrac{13\pi}{12} \qquad\qquad x = \dfrac{5\pi}{12}, \dfrac{17\pi}{12}$

**13.**
$$\cos 2x = \cos x$$
$$\cos^2 x - \sin^2 x = \cos x$$
$$\cos^2 x - (1 - \cos^2 x) - \cos x = 0$$
$$2\cos^2 x - \cos x - 1 = 0$$
$$(2\cos x + 1)(\cos x - 1) = 0$$

$$2\cos x + 1 = 0 \qquad \text{or} \quad \cos x - 1 = 0$$

$$\cos x = -\frac{1}{2} \qquad\qquad \cos x = 1$$

$$x = \frac{2\pi}{3}, \frac{4\pi}{3} \qquad\qquad x = 0$$

**15.**
$$\tan 2x - \cot x = 0$$
$$\frac{2\tan x}{1 - \tan^2 x} = \cot x$$
$$2\tan x = \cot x(1 - \tan^2 x)$$
$$2\tan x = \cot x - \cot x \tan^2 x$$
$$2\tan x = \cot x - \tan x$$
$$3\tan x = \cot x$$
$$3\tan x - \cot x = 0$$
$$3\tan x - \frac{1}{\tan x} = 0$$
$$\frac{3\tan^2 x - 1}{\tan x} = 0$$
$$\frac{1}{\tan x}(3\tan^2 x - 1) = 0$$
$$\cot x(3\tan^2 x - 1) = 0$$
$$\cot x = 0 \qquad \text{or} \quad 3\tan^2 x - 1 = 0$$

$$x = \frac{\pi}{2}, \frac{3\pi}{2} \qquad\qquad \tan^2 x = \frac{1}{3}$$

$$\tan x = \pm\frac{\sqrt{3}}{3}$$

$$x = \frac{\pi}{6}, \frac{5\pi}{6}, \frac{7\pi}{6}, \frac{11\pi}{6}$$

$$x = \frac{\pi}{6}, \frac{\pi}{2}, \frac{5\pi}{6}, \frac{7\pi}{6}, \frac{3\pi}{2}, \frac{11\pi}{6}$$

**17.**
$$\sin 4x = -2\sin 2x$$
$$\sin 4x + 2\sin 2x = 0$$
$$2\sin 2x \cos 2x + 2\sin 2x = 0$$
$$2\sin 2x(\cos 2x + 1) = 0$$

$$2\sin 2x = 0 \qquad \text{or} \qquad \cos 2x + 1 = 0$$

$$\sin 2x = 0 \qquad\qquad\qquad \cos 2x = -1$$

$$2x = n\pi \qquad\qquad\qquad 2x = \pi + 2n\pi$$

$$x = \frac{n}{2}\pi \qquad\qquad\qquad x = \frac{\pi}{2} + n\pi$$

$$x = 0, \frac{\pi}{2}, \pi, \frac{3\pi}{2} \qquad\qquad x = \frac{\pi}{2}, \frac{3\pi}{2}$$

**19.** $f(x) = 6\sin x \cos x$
$$= 3(2\sin x \cos x)$$
$$= 3\sin 2x$$

**21.** $g(x) = 4 - 8\sin^2 x$
$$= 4(1 - 2\sin^2 x)$$
$$= 4\cos 2x$$

**23.** $\sin u = \dfrac{3}{5}, \ 0 < u < \dfrac{\pi}{2} \implies \cos u = \dfrac{4}{5}$

$\sin 2u = 2 \sin u \cos u = 2 \cdot \dfrac{3}{5} \cdot \dfrac{4}{5} = \dfrac{24}{25}$

$\cos 2u = \cos^2 u - \sin^2 u = \dfrac{16}{25} - \dfrac{9}{25} = \dfrac{7}{25}$

$\tan 2u = \dfrac{2 \tan u}{1 - \tan^2 u} = \dfrac{2(3/4)}{1 - (9/16)} = \dfrac{24}{7}$

**25.** $\tan u = \dfrac{1}{2}, \ \pi < u < \dfrac{3\pi}{2} \implies \sin u = -\dfrac{1}{\sqrt{5}}$ and

$\cos u = -\dfrac{2}{\sqrt{5}}$

$\sin 2u = 2 \sin u \cos u = 2\left(-\dfrac{1}{\sqrt{5}}\right)\left(-\dfrac{2}{\sqrt{5}}\right) = \dfrac{4}{5}$

$\cos 2u = \cos^2 u - \sin^2 u = \left(-\dfrac{2}{\sqrt{5}}\right)^2 - \left(-\dfrac{1}{\sqrt{5}}\right)^2 = \dfrac{3}{5}$

$\tan 2u = \dfrac{2 \tan u}{1 - \tan^2 u} = \dfrac{2(1/2)}{1 - (1/4)} = \dfrac{4}{3}$

**27.** $\sec u = -\dfrac{5}{2}, \ \dfrac{\pi}{2} < u < \pi \implies \sin u = \dfrac{\sqrt{21}}{5}$ and $\cos u = -\dfrac{2}{5}$

$\sin 2u = 2 \sin u \cos u = 2\left(\dfrac{\sqrt{21}}{5}\right)\left(-\dfrac{2}{5}\right) = -\dfrac{4\sqrt{21}}{25}$

$\cos 2u = \cos^2 u - \sin^2 u = \left(-\dfrac{2}{5}\right)^2 - \left(\dfrac{\sqrt{21}}{5}\right)^2 = -\dfrac{17}{25}$

$\tan 2u = \dfrac{2 \tan u}{1 - \tan^2 u} = \dfrac{2(-\sqrt{21}/2)}{1 - (-\sqrt{21}/2)^2}$

$= \dfrac{-\sqrt{21}}{1 - 21/4} = \dfrac{4\sqrt{21}}{17}$

**29.** $\cos^4 x = (\cos^2 x)(\cos^2 x) = \left(\dfrac{1 + \cos 2x}{2}\right)\left(\dfrac{1 + \cos 2x}{2}\right) = \dfrac{1 + 2\cos 2x + \cos^2 2x}{4}$

$= \dfrac{1 + 2\cos 2x + (1 + \cos 4x)/2}{4}$

$= \dfrac{2 + 4\cos 2x + 1 + \cos 4x}{8}$

$= \dfrac{3 + 4\cos 2x + \cos 4x}{8}$

$= \dfrac{1}{8}(3 + 4\cos 2x + \cos 4x)$

**31.** $(\sin^2 x)(\cos^2 x) = \left(\dfrac{1 - \cos 2x}{2}\right)\left(\dfrac{1 + \cos 2x}{2}\right)$

$= \dfrac{1 - \cos^2 2x}{4}$

$= \dfrac{1}{4}\left(1 - \dfrac{1 + \cos 4x}{2}\right)$

$= \dfrac{1}{8}(2 - 1 - \cos 4x)$

$= \dfrac{1}{8}(1 - \cos 4x)$

**33.** $\sin^2 x \cos^4 x = \sin^2 x \cos^2 x \cos^2 x = \left(\dfrac{1 - \cos 2x}{2}\right)\left(\dfrac{1 + \cos 2x}{2}\right)\left(\dfrac{1 + \cos 2x}{2}\right)$

$$= \frac{1}{8}(1 - \cos 2x)(1 + \cos 2x)(1 + \cos 2x)$$

$$= \frac{1}{8}(1 - \cos^2 2x)(1 + \cos 2x)$$

$$= \frac{1}{8}(1 + \cos 2x - \cos^2 2x - \cos^3 2x)$$

$$= \frac{1}{8}\left[1 + \cos 2x - \left(\frac{1 + \cos 4x}{2}\right) - \cos 2x\left(\frac{1 + \cos 4x}{2}\right)\right]$$

$$= \frac{1}{16}[2 + 2\cos 2x - 1 - \cos 4x - \cos 2x - \cos 2x \cos 4x]$$

$$= \frac{1}{16}\left[1 + \cos 2x - \cos 4x - \left(\frac{1}{2}\cos 2x + \frac{1}{2}\cos 6x\right)\right]$$

$$= \frac{1}{32}(2 + 2\cos 2x - 2\cos 4x - \cos 2x - \cos 6x)$$

$$= \frac{1}{32}(2 + \cos 2x - 2\cos 4x - \cos 6x)$$

**Figure for Exercises 35 – 39**

$$\sin \theta = \tfrac{5}{13}$$
$$\cos \theta = \tfrac{12}{13}$$

**35.** $\cos \dfrac{\theta}{2} = \sqrt{\dfrac{1 + \cos \theta}{2}} = \sqrt{\dfrac{1 + 12/13}{2}} = \sqrt{\dfrac{25}{26}} = \dfrac{5}{\sqrt{26}}$

**37.** $\tan \dfrac{\theta}{2} = \dfrac{\sin \theta}{1 + \cos \theta} = \dfrac{5/13}{1 + 12/13} = \dfrac{5}{25} = \dfrac{1}{5}$

**39.** $\csc \dfrac{\theta}{2} = \dfrac{1}{\sin(\theta/2)} = \dfrac{1}{\sqrt{(1 - \cos \theta)/2}} = \dfrac{1}{\sqrt{\left(1 - \frac{12}{13}\right)/2}} = \dfrac{1}{\sqrt{1/26}} = \sqrt{26}$

**41.** $\sin 105° = \sin\left(\dfrac{1}{2} \cdot 210°\right) = \sqrt{\dfrac{1 - \cos 210°}{2}} = \sqrt{\dfrac{1 + (\sqrt{3}/2)}{2}} = \dfrac{1}{2}\sqrt{2 + \sqrt{3}}$

$\cos 105° = \cos\left(\dfrac{1}{2} \cdot 210°\right) = -\sqrt{\dfrac{1 + \cos 210°}{2}} = -\sqrt{\dfrac{1 - (\sqrt{3}/2)}{2}} = -\dfrac{1}{2}\sqrt{2 - \sqrt{3}}$

$\tan 105° = \tan\left(\dfrac{1}{2} \cdot 210°\right) = \dfrac{\sin 210°}{1 + \cos 210°} = \dfrac{-1/2}{1 - (\sqrt{3}/2)} = -2 - \sqrt{3}$

**43.** $\sin 112°\,30' = \sin\left(\frac{1}{2}\cdot 225°\right) = \sqrt{\dfrac{1-\cos 225°}{2}} = \sqrt{\dfrac{1+(\sqrt{2}/2)}{2}} = \frac{1}{2}\sqrt{2+\sqrt{2}}$

$\cos 112°\,30' = \cos\left(\frac{1}{2}\cdot 225°\right) = -\sqrt{\dfrac{1+\cos 225°}{2}} = -\sqrt{\dfrac{1-(\sqrt{2}/2)}{2}} = -\frac{1}{2}\sqrt{2-\sqrt{2}}$

$\tan 112°\,30' = \tan\left(\frac{1}{2}\cdot 225°\right) = \dfrac{\sin 225°}{1+\cos 225°} = \dfrac{-\sqrt{2}/2}{1-(\sqrt{2}/2)} = -1-\sqrt{2}$

**45.** $\sin\dfrac{\pi}{8} = \sin\left[\frac{1}{2}\left(\frac{\pi}{4}\right)\right] = \sqrt{\dfrac{1-\cos(\pi/4)}{2}} = \frac{1}{2}\sqrt{2-\sqrt{2}}$

$\cos\dfrac{\pi}{8} = \cos\left[\frac{1}{2}\left(\frac{\pi}{4}\right)\right] = \sqrt{\dfrac{1+\cos(\pi/4)}{2}} = \frac{1}{2}\sqrt{2+\sqrt{2}}$

$\tan\dfrac{\pi}{8} = \tan\left[\frac{1}{2}\left(\frac{\pi}{4}\right)\right] = \dfrac{\sin(\pi/4)}{1+\cos(\pi/4)} = \dfrac{\sqrt{2}/2}{1+(\sqrt{2}/2)} = \sqrt{2}-1$

**47.** $\sin u = \dfrac{5}{13},\ \dfrac{\pi}{2}<u<\pi \implies \cos u = -\dfrac{12}{13}$

$\sin\left(\dfrac{u}{2}\right) = \sqrt{\dfrac{1-\cos u}{2}} = \sqrt{\dfrac{1+(12/13)}{2}} = \dfrac{5\sqrt{26}}{26}$

$\cos\left(\dfrac{u}{2}\right) = \sqrt{\dfrac{1+\cos u}{2}} = \sqrt{\dfrac{1-(12/13)}{2}} = \dfrac{\sqrt{26}}{26}$

$\tan\left(\dfrac{u}{2}\right) = \dfrac{\sin u}{1+\cos u} = \dfrac{5/13}{1-(12/13)} = \dfrac{5}{1} = 5$

**49.** $\tan u = -\dfrac{5}{8},\ \dfrac{3\pi}{2}<u<2\pi \implies \sin u = -\dfrac{5}{\sqrt{89}}$ and $\cos u = \dfrac{8}{\sqrt{89}}$

$\sin\left(\dfrac{u}{2}\right) = \sqrt{\dfrac{1-\cos u}{2}} = \sqrt{\dfrac{1-(8/\sqrt{89})}{2}}\sqrt{\dfrac{\sqrt{89}-8}{2\sqrt{89}}} = \sqrt{\dfrac{89-8\sqrt{89}}{178}}$

$\cos\left(\dfrac{u}{2}\right) = -\sqrt{\dfrac{1+\cos u}{2}} = -\sqrt{\dfrac{1+(8/\sqrt{89})}{2}} = -\sqrt{\dfrac{\sqrt{89}+8}{2\sqrt{89}}} = -\sqrt{\dfrac{89+8\sqrt{89}}{178}}$

$\tan\left(\dfrac{u}{2}\right) = \dfrac{1-\cos u}{\sin u} = \dfrac{1-(8/\sqrt{89})}{-5/\sqrt{89}} = \dfrac{8-\sqrt{89}}{5}$

**51.** $\csc u = -\dfrac{5}{3},\ \pi<u<\dfrac{3\pi}{2} \implies \sin u = -\dfrac{3}{5}$ and $\cos u = -\dfrac{4}{5}$

$\sin\left(\dfrac{u}{2}\right) = \sqrt{\dfrac{1-\cos u}{2}} = \sqrt{\dfrac{1+(4/5)}{2}} = \dfrac{3\sqrt{10}}{10}$

$\cos\left(\dfrac{u}{2}\right) = -\sqrt{\dfrac{1+\cos u}{2}} = -\sqrt{\dfrac{1-(4/5)}{2}} - \dfrac{\sqrt{10}}{10}$

$\tan\left(\dfrac{u}{2}\right) = \dfrac{1-\cos u}{\sin u} = \dfrac{1+(4/5)}{-3/5} = -3$

**53.** $\sqrt{\dfrac{1 - \cos 6x}{2}} = |\sin 3x|$

**55.** $-\sqrt{\dfrac{1 - \cos 8x}{1 + \cos 8x}} = -\dfrac{\sqrt{(1 - \cos 8x)/2}}{\sqrt{(1 + \cos 8x)/2}}$

$$= -\left|\dfrac{\sin 4x}{\cos 4x}\right|$$

$$= -|\tan 4x|$$

**57.**   $\sin \dfrac{x}{2} + \cos x = 0$

$$\pm\sqrt{\dfrac{1 - \cos x}{2}} = -\cos x$$

$$\dfrac{1 - \cos x}{2} = \cos^2 x$$

$$0 = 2\cos^2 x + \cos x - 1$$

$$= (2\cos x - 1)(\cos x + 1)$$

$$\cos x = \dfrac{1}{2} \quad \text{or} \quad \cos x = -1$$

$$x = \dfrac{\pi}{3}, \dfrac{5\pi}{3} \qquad x = \pi$$

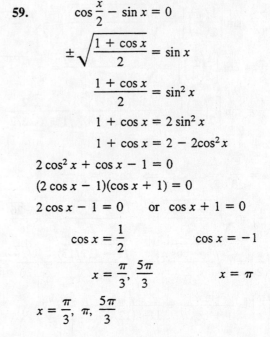

By checking these values in the original equation, we see that $x = \pi/3$ and $x = 5\pi/3$ are extraneous, and $x = \pi$ is the only solution.

**59.**   $\cos \dfrac{x}{2} - \sin x = 0$

$$\pm\sqrt{\dfrac{1 + \cos x}{2}} = \sin x$$

$$\dfrac{1 + \cos x}{2} = \sin^2 x$$

$$1 + \cos x = 2\sin^2 x$$

$$1 + \cos x = 2 - 2\cos^2 x$$

$$2\cos^2 x + \cos x - 1 = 0$$

$$(2\cos x - 1)(\cos x + 1) = 0$$

$$2\cos x - 1 = 0 \quad \text{or} \quad \cos x + 1 = 0$$

$$\cos x = \dfrac{1}{2} \qquad\qquad \cos x = -1$$

$$x = \dfrac{\pi}{3}, \dfrac{5\pi}{3} \qquad\qquad x = \pi$$

$$x = \dfrac{\pi}{3}, \ \pi, \ \dfrac{5\pi}{3}$$

$\pi/3$, $\pi$, and $5\pi/3$ are all solutions to the equation.

**61.** $6 \sin \dfrac{\pi}{4} \cos \dfrac{\pi}{4} = 6 \cdot \dfrac{1}{2}\left[\sin\left(\dfrac{\pi}{4} + \dfrac{\pi}{4}\right) + \sin\left(\dfrac{\pi}{4} - \dfrac{\pi}{4}\right)\right] = 3\left(\sin\dfrac{\pi}{2} + \sin 0\right)$

**63.** $\sin 5\theta \cos 3\theta = \dfrac{1}{2}\left[\sin(5\theta + 3\theta) + \sin(5\theta - 3\theta)\right] = \dfrac{1}{2}(\sin 8\theta + \sin 2\theta)$

**65.** $5\cos(-5\beta)\cos 3\beta = 5 \cdot \dfrac{1}{2}\left[\cos(-5\beta - 3\beta) + \cos(-5\beta + 3\beta)\right] = \dfrac{5}{2}\left[\cos(-8\beta) + \cos(-2\beta)\right]$

$$= \dfrac{5}{2}(\cos 8\beta + \cos 2\beta)$$

**67.** $\sin(x + y)\sin(x - y) = \dfrac{1}{2}(\cos 2y - \cos 2x)$

**69.** $\sin(\theta + \pi)\cos(\theta - \pi) = \dfrac{1}{2}(\sin 2\theta + \sin 2\pi)$

**71.** $\sin 60° + \sin 30° = 2 \sin\left(\dfrac{60° + 30°}{2}\right)\cos\left(\dfrac{60° - 30°}{2}\right) = 2 \sin 45° \cos 15°$

**73.** $\cos \dfrac{3\pi}{4} - \cos \dfrac{\pi}{4} = -2 \sin\left(\dfrac{(3\pi/4) + (\pi/4)}{2}\right)\sin\left(\dfrac{(3\pi/4) - (\pi/4)}{2}\right) = -2 \sin \dfrac{\pi}{2} \sin \dfrac{\pi}{4}$

**75.** $\cos 6x + \cos 2x = 2 \cos\left(\dfrac{6x + 2x}{2}\right)\cos\left(\dfrac{6x - 2x}{2}\right) = 2 \cos 4x \cos 2x$

**77.** $\sin(\alpha + \beta) - \sin(\alpha - \beta) = 2 \cos\left(\dfrac{\alpha + \beta + \alpha - \beta}{2}\right) \sin\left(\dfrac{\alpha + \beta - \alpha + \beta}{2}\right) = 2 \cos \alpha \sin \beta$

**79.** $\cos(\phi + 2\pi) + \cos \phi = 2 \cos\left(\dfrac{\phi + 2\pi + \phi}{2}\right) \cos\left(\dfrac{\phi + 2\pi - \phi}{2}\right) = 2 \cos(\phi + \pi) \cos \pi$

**81.**
$$\sin 6x + \sin 2x = 0$$
$$2 \sin\left(\frac{6x + 2x}{2}\right) \cos\left(\frac{6x - 2x}{2}\right) = 0$$
$$2(\sin 4x) \cos 2x = 0$$

$$\sin 4x = 0 \quad \text{or} \quad \cos 2x = 0$$
$$4x = n\pi \qquad\qquad 2x = \frac{\pi}{2} + n\pi$$
$$x = \frac{n\pi}{4} \qquad\qquad x = \frac{\pi}{4} + \frac{n\pi}{2}$$

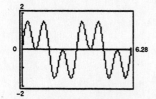

In the interval we have
$$x = 0, \frac{\pi}{4}, \frac{\pi}{2}, \frac{3\pi}{4}, \pi, \frac{5\pi}{4}, \frac{3\pi}{2}, \frac{7\pi}{4}.$$

**83.** $\dfrac{\cos 2x}{\sin 3x - \sin x} - 1 = 0$

$$\frac{\cos 2x}{\sin 3x - \sin x} = 1$$
$$\frac{\cos 2x}{2 \cos 2x \sin x} = 1$$
$$2 \sin x = 1$$
$$\sin x = \frac{1}{2}$$
$$x = \frac{\pi}{6}, \frac{5\pi}{6}$$

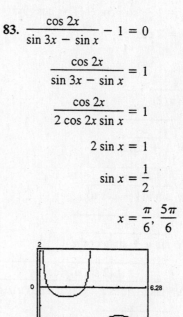

**85.** $\sin^2 \alpha = \left(\frac{5}{13}\right)^2 = \frac{25}{169}$

$\sin^2 \alpha = 1 - \cos^2 \alpha = 1 - \left(\frac{12}{13}\right)^2 = 1 - \frac{144}{169} = \frac{25}{169}$

**87.** $\sin \alpha \cos \beta = \left(\frac{5}{13}\right)\left(\frac{4}{5}\right) = \frac{4}{13}$

$\sin \alpha \cos \beta = \cos\left(\frac{\pi}{2} - \alpha\right) \sin\left(\frac{\pi}{2} - \beta\right) = \left(\frac{5}{13}\right)\left(\frac{4}{5}\right) = \frac{4}{13}$

**89.** $\csc 2\theta = \dfrac{1}{\sin 2\theta}$

$$= \frac{1}{2 \sin \theta \cos \theta}$$
$$= \frac{1}{\sin \theta} \cdot \frac{1}{2 \cos \theta}$$
$$= \frac{\csc \theta}{2 \cos \theta}$$

**91.** $\cos^2 2\alpha - \sin^2 2\alpha = \cos\left[2(2\alpha)\right]$
$$= \cos 4\alpha$$

**93.** $(\sin x + \cos x)^2 = \sin^2 x + 2\sin x \cos x + \cos^2 x$
$$= (\sin^2 x + \cos^2 x) + 2\sin x \cos x$$
$$= 1 + \sin 2x$$

**95.** $1 + \cos 10y = 1 + \cos^2 5y - \sin^2 5y$
$$= 1 + \cos^2 5y - (1 - \cos^2 5y)$$
$$= 2\cos^2 5y$$

**97.** $\sec \dfrac{u}{2} = \dfrac{1}{\cos(u/2)}$

$$= \pm\sqrt{\dfrac{2}{1 + \cos u}}$$

$$= \pm\sqrt{\dfrac{2\sin u}{\sin u(1 + \cos u)}}$$

$$= \pm\sqrt{\dfrac{2\sin u}{\sin u + \sin u \cos u}}$$

$$= \pm\sqrt{\dfrac{(2\sin u)/(\cos u)}{(\sin u)/(\cos u) + (\sin u \cos u)/(\cos u)}}$$

$$= \pm\sqrt{\dfrac{2\tan u}{\tan u + \sin u}}$$

**99.** $\dfrac{\cos 4x + \cos 2x}{\sin 4x + \sin 2x} = \dfrac{2\cos\left(\dfrac{4x + 2x}{2}\right)\cos\left(\dfrac{4x - 2x}{2}\right)}{2\sin\left(\dfrac{4x + 2x}{2}\right)\cos\left(\dfrac{4x - 2x}{2}\right)}$

$$= \dfrac{2\cos 3x \cos x}{2\sin 3x \cos x}$$

$$= \cot 3x$$

**101.** $\dfrac{\cos t + \cos 3t}{\sin 3t - \sin t} = \dfrac{2\cos\left(\dfrac{4t}{2}\right)\cos\left(-\dfrac{2t}{2}\right)}{2\cos\left(\dfrac{4t}{2}\right)\sin\left(\dfrac{2t}{2}\right)}$

$$= \dfrac{\cos(-t)}{\sin(t)}$$

$$= \dfrac{\cos(t)}{\sin(t)}$$

$$= \cot t$$

**103.** $\cos 3\beta = \cos(2\beta + \beta)$
$$= \cos 2\beta \cos\beta - \sin 2\beta \sin\beta$$
$$= (\cos^2\beta - \sin^2\beta)\cos\beta - 2\sin\beta\cos\beta\sin\beta$$
$$= \cos^3\beta - \sin^2\beta\cos\beta - 2\sin^2\beta\cos\beta$$
$$= \cos^3\beta - 3\sin^2\beta\cos\beta$$

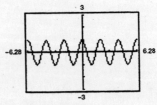

**105.** $\dfrac{\cos 4x - \cos 2x}{2 \sin 3x} = \dfrac{-2 \sin\left(\dfrac{4x + 2x}{2}\right) \sin\left(\dfrac{4x - 2x}{2}\right)}{2 \sin 3x}$

$= \dfrac{-2 \sin 3x \sin x}{2 \sin 3x}$

$= -\sin x$

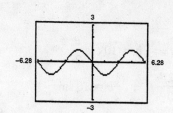

**107.** $\sin^2 x = \dfrac{1 - \cos 2x}{2} = \dfrac{1}{2} - \dfrac{\cos 2x}{2}$

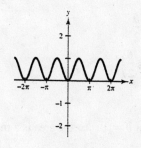

**109.** (a) $y = 4 \sin \dfrac{x}{2} + \cos x$

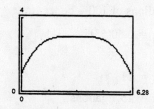

Maximum: $(\pi, 3)$

(b)    $2 \cos \dfrac{x}{2} - \sin x = 0$

$2\left(\pm\sqrt{\dfrac{1 + \cos x}{2}}\right) = \sin x$

$4\left(\dfrac{1 + \cos x}{2}\right) = \sin^2 x$

$2(1 + \cos x) = 1 - \cos^2 x$

$\cos^2 x + 2 \cos x + 1 = 0$

$(\cos x + 1)^2 = 0$

$\cos x = -1$

$x = \pi$

**111.** $f(x) = 2 \sin x \left[2 \cos^2(x/2) - 1\right]$

(a)

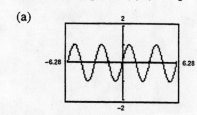

(b) and (c)

$2 \sin x [2 \cos^2 (x/2) - 1]$

$= 2 \sin x [(1 + \cos x) - 1]$

$= 2 \sin x (\cos x)$

$= 2 \sin x \cos x$

**113.** $\sin(2 \arcsin x) = 2 \sin(\arcsin x) \cos(\arcsin x) = 2x\sqrt{1 - x^2}$

**115.** (a)    $A = \dfrac{1}{2}bh$

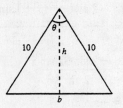

$$\cos\dfrac{\theta}{2} = \dfrac{h}{10} \implies h = 10\cos\dfrac{\theta}{2}$$

$$\sin\dfrac{\theta}{2} = \dfrac{(1/2)b}{10} \implies \dfrac{1}{2}b = 10\sin\dfrac{\theta}{2}$$

$$A = 10\sin\dfrac{\theta}{2}10\cos\dfrac{\theta}{2} \implies A = 100\sin\dfrac{\theta}{2}\cos\dfrac{\theta}{2}$$

(b) $A = 100\sin\dfrac{\theta}{2}\cos\dfrac{\theta}{2}$

$A = 50\left(2\sin\dfrac{\theta}{2}\cos\dfrac{\theta}{2}\right)$

$A = 50\sin\theta$

When $\theta = \pi/2$, $\sin\theta = 1 \implies$ the area is a maximum.

$A = 50\sin\dfrac{\pi}{2} = 50(1) = 50$ square feet

**117.** Let $x = $ profit for September,
then $x + 0.16x = $ profit for October.

$x + (x + 0.16x) = 507{,}600$

$2.16x = 507{,}600$

$x = 235{,}000$

$x + 0.16x = 272{,}600$

Profit for September: \$235,000

Profit for October: \$276,600

**119.** Let $x = $ number of gallons of 100% concentrate.

$0.30(55 - x) + 1.00x = 0.50(55)$

$16.50 - 0.30x + x = 27.50$

$0.70x = 11$

$x \approx 15.7$ gallons

# ❑ Review Exercises for Chapter 2

### Solutions to Odd-Numbered Exercises

**1.** $\dfrac{1}{\cot^2 x + 1} = \dfrac{1}{\csc^2 x} = \sin^2 x$

**3.** $\dfrac{\sin^2 \alpha - \cos^2 \alpha}{\sin^2 \alpha - \sin \alpha \cos \alpha} = \dfrac{(\sin \alpha + \cos \alpha)(\sin \alpha - \cos \alpha)}{\sin \alpha(\sin \alpha - \cos \alpha)}$

$\qquad\qquad\qquad = \dfrac{\sin \alpha + \cos \alpha}{\sin \alpha}$

$\qquad\qquad\qquad = 1 + \cot \alpha$

**5.** $\tan^2 \theta(\csc^2 \theta - 1) = \tan^2 \theta(\cot^2 \theta)$

$\qquad\qquad\qquad = \tan^2 \theta\left(\dfrac{1}{\tan^2 \theta}\right)$

$\qquad\qquad\qquad = 1$

**7.** $\dfrac{2 \tan(x + 1)}{1 - \tan^2(x + 1)} = \tan[2(x + 1)]$

$\qquad\qquad\qquad = \tan(2x + 2)$

**9.** $\tan x(1 - \sin^2 x) = \tan x \cos^2 x$

$\qquad\qquad\qquad = \dfrac{\sin x}{\cos x} \cdot \cos^2 x$

$\qquad\qquad\qquad = \sin x \cos x$

$\qquad\qquad\qquad = \dfrac{1}{2}(2 \sin x \cos x)$

$\qquad\qquad\qquad = \dfrac{1}{2} \sin 2x$

**11.** $\sec^2 x \cot x - \cot x = \cot x(\sec^2 x - 1)$

$\qquad\qquad\qquad = \cot x \tan^2 x$

$\qquad\qquad\qquad = \dfrac{1}{\tan x} \tan^2 x$

$\qquad\qquad\qquad = \tan x$

**13.** $\sin^5 x \cos^2 x = \sin^4 x \cos^2 x \sin x$

$\qquad\qquad\qquad = (1 - \cos^2 x)^2 \cos^2 x \sin x$

$\qquad\qquad\qquad = (1 - 2 \cos^2 x + \cos^4 x) \cos^2 x \sin x$

$\qquad\qquad\qquad = (\cos^2 x - 2 \cos^4 x + \cos^6 x)\sin x$

**15.** $\sin 3\theta \sin \theta = \tfrac{1}{2}[\cos(3\theta - \theta) - \cos(3\theta + \theta)]$

$\qquad\qquad\qquad = \tfrac{1}{2}(\cos 2\theta - \cos 4\theta)$

**17.** $\sqrt{\dfrac{1 - \sin \theta}{1 + \sin \theta}} = \sqrt{\dfrac{1 - \sin \theta}{1 + \sin \theta} \cdot \dfrac{1 - \sin \theta}{1 - \sin \theta}}$

$\qquad = \sqrt{\dfrac{(1 - \sin \theta)^2}{1 - \sin^2 \theta}} = \sqrt{\dfrac{(1 - \sin \theta)^2}{\cos^2 \theta}} = \dfrac{|1 - \sin \theta|}{|\cos \theta|} = \dfrac{1 - \sin \theta}{|\cos \theta|}$

Note: We can drop the absolute value on $1 - \sin \theta$ since it is always nonnegative.

**19.** $\cos 3x = \cos(2x + x)$

$\qquad = \cos 2x \cos x - \sin 2x \sin x$

$\qquad = (\cos^2 x - \sin^2 x) \cos x - 2 \sin x \cos x \sin x$

$\qquad = \cos^3 x - 3 \sin^2 x \cos x$

$\qquad = \cos^3 x - 3 \cos x(1 - \cos^2 x)$

$\qquad = \cos^3 x - 3 \cos x + 3 \cos^3 x$

$\qquad = 4 \cos^3 x - 3 \cos x$

**21.** $\cot\left(\dfrac{\pi}{2} - x\right) = \dfrac{\cos[(\pi/2) - x]}{\sin[(\pi/2) - x]}$

$\qquad = \dfrac{\cos(\pi/2) \cos x + \sin(\pi/2)\sin x}{\sin(\pi/2) \cos x - \sin x \cos(\pi/2)}$

$\qquad = \dfrac{\sin x}{\cos x}$

$\qquad = \tan x$

**23.** $\dfrac{\sec x - 1}{\tan x} = \dfrac{(1/\cos x) - 1}{\sin x/\cos x} = \dfrac{1 - \cos x}{\sin x} = \tan \dfrac{x}{2}$

**25.** $2 \sin y \cos y \sec 2y = (\sin 2y)(\sec 2y)$

$\qquad = \dfrac{\sin 2y}{\cos 2y}$

$\qquad = \tan 2y$

**27.** $\sin\left(x - \dfrac{3\pi}{2}\right) = \sin x \cos \dfrac{3\pi}{2} - \sin \dfrac{3\pi}{2} \cos x$

$\qquad = (\sin x)(0) - (-1)(\cos x)$

$\qquad = \cos x$

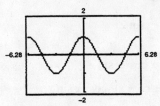

**29.** $\dfrac{1 - \cos 2x}{1 + \cos 2x} = \dfrac{1 - (1 - 2 \sin^2 x)}{1 + (2 \cos x^2 - 1)}$

$\qquad = \dfrac{2 \sin^2 x}{2 \cos^2 x}$

$\qquad = \tan^2 x$

**31.** $\sin \dfrac{5\pi}{12} = \sin\left(\dfrac{2\pi}{3} - \dfrac{\pi}{4}\right)$

$\qquad = \sin\left(\dfrac{2\pi}{3}\right) \cos\left(\dfrac{\pi}{4}\right) - \cos\dfrac{2\pi}{3} \sin\left(\dfrac{\pi}{4}\right)$

$\qquad = \left(\dfrac{\sqrt{3}}{2}\right)\left(\dfrac{\sqrt{2}}{2}\right) - \left(-\dfrac{1}{2}\right)\left(\dfrac{\sqrt{2}}{2}\right)$

$\qquad = \dfrac{\sqrt{2}}{4}\left(\sqrt{3} + 1\right)$

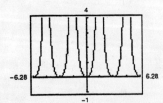

**33.** $\cos(157°\,30') = \cos \dfrac{315°}{2} = -\sqrt{\dfrac{1 + \cos 315°}{2}} = -\sqrt{\dfrac{1 + \cos 45°}{2}}$

$\qquad = -\sqrt{\dfrac{1 + \sqrt{2}/2}{2}} = -\sqrt{\dfrac{2 + \sqrt{2}}{4}} = -\dfrac{\sqrt{2 + \sqrt{2}}}{2}$

**For Exercises 35–39**

$$\sin u = \frac{3}{4}, \ u \text{ in Quadrant II} \implies \cos u = -\frac{\sqrt{7}}{4}$$

$$\cos v = -\frac{5}{13}, \ v \text{ in Quadrant II} \implies \sin v = \frac{12}{13}$$

**35.** $\sin(u + v) = \sin u \cos v + \cos u \sin v$

$$= \left(\frac{3}{4}\right)\left(-\frac{5}{13}\right) + \left(-\frac{\sqrt{7}}{4}\right)\left(\frac{12}{13}\right)$$

$$= -\frac{15}{52} - \frac{12\sqrt{7}}{52}$$

$$= \frac{-3(5 + 4\sqrt{7})}{52}$$

**37.** $\cos(u - v) = \cos u \cos v + \sin u \sin v$

$$= \left(-\frac{\sqrt{7}}{4}\right)\left(-\frac{5}{13}\right) + \left(\frac{3}{4}\right)\left(\frac{12}{13}\right)$$

$$= \frac{5\sqrt{7} + 36}{52}$$

**39.** $\cos \dfrac{u}{2} = \sqrt{\dfrac{1 + \cos u}{2}}$

$$= \sqrt{\frac{1 + (-\sqrt{7}/4)}{2}}$$

$$= \sqrt{\frac{4 - \sqrt{7}}{8}}$$

$$= \frac{1}{4}\sqrt{2(4 - \sqrt{7})}$$

**41.** If $\dfrac{\pi}{2} < \theta < \pi$, then $\cos \dfrac{\theta}{2} < 0$. False, if

$$\frac{\pi}{2} < \theta < \pi \implies \frac{\pi}{4} < \frac{\theta}{2} < \frac{\pi}{2},$$

which is in Quadrant I $\implies \cos(\theta/2) > 0.$

**43.** $4 \sin(-x) \cos(-x) = -2 \sin 2x.$ True.

$$4 \sin(-x) \cos(-x) = 4(-\sin x)(\cos x) = -4 \sin x \cos x = -2(2 \sin x \cos x) = -2 \sin 2x$$

**45.**
$$\sin x - \tan x = 0$$
$$\sin x - \frac{\sin x}{\cos x} = 0$$
$$\sin x \cos x - \sin x = 0$$
$$\sin x(\cos x - 1) = 0$$
$$\sin x = 0 \quad \text{or} \quad \cos x - 1 = 0$$
$$x = 0, \pi \qquad \cos x = 1$$
$$x = 0$$

**47.**
$$\sin 2x + \sqrt{2} \sin x = 0$$
$$2 \sin x \cos x + \sqrt{2} \sin x = 0$$
$$\sin x(2 \cos x + \sqrt{2}) = 0$$
$$\sin x = 0 \quad \text{or} \quad 2 \cos x + \sqrt{2} = 0$$
$$x = 0, \pi \qquad \cos x = -\frac{\sqrt{2}}{2}$$
$$x = \frac{3\pi}{4}, \frac{5\pi}{4}$$

**49.**
$$\cos^2 x + \sin x = 1$$
$$1 - \sin^2 x + \sin x = 1$$
$$\sin x(\sin x - 1) = 0$$
$$\sin x = 0 \quad \text{or} \quad \sin x = 1$$
$$x = 0, \pi \qquad x = \frac{\pi}{2}$$

**51.** $\dfrac{1 + \sin x}{\cos x} + \dfrac{\cos x}{1 + \sin x} = 4$

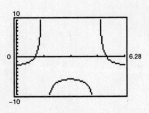

$$(1 + \sin x)^2 + \cos^2 x = 4 \cos x(1 + \sin x)$$

$$1 + 2 \sin x + \sin^2 x + \cos^2 x = 4 \cos x(1 + \sin x)$$

$$2 + 2 \sin x - 4 \cos x(1 + \sin x) = 0$$

$$2(1 + \sin x)(1 - 2 \cos x) = 0$$

$$1 + \sin x = 0 \quad \text{or} \quad 1 - 2 \cos x = 0$$

$$\sin x = -1 \qquad\qquad \cos x = \dfrac{1}{2}$$

$$x = \dfrac{3\pi}{2} \qquad\qquad x = \dfrac{\pi}{3}, \dfrac{5\pi}{3}$$

(extraneous solution)

**53.**   $\tan^3 x - \tan^2 x + 3 \tan x - 3 = 0$

$$\tan^2 x(\tan x - 1) + 3(\tan x - 1) = 0$$

$$(\tan^2 x + 3)(\tan x - 1) = 0$$

$$\tan^2 x + 3 = 0 \quad \text{or} \quad \tan x - 1 = 0$$

(No solution)   or   $\tan x = 0$

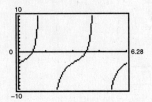

$$x = \dfrac{\pi}{4}, \dfrac{5\pi}{4}$$

**55.** False, $\sin \theta = \frac{1}{2}$ has an infinite number of solutions but is not an identity.

**57.** $\cos 3\theta + \cos 2\theta = 2 \cos\left(\dfrac{3\theta + 2\theta}{2}\right) \cos\left(\dfrac{3\theta - 2\theta}{2}\right)$

$$= 2 \cos \dfrac{5\theta}{2} \cos \dfrac{\theta}{2}$$

**59.** $\sin 3\alpha \sin 2\alpha = \frac{1}{2}[\cos(3\alpha - 2\alpha) - \cos(3\alpha + 2\alpha)]$

$$= \tfrac{1}{2}(\cos \alpha - \cos 5\alpha)$$

**61.** $\cos(2 \arccos 2x) = \cos 2\theta$

$$= \cos^2 \theta - \sin^2 \theta$$

$$= (2x)^2 - \left(\sqrt{1 - 4x^2}\right)^2$$

$$= 4x^2 - (1 - 4x^2)$$

$$= 8x^2 - 1$$

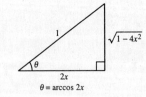

$\theta = \arccos 2x$

**63.** $\sin^{-1/2} \cos x = \dfrac{\cos x}{\sin^{1/2} x}$

$$= \dfrac{\cos x}{\sqrt{\sin x}} \cdot \dfrac{\sqrt{\sin x}}{\sqrt{\sin x}}$$

$$= \dfrac{\cos x}{\sin x} \sqrt{\sin x}$$

$$= \cot x \sqrt{\sin x}$$

**65.** $y = 1.5 \sin 8t - 0.5 \cos 8t$

(a) $a = \dfrac{3}{2}, \ b = -\dfrac{1}{2}, \ B = 8, \ c = \arctan\left(-\dfrac{1/2}{3/2}\right)$

$$y = \sqrt{(3/2)^2 + (-1/2)^2} \sin\left(8t + \arctan -\dfrac{1}{3}\right)$$

$$y = \dfrac{1}{2}\sqrt{10} \sin\left(8t - \arctan \dfrac{1}{3}\right)$$

(b) Amplitude: $\dfrac{1}{2}\sqrt{10}$ sin feet

(c) Frequency: $\dfrac{8}{2\pi} = \dfrac{4}{\pi}$ cycles per second

# ❑ Practice Test for Chapter 2

**1** Find the value of the other five trigonometric functions, given $\tan x = \frac{4}{11}$, $\sec x < 0$.

**2.** Simplify $\dfrac{\sec^2 x + \csc^2 x}{\csc^2 x(1 + \tan^2 x)}$.

**3.** Rewrite as a single logarithm and simplify $\ln|\tan \theta| - \ln|\cot \theta|$.

**4.** True or false:

$$\cos\left(\frac{\pi}{2} - x\right) = \frac{1}{\csc x}$$

**5.** Factor and simplify: $\sin^4 x + (\sin^2 x)\cos^2 x$

**6.** Multiply and simplify: $(\csc x + 1)(\csc x - 1)$

**7.** Rationalize the denominator and simplify:

$$\frac{\cos^2 x}{1 - \sin x}$$

**8.** Verify:

$$\frac{1 + \cos \theta}{\sin \theta} + \frac{\sin \theta}{1 + \cos \theta} = 2 \csc \theta$$

**9.** Verify:

$$\tan^4 x + 2 \tan^2 x + 1 = \sec^4 x$$

**10.** Use the sum or difference formulas to determine:

(a) $\sin 105°$        (b) $\tan 15°$

**11.** Simplify: $(\sin 42°)\cos 38° - (\cos 42°)\sin 38°$

**12.** Verify $\tan\left(\theta + \dfrac{\pi}{4}\right) = \dfrac{1 + \tan \theta}{1 - \tan \theta}$.

**13.** Write $\sin(\arcsin x - \arccos x)$ as an algebraic expression in $x$.

**14.** Use the double-angle formulas to determine:

(a) $\cos 120°$        (b) $\tan 300°$

**15.** Use the half-angle formulas to determine:

(a) $\sin 22.5°$        (d) $\tan \dfrac{\pi}{12}$

**16.** Given $\sin = 4/5$, $\theta$ lies in Quadrant II, find $\cos(\theta/2)$.

**17.** Use the power-reducing identities to write $(\sin^2 x)\cos^2 x$ in terms of the first power of cosine.

**18.** Rewrite as a sum: $6(\sin 5\theta)\cos 2\theta$.

**19.** Rewrite as a product:

$\sin(x + \pi) + \sin(x - \pi)$.

**20.** Verify $\dfrac{\sin 9x + \sin 5x}{\cos 9x - \cos 5x} = -\cot 2x$.

**21.** Verify:

$(\cos u)\sin v = \frac{1}{2}[\sin(u + v) - \sin(u - v)]$.

**22.** Find all solutions in the interval $[0, 2\pi)$:

$4 \sin^2 x = 1$

**23.** Find all solutions in the interval $[0, 2\pi)$:

$\tan^2 \theta + \left(\sqrt{3} - 1\right)\tan\theta - \sqrt{3} = 0$

**24.** Find all solutions in the interval $[0, 2\pi)$:

$\sin 2x = \cos x$

**25.** Use the quadratic formula to find all solutions in the interval $[0, 2\pi)$:

$\tan^2 x - 6 \tan x + 4 = 0$

# C H A P T E R 3
## Additional Topics in Trigonometry

# TER 3

## al Topics in Trigonometry

### .1   Law of Sines

is any oblique triangle with sides $a$, $b$, and $c$, then

$$\frac{}{A} = \frac{b}{\sin B} = \frac{c}{\sin C}.$$

ould be able to use the Law of Sines to solve an oblique triangle for the remaining three parts, given:

o angles and any side (AAS or ASA)

vo sides and an angle opposite one of them (SSA)

. If $A$ is acute and $h = b \sin A$:

   (a) $a < h$, no triangle is possible.

   (b) $a = h$ or $a > b$, one triangle is possible.

   (c) $h < a < b$, two triangles are possible.

2. If $A$ is obtuse and $h = b \sin A$:

   (a) $a \le b$, no triangle is possible.

   (b) $a > b$, one triangle is possible.

area of any triangle equals one-half the product of the lengths of two sides times the sine of their luded angle.

$$A = \tfrac{1}{2}ab \sin C = \tfrac{1}{2}ac \sin B = \tfrac{1}{2}bc \sin A$$

n: $A = 30°$, $B = 45°$, $a = 20$

$180° - A - B = 105°$

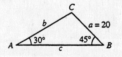

$$\frac{a}{\sin A}(\sin B) = \frac{20 \sin 45°}{\sin 30°} = 20\sqrt{2} \approx 28.28$$

$$= \frac{a}{\sin A}(\sin C) = \frac{20 \sin 105°}{\sin 30°} \approx 38.64$$

ven: $A = 10°$, $B = 60°$, $a = 7.5$

$= 180° - A - B = 110°$

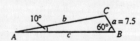

$$= \frac{a}{\sin A}(\sin B) = \frac{7.5}{\sin 10°}(\sin 60°) \approx 37.40$$

$$= \frac{a}{\sin A}(\sin C) = \frac{7.5}{\sin 10°}(\sin 110°) \approx 40.59$$

**5.** Given: $A = 36°$, $a = 8$, $b = 5$

$$\sin B = \frac{b \sin A}{a} = \frac{5 \sin 36°}{8} \approx 0.36737 \implies B \approx 21.55°$$

$$C = 180° - A - B \approx 180° - 36° - 21.55 = 122.45°$$

$$c = \frac{a}{\sin A}(\sin C) = \frac{8}{\sin 36°}(\sin 122.45°) \approx 11.49$$

**7.** Given: $A = 150°$, $C = 20°$, $a = 200$

$$B = 180° - A - C = 180° - 150° - 20° = 10°$$

$$b = \frac{a}{\sin A}(\sin B) = \frac{200}{\sin 150°}(\sin 10°) \approx 69.46$$

$$c = \frac{a}{\sin A}(\sin C) = \frac{200}{\sin 150°}(\sin 20°) \approx 136.81$$

**9.** Given: $A = 83°20'$, $C = 54.6°$, $c = 18.1$

$$B = 180° - A - C = 180° - 80°20' - 54°36' = 42°4'$$

$$a = \frac{c}{\sin C}(\sin A) = \frac{18.1}{\sin 54.6°}(\sin 83°20') \approx 22.05$$

$$b = \frac{c}{\sin C}(\sin B) = \frac{18.1}{\sin 54.6°}(\sin 42°4') \approx 14.88$$

**11.** Given: $B = 15°30'$, $a = 4.5$, $b = 6.8$

$$\sin A = \frac{a \sin B}{b} = \frac{4.5 \sin 15°30'}{6.8} \approx 0.17685 \implies A \approx 10°11'$$

$$C = 180° - A - B \approx 180° - 10°11' - 15°30' = 154°19'$$

$$c = \frac{b}{\sin B}(\sin C) = \frac{6.8}{\sin 15°30'}(\sin 154°19') \approx 11.03$$

**13.** Given: $C = 145°$, $b = 4$, $c = 14$

$$\sin B = \frac{b \sin C}{c} = \frac{4 \sin 145°}{14} \approx 0.1639 \implies B \approx 9.43°$$

$$A = 180° - B - C \approx 180° - 9.43° - 145° = 25.57°$$

$$a = \frac{c}{\sin C}(\sin A) \approx \frac{14}{\sin 145°}(\sin 25.57°) \approx 10.5$$

**15.** Given: $B = 110°15'$, $a = 48$, $b = 16$

$$\sin B = \frac{b \sin A}{a} = \frac{16 \sin 110°15'}{48} \approx 0.31273 \implies B \approx 18°13'$$

$$C = 180° - A - B \approx 180° - 110°15' - 18°13' = 51°32'$$

$$c = \frac{a}{\sin A}(\sin C) = \frac{48}{\sin 110°15'}(\sin 51°32') \approx 40.06$$

**17.** Given: $a = 4.5$, $b = 12.8$, $A = 58°$

$h = 12.8 \sin 58° \approx 10.86$

Since $a < h$, no triangle is formed.

**19.** Given: $a = 4.5$, $b = 5$, $A = 58°$

$\sin B = \dfrac{b \sin A}{a} = \dfrac{5 \sin 58°}{4.5} \approx 0.9423 \implies B = 70.4°$ or $B = 109.6°$

Case 1

$B \approx 70.4°$

$C \approx 180° - 70.4° - 58° = 51.6°$

$c \approx \dfrac{4.5}{\sin 58°} (\sin 51.6°) \approx 4.16$

Case 2

$B \approx 109.6°$

$C \approx 180° - 109.6° - 58° = 12.4°$

$c \approx \dfrac{4.5}{\sin 58°} (\sin 12.4°) \approx 1.14$

**21.** Given: $a = 125$, $b = 200$, $A = 110°$

No triangle is formed because $A$ is obtuse and $a < b$.

**23.** Given: $A = 36°$, $a = 5$

(a) One solution if $b \le 5$ or $b = \dfrac{5}{\sin 36°}$

(b) Two solutions if $5 < b < \dfrac{5}{\sin 36°}$

(c) No solution is $b > \dfrac{5}{\sin 36°}$

**25.** (a)

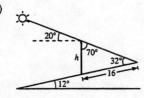

(b) $\dfrac{16}{\sin 70°} = \dfrac{h}{\sin 32°}$

(c) $h = \dfrac{16 \sin 32°}{\sin 70°} \approx 9$ meters

**27.** $\dfrac{\sin(42° - \theta)}{10} = \dfrac{\sin 48°}{17}$

$\sin(42° - \theta) \approx 0.43714$

$\theta \approx 16.1°$

**29.** Given: $A = 74° - 28° = 46°$,

$B = 180° - 41° - 74° = 65°$, $c = 100$

$C = 180° - 46° - 65° = 69°$

$a = \dfrac{c}{\sin C} (\sin A) = \dfrac{100}{\sin 69°} (\sin 46°) \approx 77$ meters

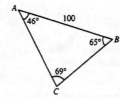

**31. (a)**

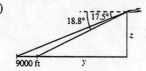

(b) $\dfrac{x}{\sin 17.5°} = \dfrac{9000}{\sin 1.3°}$

$x \approx 119{,}289.1261 \text{ feet } \approx 22.6 \text{ miles}$

(c) $\dfrac{y}{\sin 71.2°} = \dfrac{x}{\sin 90°}$

$y = x \sin 71.2° \approx 119{,}289.1261 \sin 71.2°$

$\approx 112{,}924.963 \text{ feet } \approx 21.4 \text{ miles}$

(d) $z = x \sin 18.8° \approx 119{,}289.1261 \sin 18.8° \approx 37{,}443 \text{ feet}$

**33.** $A = 65° - 28° = 37°$

$c = 30$

$B = 180° - 16.5° - 65° = 98.5°$

$C = 180° - 37° - 98.5° = 44.5°$

$a = \dfrac{c}{\sin C}(\sin A) = \dfrac{30}{\sin 44.5°}(\sin 37°) \approx 25.8 \text{ km to } B$

$b = \dfrac{c}{\sin C}(\sin B) = \dfrac{30}{\sin 44.5°}(\sin 98.5°) \approx 42.3 \text{ km to } A$

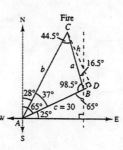

**35.** $A = 90° - 62° = 28°$

$B = 90° + 38° = 128°, \quad c = 5$

$C = 180° - 128° - 28° = 24°$

$a = \dfrac{c}{\sin C}(\sin A) = \dfrac{5}{\sin 24°}(\sin 28°) \approx 5.77$

$d = a \sin(90° - 38°) \approx 5.77 \sin 52° \approx 4.55 \text{ miles}$

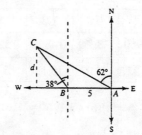

**37.** (a) $\dfrac{\sin \alpha}{9} = \dfrac{\sin \beta}{18}$

$\sin \alpha = 0.5 \sin \beta$

$\alpha = \arcsin(0.5 \sin \beta)$

(b)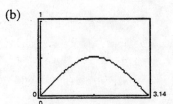

Domain: $0 < \beta < \pi$

Range: $0 < \alpha \le \dfrac{\pi}{6}$

(c) $\gamma = \pi - \alpha - \beta = \pi - \beta - \arcsin(0.5 \sin \beta)$

$\dfrac{c}{\sin \gamma} = \dfrac{18}{\sin \beta}$

$c = \dfrac{18 \sin \gamma}{\sin \beta} = \dfrac{18 \sin \left[ \pi - \beta - \arcsin(0.5 \sin \beta) \right]}{\sin \beta}$

(d)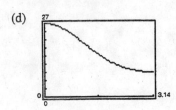

Domain: $0 < \beta < \pi$

Range: $9 < \alpha < 27$

(e)

| $\beta$ | 0 | 0.4 | 0.8 | 1.2 | 1.6 | 2.0 | 2.4 | 2.8 |
|---------|---|-----|-----|-----|-----|-----|-----|-----|
| $\alpha$ | 0 | 0.1960 | 0.3669 | 0.4848 | 0.5234 | 0.4720 | 0.3445 | 0.1683 |
| $c$ | 27 | 25.92 | 27.07 | 19.19 | 15.33 | 12.29 | 10.31 | 9.27 |

As $\beta \rightarrow 0$, $c \rightarrow 27$

As $\beta \rightarrow \pi$, $c \rightarrow 9$

**39.** Area $= \frac{1}{2} ab \sin C = \frac{1}{2}(4)(6) \sin 120° \approx 10.4$

**41.** Area $= \frac{1}{2} bc \sin A = \frac{1}{2}(57)(85) \sin 43°45' \approx 1675.2$

**43.** Area $= \frac{1}{2} ac \sin B = \frac{1}{2}(62)(20) \sin 130° \approx 474.9$

**45.** (a) $A = \dfrac{1}{2}(30)(20) \sin\left(\theta + \dfrac{\theta}{2}\right) - \dfrac{1}{2}(8)(20) \sin \dfrac{\theta}{2} - \dfrac{1}{2}(8)(30) \sin \theta$

$= 300 \sin \dfrac{3\theta}{2} - 80 \sin \dfrac{\theta}{2} - 120 \sin \theta$

$= 20\left[ 15 \sin \dfrac{3\theta}{2} - 4 \sin \dfrac{\theta}{2} - 6 \sin \theta \right]$

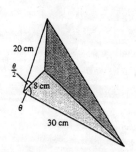

(c) Domain: $0 \le \theta \le 1.6690$

The domain would increase in length and the area would increase if the 8-centimeter line segment were decreased.

(b) 

170

1.7

# Section 3.2    Law of Cosines

- If $ABC$ is any oblique triangle with sides $a$, $b$, and $c$, the following equations are valid.

    (a) $a^2 = b^2 + c^2 - 2bc \cos A$   or   $\cos A = \dfrac{b^2 + c^2 - a^2}{2bc}$

    (b) $b^2 = a^2 + c^2 - 2ac \cos B$   or   $\cos B = \dfrac{a^2 + c^2 - b^2}{2ac}$

    (c) $c^2 = a^2 + b^2 - 2ab \cos C$   or   $\cos C = \dfrac{a^2 + b^2 - c^2}{2ab}$

- You should be able to use the Law of Cosines to solve an oblique triangle for the remaining three parts, given:

    (a) Three sides (SSS)

    (b) Two sides and their included angle (SAS)

- Given any triangle with sides of length $a$, $b$, and $c$, then the area of the triangle is

$$\text{Area} = \sqrt{s(s - a)(s - b)(s - c)}, \text{ where } s = \frac{a + b + c}{2}. \quad \text{(Heron's Formula)}$$

**Solutions to Odd-Numbered Exercises**

1. Given: $a = 6$, $b = 8$, $c = 12$

$$\cos A = \frac{b^2 + c^2 - a^2}{2bc} = \frac{64 + 144 - 36}{2(8)(12)} \approx 0.8958 \implies A \approx 26.4°$$

$$\sin B = \frac{b \sin A}{a} \approx \frac{8 \sin 26.4°}{6} \approx 0.5928 \implies B \approx 36.3°$$

$$C \approx 180° - 26.4° - 36.3° = 117.3°$$

3. Given: $A = 30°$, $b = 15$, $c = 30$

$$a^2 = b^2 + c^2 - 2bc \cos A$$

$$= 225 + 900 - 2(15)(30) \cos 30° \approx 18.5897$$

$$a \approx 18.6$$

$$\cos B = \frac{a^2 + c^2 - b^2}{2ac} \approx \frac{(18.6)^2 + 900 - 225}{2(18.6)(30)} \approx 0.9148$$

$$B \approx 23.8°$$

$$C \approx 180° - 30° - 23.8° = 126.2°$$

5. Given: $a = 9$, $b = 12$, $c = 15$

$$\cos C = \frac{a^2 + b^2 - c^2}{2ab} = \frac{81 + 144 - 225}{2(9)(12)} = 0 \implies C = 90°$$

$$\sin A = \frac{9}{15} = \frac{3}{5} \implies A \approx 36.9°$$

$$B \approx 180° - 90° - 36.9° = 53.1°$$

**7.** Given: $a = 75.4$, $b = 52$, $c = 52$

$$\cos A = \frac{b^2 + c^2 - a^2}{2bc} = \frac{52^2 + 52^2 - 75.4^2}{2(52)(52)} = -0.05125 \implies A \approx 92.94°$$

$$\sin B = \frac{b \sin A}{a} \approx \frac{52(0.9987)}{75.4} \approx 0.68875 \implies B \approx 43.53°$$

$$C = B \approx 43.53°$$

**9.** Given: $A = 120°$, $b = 3$, $c = 10$

$$a^2 = b^2 + c^2 - 2bc \cos A = 9 + 100 - 60 \cos 120° = 139 \implies a \approx 11.79$$

$$\sin B = \frac{b \sin A}{a} \approx \frac{3 \sin 120°}{11.79} \approx 0.2204 \implies B \approx 12.7°$$

$$C = 180° - 120° - 12.7° = 47.3°$$

**11.** Given: $B = 8°45'$, $a = 25$, $c = 15$

$$b^2 = a^2 + c^2 - 2ac \cos B \approx 625 + 225 - 2(25)(15)(0.9884) \approx 108.7 \implies b \approx 10.4$$

$$\sin C = \frac{c \sin B}{b} \approx \frac{15(0.1521)}{10.43} \approx 0.2188 \implies C \approx 12.64° \approx 12°38'$$

$$A \approx 180° - 8°45' - 12°38' = 158°37'$$

**13.** Given: $C = 125°40'$, $a = 32$, $b = 32$

$$c^2 = a^2 + b^2 - 2ab \cos C \approx 32^2 + 32^2 - 2(32)(32)(-0.5831) \approx 3242.1 \implies c \approx 56.9$$

$$A = B \implies 2A = 180° - 125°40' = 54°20' \implies A = B = 27°10'$$

**15.**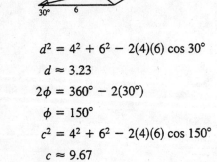

$$d^2 = 4^2 + 6^2 - 2(4)(6) \cos 30°$$

$$d \approx 3.23$$

$$2\phi = 360° - 2(30°)$$

$$\phi = 150°$$

$$c^2 = 4^2 + 6^2 - 2(4)(6) \cos 150°$$

$$c \approx 9.67$$

**17.**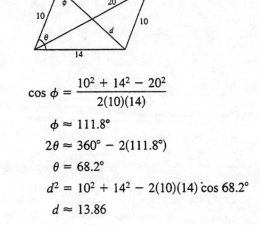

$$\cos \phi = \frac{10^2 + 14^2 - 20^2}{2(10)(14)}$$

$$\phi \approx 111.8°$$

$$2\theta \approx 360° - 2(111.8°)$$

$$\theta = 68.2°$$

$$d^2 = 10^2 + 14^2 - 2(10)(14) \cos 68.2°$$

$$d \approx 13.86$$

**19.** $\cos \alpha = \dfrac{(9)^2 + (10)^2 - (6)^2}{2(9)(10)}$

$\alpha \approx 36.3°$

$\cos \beta = \dfrac{6^2 + 10^2 - 9^2}{2(6)(10)}$

$\beta \approx 62.7°$

$z = 180° - \alpha - \beta \approx 81.0°$

$\mu = 180° - z \approx 99.0°$

$b^2 = 9^2 + 6^2 - 2(9)(6)(\cos 99.0°)$

$b \approx 11.58$

$\cos \omega \approx \dfrac{9^2 + 11.58^2 - 6^2}{2(9)(11.58)}$

$\omega \approx 30.8°$

$\theta = \alpha + \omega \approx 67.1°$

$\cos x \approx \dfrac{6^2 + 11.58^2 - 9^2}{2(6)(11.58)}$

$x \approx 50.2°$

$\phi = \beta + x \approx 112.9°$

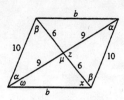

**21.** $\cos B = \dfrac{1100^2 + 2500^2 - 2000^2}{2(1100)(2500)}$

$B \approx 51° \implies$  Bearing of N 39° E

$\cos C = \dfrac{1100^2 + 2000^2 - 2500^2}{2(1100)(2000)}$

$C \approx 103.7°$

$\alpha = 180° - 51° - 103.7° = 25.3° \implies$  Bearing of S 64.7° E

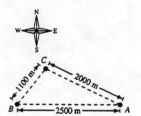

**23.**

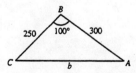

$b^2 = 250^2 + 300^2 - 2(250)(300) \cos 100°$

$b \approx 422.5$ meters

**25.**

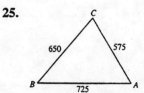

The largest angle is across from the largest side.

$\cos C = \dfrac{650^2 + 575^2 - 725^2}{2(650)(575)}$

$C \approx 72.3°$

**27.** $C = 180° - 53° - 67° = 60°$

$c^2 = a^2 + b^2 - 2ab \cos C$

$\quad = 36^2 + 48^2 - 2(36)(48)(0.5)$

$\quad = 1872$

$\quad c \approx 43.3$ mi

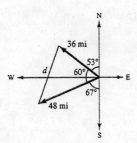

**29.** (a) $\cos \theta = \dfrac{273^2 + 178^2 - 235^2}{2(273)(178)}$

$\qquad \theta \approx 58.4°$

Bearing: N 58.4° W

(b) $\cos \phi = \dfrac{235^2 + 178^2 - 273^2}{2(235)(178)}$

$\qquad \phi \approx 81.5°$

Bearing: S 81.5° W

**31.** $d^2 = 60.5^2 + 90^2 - 2(60.5)(90) \cos 45° \approx 4059.9 \implies d \approx 63.7$ ft

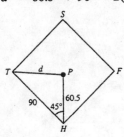

**33.** $\overline{RS} = \sqrt{8^2 + 10^2} = \sqrt{164} = 2\sqrt{41} \approx 12.8$ ft

$\overline{PQ} = \frac{1}{2}\sqrt{16^2 + 10^2} = \frac{1}{2}\sqrt{356} = \sqrt{89} \approx 9.4$ ft

$\tan P = \frac{10}{16}$

$\quad P = \arctan \frac{5}{8} \approx 32.0°$

$\overline{QS} \approx \sqrt{8^2 + 9.4^2 - 2(8)(9.4) \cos 32°} \approx \sqrt{24.81} \approx 5.0$ ft

**35.** (a)  $7^2 = 1.5^2 + x^2 - 2(1.5)(x) \cos \theta$

$49 = 2.25 + x^2 - 3x \cos \theta$

(b) $\qquad x^2 - 3x \cos \theta = 46.75$

$$x^2 - 3x \cos \theta + \left(\frac{3 \cos \theta}{2}\right)^2 = 46.75 + \left(\frac{3 \cos \theta}{2}\right)^2$$

$$\left[x - \frac{3 \cos \theta}{2}\right]^2 = \frac{187}{4} - \frac{9 \cos^2 \theta}{4}$$

$$x - \frac{3 \cos \theta}{2} = \pm \sqrt{\frac{187 + 9 \cos^2 \theta}{4}}$$

Choosing the positive values of $x$, we have $x = \frac{1}{2}\left[3 \cos \theta + \sqrt{9 \cos^2 \theta + 187}\right]$.

(c)

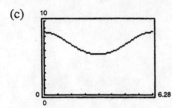

(d) When $\theta = 2\pi$, the piston has moved $2(2)(1.5) = 6$ inches.

**37.**  $A = 180° - 40° - 20° = 120°$

$$\frac{x}{\sin 20°} = \frac{7}{\sin 120°}$$

$$x = \frac{7 \sin 20°}{\sin 120°}$$

$$x = 2.76 \text{ feet}$$

**39.**  $a = 25,\ b = 55,\ c = 72,\ s = \dfrac{25 + 55 + 72}{2} = 76$

(a) $A = \sqrt{(76)(76 - 25)(76 - 55)(76 - 72)} \approx 570.60$

(b) $\cos C = \dfrac{a^2 + b^2 - c^2}{2ab} = \dfrac{25^2 + 55^2 - 72^2}{2(25)(55)} \implies C \approx 123.905°$

$\qquad 2R = \dfrac{c}{\sin C} \approx \dfrac{72}{\sin 123.905°} \implies R \approx 43.3754$

$\qquad A = \pi R^2 \approx 5910.68$

(c) $r = \sqrt{\dfrac{(s - a)(s - b)(s - c)}{s}} = \sqrt{\dfrac{(51)(21)(4)}{76}} \approx 7.5079$

$\qquad A = \pi r^2 \approx 177.09$

**41.**  $a = 5,\ b = 7,\ c = 10 \implies s = \dfrac{a + b + c}{2} = 11$

Area $= \sqrt{s(s - a)(s - b)(s - c)} = \sqrt{11(6)(4)(1)} \approx 16.25$

**43.** $a = 12$, $b = 15$, $c = 9$ $\Rightarrow$ $s = \dfrac{12 + 15 + 9}{2} = 18$

Area $= \sqrt{18(6)(3)(9)} = 54$

**45.** $a = 20$, $b = 20$, $c = 10$ $\Rightarrow$ $s = \dfrac{20 + 20 + 10}{2} = 25$

Area $= \sqrt{25(5)(5)(15)} \approx 96.82$

**47.** $a = 200$, $b = 500$, $c = 600 \Rightarrow s = \dfrac{200 + 500 + 600}{2} = 650$

Area $= \sqrt{650(450)(150)(50)} \approx 46{,}837.5$ sq ft

**49.**
$$\frac{1}{2}bc(1 + \cos A) = \frac{1}{2}bc\left[1 + \frac{b^2 + c^2 - a^2}{2bc}\right]$$
$$= \frac{1}{2}bc\left[\frac{2bc + b^2 + c^2 - a^2}{2bc}\right]$$
$$= \frac{1}{4}\left[(b + c)^2 - a^2\right]$$
$$= \frac{1}{4}\left[(b + c) + a\right]\left[(b + c) - a\right]$$
$$= \frac{b + c + a}{2} \cdot \frac{b + c - a}{2}$$
$$= \frac{a + b + c}{2} \cdot \frac{-a + b + c}{2}$$

**51.** Let $\theta = \arcsin 2x$, then

$\sin \theta = 2x = \dfrac{2x}{1}$ and $\sec \theta = \dfrac{1}{\sqrt{1 - 4x^2}}$.

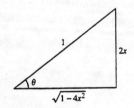

**53.** Let $\theta = \arctan(x - 2)$, then

$\tan \theta = x - 2 = \dfrac{x - 2}{1}$ and $\cot \theta = \dfrac{1}{x - 2}$.

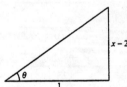

# Section 3.3    Vectors in the Plane

<table>
<tr><td>■</td><td>A vector v is the collection of all directed line segments that are equivalent to a given directed line segment $\overrightarrow{PQ}$.</td></tr>
</table>

■ A vector **v** is the collection of all directed line segments that are equivalent to a given directed line segment $\overrightarrow{PQ}$.

■ You should be able to *geometrically* perform the operations of vector addition and scalar multiplication.

■ The component form of the vector with initial point $P = (p_1, p_2)$ and terminal point $Q = (q_1, q_2)$ is

$$\overrightarrow{PQ} = \langle q_1 - p_1, q_2 - p_2 \rangle = \langle v_1, v_2 \rangle = \mathbf{v}.$$

■ The magnitude of $\mathbf{v} = \langle v_1, v_2 \rangle$ is given by $\|\mathbf{v}\| = \sqrt{v_1{}^2 + v_2{}^2}$.

■ You should be able to perform the operations of scalar multiplication and vector addition in component form.

■ You should know the following properties of vector addition and scalar multiplication.

(a) $\mathbf{u} + \mathbf{v} = \mathbf{v} + \mathbf{u}$

(b) $(\mathbf{u} + \mathbf{v}) + \mathbf{w} = \mathbf{u} + (\mathbf{v} + \mathbf{w})$

(c) $\mathbf{u} + \phi = \mathbf{u}$

(d) $\mathbf{u} + (-\mathbf{u}) = \phi$

(e) $c(d\mathbf{u}) = (cd)\mathbf{u}$

(f) $(c + d)\mathbf{u} = c\mathbf{u} + d\mathbf{u}$

(g) $c(\mathbf{u} + \mathbf{v}) = c\mathbf{u} + c\mathbf{v}$

(h) $1(\mathbf{u}) = \mathbf{u}, 0\mathbf{u} = \phi$

(i) $\|c\mathbf{v}\| = |c|\,\|\mathbf{v}\|$

■ A unit vector in the direction of **v** is given $\mathbf{u} = \dfrac{\mathbf{v}}{\|\mathbf{v}\|}$.

■ The standard unit vectors are $\mathbf{i} = \langle 1, 0 \rangle$ and $\mathbf{j} = \langle 0, 1 \rangle$. $\mathbf{v} = \langle v_1, v_2 \rangle$ can be written as $\mathbf{v} = v_1\mathbf{i} + v_2\mathbf{j}$.

■ A vector **v** with magnitude $\|\mathbf{v}\|$ and direction $\theta$ can be written as $\mathbf{v} = a\mathbf{i} + b\mathbf{j} = \|\mathbf{v}\|(\cos\theta)\mathbf{i} + \|\mathbf{v}\|(\sin\theta)\mathbf{j}$ where $\tan\theta = b/a$.

## Solutions to Odd-Numbered Exercises

**1.** Initial point: $(0, 0)$

Terminal point: $(4, 3)$

$\mathbf{v} = \langle 4 - 0, 3 - 0 \rangle = \langle 4, 3 \rangle$

$\|\mathbf{v}\| = \sqrt{4^2 + 3^2} = 5$

**3.** Initial point: $(2, 2)$

Terminal point: $(-1, 4)$

$\mathbf{v} = \langle -1 - 2, 4 - 2 \rangle = \langle -3, 2 \rangle$

$\|\mathbf{v}\| = \sqrt{(-3)^2 + 2^2} = \sqrt{13}$

**5.** Initial point: $(3, -2)$

Terminal point: $(3, 3)$

$\mathbf{v} = \langle 3 - 3, 3 - (-2) \rangle = \langle 0, 5 \rangle$

$\|\mathbf{v}\| = \sqrt{0^2 + 5^2} = \sqrt{25} = 5$

**7.** Initial point: $(-1, 5)$

Terminal point: $(15, 12)$

$\mathbf{v} = \langle 15 - (-1), 12 - 5 \rangle = \langle 16, 7 \rangle$

$\|\mathbf{v}\| = \sqrt{16^2 + 7^2} = \sqrt{305}$

**9.** Initial point: $(-3, -5)$

Terminal point: $(5, 1)$

$\mathbf{v} = \langle 5 - (-3), 1 - (-5) \rangle = \langle 8, 6 \rangle$

$\|\mathbf{v}\| = \sqrt{8^2 + 6^2} = \sqrt{100} = 10$

**11.**

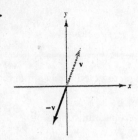

**13.**

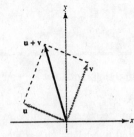

**15.**

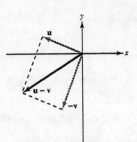

**17.** $\mathbf{u} = \langle 1, 2 \rangle$, $\mathbf{v} = \langle 3, 1 \rangle$

   (a) $\mathbf{u} + \mathbf{v} = \langle 4, 3 \rangle$

   (b) $\mathbf{u} - \mathbf{v} = \langle -2, 1 \rangle$

   (c) $2\mathbf{u} - 3\mathbf{v} = \langle 2, 4 \rangle - \langle 9, 3 \rangle = \langle -7, 1 \rangle$

**19.** $\mathbf{u} = \langle 4, -2 \rangle$, $\mathbf{v} = \langle 0, 0 \rangle$

   (a) $\mathbf{u} + \mathbf{v} = \langle 4, -2 \rangle$

   (b) $\mathbf{u} - \mathbf{v} = \langle 4, -2 \rangle$

   (c) $2\mathbf{u} - 3\mathbf{v} = \langle 8, -4 \rangle - \langle 0, 0 \rangle = \langle 8, -4 \rangle$

**21.** $\mathbf{u} = \mathbf{i} + \mathbf{j}$, $\mathbf{v} = 2\mathbf{i} - 3\mathbf{j}$

   (a) $\mathbf{u} + \mathbf{v} = 3\mathbf{i} - 2\mathbf{j}$

   (b) $\mathbf{u} - \mathbf{v} = -\mathbf{i} + 4\mathbf{j}$

   (c) $2\mathbf{u} - 3\mathbf{v} = (2\mathbf{i} + 2\mathbf{j}) - (6\mathbf{i} - 9\mathbf{j})$
$$= -4\mathbf{i} + 11\mathbf{j}$$

**23.** $\mathbf{u} = 2\mathbf{i}$, $\mathbf{v} = \mathbf{j}$

   (a) $\mathbf{u} + \mathbf{v} = 2\mathbf{i} + \mathbf{j}$

   (b) $\mathbf{u} - \mathbf{v} = 2\mathbf{i} - \mathbf{j}$

   (c) $2\mathbf{u} - 3\mathbf{v} = 4\mathbf{i} - 3\mathbf{j}$

**25.** $\mathbf{u} = \dfrac{1}{\|\mathbf{v}\|} \mathbf{v}$

$$= \frac{1}{\sqrt{5^2 + 0^2}} \langle 5, 0 \rangle$$

$$= \frac{1}{5} \langle 5, 0 \rangle$$

$$= \langle 1, 0 \rangle$$

**27.** $\mathbf{u} = \dfrac{1}{\|\mathbf{v}\|} \mathbf{v}$

$$= \frac{1}{\sqrt{(-2)^2 + 2^2}} \langle -2, 2 \rangle$$

$$= \frac{1}{2\sqrt{2}} \langle -2, 2 \rangle$$

$$= \left\langle -\frac{1}{\sqrt{2}}, \frac{1}{\sqrt{2}} \right\rangle$$

**29.** $\mathbf{u} = \dfrac{1}{\|\mathbf{v}\|} \mathbf{v}$

$$= \frac{1}{\sqrt{16 + 9}} (4\mathbf{i} - 3\mathbf{j}) = \frac{1}{5} (4\mathbf{i} - 3\mathbf{j})$$

$$= \frac{4}{5}\mathbf{i} - \frac{3}{5}\mathbf{j}$$

**31.** $\mathbf{u} = \dfrac{1}{\|\mathbf{v}\|} \mathbf{v} = \dfrac{1}{2} (2\mathbf{j}) = \mathbf{j}$

**33.** $5\left(\dfrac{1}{\|\mathbf{v}\|} \mathbf{v}\right) = 5\left(\dfrac{1}{\sqrt{3^2 + 3^2}} \langle 3, 3 \rangle\right)$

$$= 5\left(\frac{1}{3\sqrt{2}} \langle 3, 3 \rangle\right)$$

$$= \left\langle \frac{5}{\sqrt{2}}, \frac{5}{\sqrt{2}} \right\rangle$$

**35.** $7\left(\dfrac{1}{\|\mathbf{v}\|} \mathbf{v}\right) = 7\left(\dfrac{1}{\sqrt{(-3)^2 + 4^2}} \langle -3, 4 \rangle\right)$

$$= \frac{7}{5} \langle -3, 4 \rangle$$

$$= \left\langle -\frac{21}{5}, \frac{28}{5} \right\rangle$$

**37.** $\mathbf{v} = \frac{3}{2}\mathbf{u}$

$\quad = \frac{3}{2}(2\mathbf{i} - \mathbf{j})$

$\quad = 3\mathbf{i} - \frac{3}{2}\mathbf{j} = \langle 3, -\frac{3}{2} \rangle$

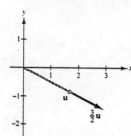

**39.** $\mathbf{v} = \mathbf{u} + 2\mathbf{w}$

$\quad = (2\mathbf{i} - \mathbf{j}) + 2(\mathbf{i} + 2\mathbf{j})$

$\quad = 4\mathbf{i} + 3\mathbf{j} = \langle 4, 3 \rangle$

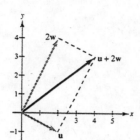

**41.** $\mathbf{v} = \frac{1}{2}(3\mathbf{u} + \mathbf{w})$

$\quad = \frac{1}{2}(6\mathbf{i} - 3\mathbf{j} + \mathbf{i} + 2\mathbf{j})$

$\quad = \frac{7}{2}\mathbf{i} - \frac{1}{2}\mathbf{j} = \langle \frac{7}{2}, -\frac{1}{2} \rangle$

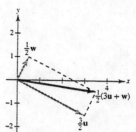

**43.** $\mathbf{v} = 5(\cos 30°\mathbf{i} + \sin 30°\mathbf{j})$

$\quad \|\mathbf{v}\| = 5, \ \theta = 30°$

**45.** $\mathbf{v} = 6\mathbf{i} - 6\mathbf{j}$

$\quad \|\mathbf{v}\| = \sqrt{6^2 + (-6)^2} = \sqrt{72} = 6\sqrt{2}$

$\quad \tan \theta = \dfrac{-6}{6} = -1$

Since $\mathbf{v}$ lies in Quadrant IV, $\theta = 315°$.

**47.** $\mathbf{v} = \langle 3 \cos 0°, 3 \sin 0° \rangle$

$\quad = \langle 3, 0 \rangle$

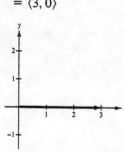

**49.** $\mathbf{v} = \langle \cos 150°, \sin 150° \rangle$

$\quad = \left\langle -\dfrac{\sqrt{3}}{2}, \dfrac{1}{2} \right\rangle$

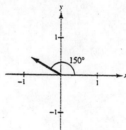

**51.** $\mathbf{v} = \langle 3\sqrt{2} \cos 150°, \ 3\sqrt{2} \sin 150° \rangle$

$\quad = \left\langle -\dfrac{3\sqrt{6}}{2}, \dfrac{3\sqrt{2}}{2} \right\rangle$

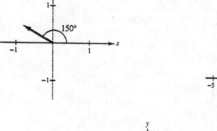

**53.** $\mathbf{v} = 2\left( \dfrac{1}{\sqrt{3^2 + 1^2}} \right)(\mathbf{i} + 3\mathbf{j})$

$\quad = \dfrac{2}{\sqrt{10}}(\mathbf{i} + 3\mathbf{j})$

$\quad = \dfrac{\sqrt{10}}{5}\mathbf{i} + \dfrac{3\sqrt{10}}{5}\mathbf{j} = \left\langle \dfrac{\sqrt{10}}{5}, \dfrac{3\sqrt{10}}{5} \right\rangle$

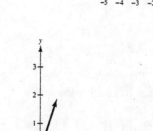

**55.** $\mathbf{u} = \langle 5 \cos 0°, 5 \sin 0° \rangle = \langle 5, 0 \rangle$

$\quad \mathbf{v} = \langle 5 \cos 90°, 5 \sin 90° \rangle = \langle 0, 5 \rangle$

$\quad \mathbf{u} + \mathbf{v} = \langle 5, 5 \rangle$

**57.** $\mathbf{u} = \langle 20 \cos 45°, 20 \sin 45° \rangle = \langle 10\sqrt{2}, 10\sqrt{2} \rangle$

$\quad \mathbf{v} = \langle 50 \cos 180°, 50 \sin 180° \rangle = \langle -50, 0 \rangle$

$\quad \mathbf{u} + \mathbf{v} = \langle 10\sqrt{2} - 50, 10\sqrt{2} \rangle$

**59.** $v = i + j$

$w = 2(i - j)$

$u = v - w = -i + 3j$

$\|v\| = \sqrt{2}$

$\|w\| = 2\sqrt{2}$

$\|v - w\| = \sqrt{10}$

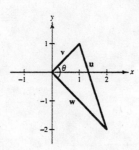

$\cos \alpha = \dfrac{\|v\|^2 + \|w\|^2 - \|v - w\|^2}{2\|v\|\,\|w\|} = \dfrac{2 + 8 - 10}{2\sqrt{2} \cdot 2\sqrt{2}} = 0$

$\alpha = 90°$

**61.** $v = i + j$

$w = 3i - j$

$u = v - w = -2i + 2j$

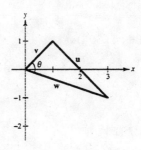

$\cos \alpha = \dfrac{\|v\|^2 + \|w\|^2 - \|v - w\|^2}{2\|v\|\,\|w\|} = \dfrac{2 + 10 - 8}{2\sqrt{2}\,\sqrt{10}} \approx 0.4472$

$\alpha = 63.4°$

**63.** Force One:  $u = 45i$

Force Two:  $v = 60 \cos \theta i + 60 \sin \theta j$

Resultant Force:  $u + v = (45 + 60 \cos \theta)i + 60 \sin \theta j$

$\|u + v\| = \sqrt{(45 + 60 \cos \theta)^2 + (60 \sin \theta)^2} = 90$

$$2025 + 5400 \cos \theta + 3600 = 8100$$

$$5400 \cos \theta = 2475$$

$$\cos \theta = \frac{2475}{5400} \approx 0.4583$$

$$\theta \approx 62.7°$$

**65.** (a)  The angle between them is 0°.

(b)  The angle between them is 180°.

(c)  No. At most it can be equal to the sum when the angle between them is 0°.

**67.** $u = 220i$

$v = (150 \cos 30°)i + (150 \sin 30°)j = 75\sqrt{3}i + 75j$

$u + v = (220 + 75\sqrt{3})i + 75j$

$\|u + v\| = \sqrt{\left(220 + 75\sqrt{3}\right)^2 + 75^2} \approx 357.85$ newtons

$\tan \theta = \dfrac{75}{220 + 75\sqrt{3}} \implies \theta \approx 12.1°$

**69.** $\mathbf{u} = (75 \cos 30°)\mathbf{i} + (75 \sin 30°)\mathbf{j} \approx 64.95\mathbf{i} + 37.5\mathbf{j}$

$\mathbf{v} = (100 \cos 45°)\mathbf{i} + (100 \sin 45°)\mathbf{j} \approx 70.71\mathbf{i} + 70.71\mathbf{j}$

$\mathbf{w} = (125 \cos 120°)\mathbf{i} + (125 \sin 120°)\mathbf{j} \approx -62.5\mathbf{i} + 108.3\mathbf{j}$

$\mathbf{u} + \mathbf{v} + \mathbf{w} \approx 73.16\mathbf{i} + 216.5\mathbf{j}$

$\|\mathbf{u} + \mathbf{v} + \mathbf{w}\| \approx 228.5$ pounds

$\tan \theta \approx \dfrac{216.5}{73.16} \approx 2.9592$

$\theta \approx 71.3°$

**71.** Horizontal component of velocity: $80 \cos 40° \approx 61.28$ ft/sec

Vertical component of velocity: $80 \sin 40° \approx 51.42$ ft/sec

**73.** Cable $\overrightarrow{AC}$: $\mathbf{u} = \|\mathbf{u}\|(\cos 50°\mathbf{i} - \sin 50°\mathbf{j})$

Cable $\overrightarrow{BC}$: $\mathbf{u} = \|\mathbf{u}\|(\cos 30°\mathbf{i} - \sin 30°\mathbf{j})$

Resultant: $\mathbf{u} + \mathbf{v} = -1000\mathbf{j}$

$\|\mathbf{u}\| \cos 50° - \|\mathbf{v}\| \cos 30° = 0$

$-\|\mathbf{u}\| \sin 50° - \|\mathbf{v}\| \sin 30° = -2000$

Solving this system of equations yields:

$T_{AC} = \|\mathbf{u}\| \approx 1758.8$ pounds

$T_{BC} = \|\mathbf{v}\| \approx 1305.4$ pounds

**75.** Towline 1: $\mathbf{u} = \|\mathbf{u}\|(\cos 18°\mathbf{i} + \sin 18°\mathbf{j})$

Towline 2: $\mathbf{v} = \|\mathbf{u}\|(\cos 18°\mathbf{i} - \sin 18°\mathbf{j})$

Resultant: $\mathbf{u} + \mathbf{v} = 6000\mathbf{i}$

$\|\mathbf{u}\| \cos 18° + \|\mathbf{u}\| \cos 18° = 6000$

$\|\mathbf{u}\| \approx 3154.4$

Therefore, the tension on each towline is

$\|\mathbf{u}\| \approx 3154.4$ pounds.

**77.** Airspeed: $\mathbf{u} = (875 \cos 32°)\mathbf{i} - (875 \sin 32°)\mathbf{j}$

Groundspeed: $\mathbf{v} = (800 \cos 40°)\mathbf{i} - (800 \sin 40°)\mathbf{j}$

Wind: $\mathbf{w} = \mathbf{v} - \mathbf{u} = (800 \cos 40° - 875 \cos 32°)\mathbf{i} + (-800 \sin 40° + 875 \sin 32°)\mathbf{j}$

$\approx -129.2065\mathbf{i} - 50.5507\mathbf{j}$

Wind speed: $\|\mathbf{w}\| \approx \sqrt{(-129.2065)^2 + (-50.5507)^2}$

$\approx 138.7$ kilometers per hour

Wind direction: $\tan \theta \approx \dfrac{-50.5507}{-129.2065}$

$\theta \approx 21.4°$

N 21.4° E

**79.** $W = FD = (85 \cos 60°)(20) = 850$ ft/lb

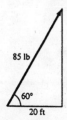

85 lb    60°    20 ft

**81.** True

**83.** False, $a = b = 0$.

**85.** Let $\mathbf{v} = (\cos\theta)\mathbf{i} + (\sin\theta)\mathbf{j}$.

$\|\mathbf{v}\| = \sqrt{\cos^2\theta + \sin^2\theta} = \sqrt{1} = 1$

Therefore, $\mathbf{v}$ is a unit vector for any value of $\theta$.

**87.** $\mathbf{u} = \langle 5 - 1, 2 - 6 \rangle = \langle 4, -4 \rangle$

$\mathbf{v} = \langle 9 - 4, 4 - 5 \rangle = \langle 5, -1 \rangle$

$\mathbf{u} - \mathbf{v} = \langle -1, -3 \rangle$ or $\mathbf{v} - \mathbf{u} = \langle 1, 3 \rangle$

**89.** Let $d$ be the distance from the bridge deck to the water level.

$$\frac{d}{\sin 55} = \frac{17.779}{\sin 90}$$

$$d \approx 14.6 \text{ meters}$$

**91.**

$$\sqrt{x^2 - 64} = \sqrt{(8\sec\theta)^2 - 64}$$
$$= \sqrt{64(\sec^2\theta - 1)}$$
$$= 8\sqrt{\tan^2\theta}$$
$$= 8\tan\theta \quad \text{for } 0 < \theta < \frac{\pi}{2}$$

**93.**

$$\sqrt{x^2 + 36} = \sqrt{(6\tan\theta)^2 + 36}$$
$$= \sqrt{36(\tan^2\theta + 1)}$$
$$= 6\sqrt{\sec^2\theta}$$
$$= 6\sec\theta \quad \text{for } 0 < \theta < \frac{\pi}{2}$$

# Section 3.4    Vintage Vectors and Dot Products

- Know the definition of the dot product of $\mathbf{u} = \langle u_1, u_2 \rangle$ and $\mathbf{v} = \langle v_1, v_2 \rangle$.

  $$\mathbf{u} \cdot \mathbf{v} = u_1 v_1 + u_2 v_2$$

- Know the following properties of the dot product:

  1. $\mathbf{u} \cdot \mathbf{v} = \mathbf{v} \cdot \mathbf{u}$
  2. $\mathbf{0} \cdot \mathbf{v} = 0$
  3. $\mathbf{u} \cdot (\mathbf{v} + \mathbf{w}) = \mathbf{u} \cdot \mathbf{v} + \mathbf{u} \cdot \mathbf{w}$
  4. $\mathbf{v} \cdot \mathbf{v} = \|\mathbf{v}\|^2$
  5. $c(\mathbf{u} \cdot \mathbf{v}) = c\mathbf{u} \cdot \mathbf{v} = \mathbf{u} \cdot c\mathbf{v}$

- If $\theta$ is the angle between two nonzero vectors $\mathbf{u}$ and $\mathbf{v}$, then

  $$\cos\theta = \frac{\mathbf{u} \cdot \mathbf{v}}{\|\mathbf{u}\|\,\|\mathbf{v}\|}.$$

- The vectors $\mathbf{u}$ and $\mathbf{v}$ are orthogonal if $\mathbf{u} \cdot \mathbf{v} = 0$.

- Know the definition of vector components.

  $\mathbf{u} = \mathbf{w}_1 + \mathbf{w}_2$ where $\mathbf{w}_1$ and $\mathbf{w}_2$ are orthogonal, and $\mathbf{w}_1$ is parallel to $\mathbf{v}$. $\mathbf{w}_1$ is called the projection of $\mathbf{u}$ onto $\mathbf{v}$

  and is denoted by $\mathbf{w}_1 = \text{proj}_{\mathbf{v}}\mathbf{u} = \left(\dfrac{\mathbf{u} \cdot \mathbf{v}}{\|\mathbf{v}\|^2}\right)\mathbf{v}$. Then we have $\mathbf{w}_2 = \mathbf{u} - \mathbf{w}_1$.

- Know the definition of work.

  1. Projection form: $w = \|\text{proj}_{\overrightarrow{PQ}}\,\mathbf{F}\|\,\|\overrightarrow{PQ}\|$
  2. Dot product form: $w = \mathbf{F} \cdot \overrightarrow{PQ}$

### Solutions to Odd-Numbered Exercises

**1.** $\mathbf{u} = \langle 3, 4 \rangle$, $\mathbf{v} = \langle 2, -3 \rangle$

$\mathbf{u} \cdot \mathbf{v} = 3(2) + 4(-3) = -6$

**3.** $\mathbf{u} = 4\mathbf{i} - 2\mathbf{j}$, $\mathbf{v} = \mathbf{i} - \mathbf{j}$

$\mathbf{u} \cdot \mathbf{v} = 4(1) + (-2)(-1) = 6$

**5.** $\mathbf{u} = \langle 2, 2 \rangle$

$\mathbf{u} \cdot \mathbf{u} = 2(2) + 2(2) = 8$

The result is a scalar.

**7.** $\mathbf{u} = \langle 2, 2 \rangle$, $\mathbf{v} = \langle -3, 4 \rangle$

$(\mathbf{u} \cdot \mathbf{v})\mathbf{v} = [(2)(-3) + 2(4)]\langle -3, 4 \rangle$

$= 2\langle -3, 4 \rangle = \langle -6, 8 \rangle$

The result is a vector.

**9.** $\mathbf{u} = \langle -5, 12 \rangle$

$\|\mathbf{u}\| = \sqrt{\mathbf{u} \cdot \mathbf{u}} = \sqrt{(-5)^2 + 12^2} = 13$

**11.** $\mathbf{u} = 20\mathbf{i} + 25\mathbf{j}$

$\|\mathbf{u}\| = \sqrt{(20)^2 + (25)^2} = \sqrt{1025} = 5\sqrt{41}$

**13.** $\mathbf{u} = \langle 1245, 2600 \rangle$, $\mathbf{v} = \langle 12.20, 8.50 \rangle$

$\mathbf{u} \cdot \mathbf{v} = 1245(12.20) + 2600(8.50) = \$37,289$

This gives the total revenue that can be earned by selling all of the units.

**15.** $\mathbf{u} = \langle 1, 0 \rangle$, $\mathbf{v} = \langle 0, -2 \rangle$

$\cos \theta = \dfrac{\mathbf{u} \cdot \mathbf{v}}{\|\mathbf{u}\| \, \|\mathbf{v}\|} = \dfrac{0}{(1)(2)} = 0$

$\theta = 90°$

**17.** $\mathbf{u} = 3\mathbf{i} + 4\mathbf{j}$, $\mathbf{v} = -2\mathbf{j}$

$\cos \theta = \dfrac{\mathbf{u} \cdot \mathbf{v}}{\|\mathbf{u}\| \, \|\mathbf{v}\|} = -\dfrac{8}{(5)(2)}$

$\theta = \arccos\left(-\dfrac{4}{5}\right)$

$\theta \approx 143.13°$

**19.** $\mathbf{u} = \left(\cos \dfrac{\pi}{3}\right)\mathbf{i} + \left(\sin \dfrac{\pi}{3}\right)\mathbf{j} = \dfrac{1}{2}\mathbf{i} + \dfrac{\sqrt{3}}{2}\mathbf{j}$

$\mathbf{v} = \left(\cos \dfrac{3\pi}{4}\right)\mathbf{i} + \left(\sin \dfrac{3\pi}{4}\right)\mathbf{j} = -\dfrac{\sqrt{2}}{2}\mathbf{i} + \dfrac{\sqrt{2}}{2}\mathbf{j}$

$\|\mathbf{u}\| = \|\mathbf{v}\| = 1$

$\cos \theta = \dfrac{\mathbf{u} \cdot \mathbf{v}}{\|\mathbf{u}\| \, \|\mathbf{v}\|} = \mathbf{u} \cdot \mathbf{v} = \left(\dfrac{1}{2}\right)\left(-\dfrac{\sqrt{2}}{2}\right) + \left(\dfrac{\sqrt{3}}{2}\right)\left(\dfrac{\sqrt{2}}{2}\right) = \dfrac{-\sqrt{2} + \sqrt{6}}{4}$

$\theta = \arccos\left(\dfrac{-\sqrt{2} + \sqrt{6}}{4}\right) = 75° = \dfrac{5\pi}{12}$

**21.** $\mathbf{u} = 3\mathbf{i} + 4\mathbf{j}$, $\mathbf{v} = -7\mathbf{i} + 5\mathbf{j}$

$\cos \theta = \dfrac{\mathbf{u} \cdot \mathbf{v}}{\|\mathbf{u}\| \, \|\mathbf{v}\|} = -\dfrac{1}{(5)(\sqrt{74})} \implies \theta \approx 91.33°$

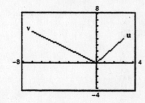

**23.** $\mathbf{u} = 5\mathbf{i} + 5\mathbf{j}$, $\mathbf{v} = -6\mathbf{i} + 6\mathbf{j}$

$\cos \theta = \dfrac{\mathbf{u} \cdot \mathbf{v}}{\|\mathbf{u}\| \, \|\mathbf{v}\|} = 0 \implies \theta = 90°$

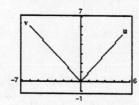

**25.** $P = (1, 2)$, $Q = (3, 4)$, $R = (2, 5)$

$\overrightarrow{PQ} = \langle 2, 2 \rangle$, $\overrightarrow{PR} = \langle 1, 3 \rangle$, $\overrightarrow{QR} = \langle -1, -1 \rangle$

$\cos \alpha = \dfrac{\overrightarrow{PQ} \cdot \overrightarrow{PR}}{\|\overrightarrow{PQ}\| \, \|\overrightarrow{PR}\|} = \dfrac{8}{(2\sqrt{2})(\sqrt{10})} \implies \alpha = \arccos \dfrac{2}{\sqrt{5}} \approx 26.6°$

$\cos \beta = \dfrac{\overrightarrow{PQ} \cdot \overrightarrow{QR}}{\|\overrightarrow{PQ}\| \, \|\overrightarrow{QR}\|} = 0 \implies \beta = 90°$. Thus, $\gamma = 180° - 26.6° - 90° = 63.4°$.

**27.** $\mathbf{u} \cdot \mathbf{v} = \|\mathbf{u}\| \, \|\mathbf{v}\| \cos \theta$

$\qquad = (4)(10) \cos \dfrac{2\pi}{3}$

$\qquad = 40\left(-\dfrac{1}{2}\right)$

$\qquad = -20$

**29.** $\mathbf{u} = \langle -12, 30 \rangle, \; \mathbf{v} = \left\langle \dfrac{1}{2}, -\dfrac{5}{4} \right\rangle$

$\mathbf{u} = -24\mathbf{v} \implies \mathbf{u}$ and $\mathbf{v}$ are parallel.

**31.** $\mathbf{u} = \frac{1}{4}(3\mathbf{i} - \mathbf{j}), \; \mathbf{v} = 5\mathbf{i} + 6\mathbf{j}$

$\mathbf{u} \neq k\mathbf{v} \implies$ Not parallel

$\mathbf{u} \cdot \mathbf{v} \neq 0 \implies$ Not orthogonal

Neither

**33.** $\mathbf{u} = 2\mathbf{i} - 2\mathbf{j}, \; \mathbf{v} = -\mathbf{i} - \mathbf{j}$

$\mathbf{u} \cdot \mathbf{v} = 0 \implies \mathbf{u}$ and $\mathbf{v}$ are orthogonal.

**35.** $\mathbf{u} = \langle 3, 4 \rangle, \; \mathbf{v} = \langle 8, 2 \rangle$

$\mathbf{w}_1 = \text{proj}_{\mathbf{v}}\mathbf{u} = \left(\dfrac{\mathbf{u} \cdot \mathbf{v}}{\|\mathbf{v}\|^2}\right)\mathbf{v} = \left(\dfrac{32}{68}\right)\mathbf{v} = \dfrac{8}{17}\langle 8, 2 \rangle = \dfrac{16}{17}\langle 4, 1 \rangle$

$\mathbf{w}_2 = \mathbf{u} - \mathbf{w}_1 = \langle 3, 4 \rangle - \dfrac{16}{17}\langle 4, 1 \rangle = \dfrac{13}{17}\langle -1, 4 \rangle$

**37.** $\mathbf{u} = \langle 0, 3 \rangle, \; \mathbf{v} = \langle 2, 15 \rangle$

$\mathbf{w}_1 = \text{proj}_{\mathbf{v}}\mathbf{u} = \left(\dfrac{\mathbf{u} \cdot \mathbf{v}}{\|\mathbf{v}\|^2}\right)\mathbf{v} = \dfrac{45}{229}\langle 2, 15 \rangle$

$\mathbf{w}_2 = \mathbf{u} - \mathbf{w}_1 = \langle 0, 3 \rangle - \dfrac{45}{229}\langle 2, 15 \rangle = \left\langle -\dfrac{90}{229}, \dfrac{12}{229} \right\rangle = \dfrac{6}{229}\langle -15, 2 \rangle$

**39.** $\mathbf{u} = \langle 3, 5 \rangle$

For $\mathbf{v}$ to be orthogonal to $\mathbf{u}$, $\mathbf{u} \cdot \mathbf{v}$ must equal 0.

Two possibilities: $\langle -5, 3 \rangle$ and $\langle 5, -3 \rangle$

**41.** $\mathbf{u} = \frac{1}{2}\mathbf{i} - \frac{2}{3}\mathbf{j}$

For $\mathbf{u}$ and $\mathbf{v}$ to be orthogonal, $\mathbf{u} \cdot \mathbf{v}$ must equal 0.

Two possibilities: $\frac{2}{3}\mathbf{i} + \frac{1}{2}\mathbf{j}$ and $-\frac{2}{3}\mathbf{i} - \frac{1}{2}\mathbf{j}$

**43. (a)** $\mathbf{F} = -36{,}000\mathbf{j}$    Gravitational force

$\mathbf{v} = (\cos 10°)\mathbf{i} + (\sin 10°)\mathbf{j}$

$\mathbf{w}_1 = \text{proj}_{\mathbf{v}}\mathbf{F} = \left(\dfrac{\mathbf{F} \cdot \mathbf{v}}{\|\mathbf{v}\|^2}\right)\mathbf{v} = (\mathbf{F} \cdot \mathbf{v})\,\mathbf{v} \approx -6251.3\mathbf{v}$

The magnitude of this force is 6251.3, therefore a force of 6251.3 pounds is needed to keep the truck from rolling down the hill.

**(b)** $\mathbf{w}_2 = \mathbf{F} - \mathbf{w}_1 = -36{,}000\mathbf{j} + 6251.3\,(\cos 10°\mathbf{i} + \sin 10°\mathbf{j})$

$\qquad = [(6251.3 \cos 10°)\mathbf{i} + (6251.3 \sin 10° - 36{,}000)\mathbf{j}]$

$\|\mathbf{w}_2\| \approx 35{,}453.1$ pounds

**45. (a)** $\mathbf{u} \cdot \mathbf{v} = 0 \implies \mathbf{u}$ and $\mathbf{v}$ are orthogonal and $\theta = \dfrac{\pi}{2}$.

**(b)** $\mathbf{u} \cdot \mathbf{v} > 0 \implies \cos \theta > 0 \implies 0 \leq \theta < \dfrac{\pi}{2}$

**(c)** $\mathbf{u} \cdot \mathbf{v} < 0 \implies \cos \theta < 0 \implies \dfrac{\pi}{2} < \theta \leq \pi$

**47.** $\mathbf{w} = (245)(3) = 735$ Newton-meters

**49.** $\mathbf{w} = (\cos 30°)(45)(20) \approx 779.4$ foot-pounds

**51.** $\mathbf{w} = \|\operatorname{proj}_{\overrightarrow{PQ}}\mathbf{v}\|\|\overrightarrow{PQ}\|$ where $\overrightarrow{PQ} = \langle 4, 7 \rangle$ and $\mathbf{v} = \langle 1, 4 \rangle$.

$$\operatorname{proj}_{\overrightarrow{PQ}}\mathbf{v} = \left( \frac{\mathbf{v} \cdot \overrightarrow{PQ}}{\|\overrightarrow{PQ}\|^2} \right)\overrightarrow{PQ} = \left( \frac{32}{65} \right)\langle 4, 7 \rangle$$

$$\mathbf{w} = \|\operatorname{proj}_{\overrightarrow{PQ}}\mathbf{v}\|\|\overrightarrow{PQ}\| = \left( \frac{32\sqrt{65}}{65} \right)\left( \sqrt{65} \right) = 32$$

**53.** In a rhombus, $\|\mathbf{u}\| = \|\mathbf{v}\|$. The diagonals are $\mathbf{u} + \mathbf{v}$ and $\mathbf{u} - \mathbf{v}$.

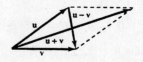

$$(\mathbf{u} + \mathbf{v}) \cdot (\mathbf{u} - \mathbf{v}) = (\mathbf{u} + \mathbf{v}) \cdot \mathbf{u} - (\mathbf{u} + \mathbf{v}) \cdot \mathbf{v}$$
$$= \mathbf{u} \cdot \mathbf{u} + \mathbf{v} \cdot \mathbf{u} - \mathbf{u} \cdot \mathbf{v} - \mathbf{v} \cdot \mathbf{v}$$
$$= \|\mathbf{u}\|^2 - \|\mathbf{v}\|^2 = 0$$

Therefore, the diagonals are orthogonal.

**55.** (a) Let $\mathbf{v} = \langle v_1, v_2 \rangle$.

$$\mathbf{0} \cdot \mathbf{v} = 0(v_1) + 0(v_2) = 0$$

(b) Let $\mathbf{u} = \langle u_1, u_2 \rangle$, $\mathbf{v} = \langle v_1, v_2 \rangle$ and $\mathbf{w} = \langle w_1, w_2 \rangle$.

$$\mathbf{u} \cdot (\mathbf{v} + \mathbf{w}) = \langle u_1, u_2 \rangle \cdot \langle v_1 + w_1, v_2 + w_2 \rangle$$
$$= u_1(v_1 + w_1) + u_2(v_2 + w_2)$$
$$= u_1 v_1 + u_1 w_1 + u_2 v_2 + u_2 w_2$$
$$= (u_1 v_1 + u_2 v_2) + (u_1 w_1 + u_2 w_2)$$
$$= \mathbf{u} \cdot \mathbf{v} + \mathbf{u} \cdot \mathbf{w}$$

(c) Let $\mathbf{u} = \langle u_1, u_2 \rangle$ and $\mathbf{v} = \langle v_1, v_2 \rangle$.

$$c(\mathbf{u} \cdot \mathbf{v}) = c(u_1 v_1 + u_2 v_2)$$
$$= c(u_1 v_1) + c(u_2 v_2)$$
$$= u_1(c v_1) + u_2(c v_2)$$
$$= \mathbf{u} \cdot (c\mathbf{v})$$

# ❑ Review Exercises for Chapter 3

**Solutions to Even-Numbered Exercises**

1. Given: $a = 5$, $b = 8$, $c = 10$

$$\cos C = \frac{a^2 + b^2 - c^2}{2ab} = \frac{25 + 64 - 100}{80} \approx -0.1375 \implies C \approx 97.9°$$

$$\sin A = \frac{a \sin C}{c} \approx \frac{5(0.9905)}{10} \approx 0.4953 \implies A \approx 29.7°$$

$$B = 180° - A - C = 180° - 29.7° - 97.9° = 52.4°$$

3. Given: $A = 12°$, $B = 58°$, $a = 5$

$$C = 180° - A - B = 180° - 12° - 58° = 110°$$

$$b = \frac{a \sin B}{\sin A} = \frac{5 \sin 58°}{\sin 12°} \approx \frac{5(0.8480)}{0.2079} \approx 20.4$$

$$c = \frac{a \sin C}{\sin A} = \frac{5 \sin 110°}{\sin 12°} \approx \frac{5(0.9397)}{0.2079} \approx 22.6$$

5. Given: $B = 110°$, $a = 4$, $c = 4$

$$b^2 = a^2 + c^2 - 2ac \cos B \approx 16 + 16 - 2(4)(4)(-0.3420) \approx 42.94 \implies b \approx 6.6$$

$$\sin A = \frac{a \sin B}{b} \approx \frac{4 \sin 110°}{6.6} \approx \frac{4(0.9397)}{6.6} \approx 0.5736 \implies A \approx 35°$$

$$c = a \implies C = A \approx 35°$$

7. Given: $A = 75°$, $a = 2.5$, $b = 16.5$

$$\sin B = \frac{b \sin A}{a} = \frac{16.5 \sin 75°}{2.5} \approx \frac{16.5(0.9659)}{2.5} \approx 5.375 \implies \text{no triangle formed}$$

No solution

9. Given: $B = 115°$, $a = 7$, $b = 14.5$

$$\sin A = \frac{a \sin B}{b} = \frac{7 \sin 115°}{14.5} \approx \frac{7(0.9063)}{14.5} \approx 0.4375 \implies A \approx 25.9°$$

$$C \approx 180° - 115° - 25.9° = 39.1°$$

$$c^2 = a^2 + b^2 - 2ab \cos C \approx 7^2 + 14.5^2 - 2(7)(14.5)(0.7760) \approx 101.7 \implies c \approx 10.1$$

11. Given: $A = 15°$, $a = 5$, $b = 10$

$$\sin B = \frac{b \sin A}{a} = \frac{10 \sin 15°}{5} \approx \frac{10(0.2588)}{5} \approx 0.5176 \implies B \approx 31.2° \text{ or } 148.8°$$

| Case 1: $B \approx 31.2°$ | Case 2: $B \approx 148.8°$ |
|---|---|
| $C \approx 180° - 15° - 31.2° = 133.8°$ | $C \approx 180° - 15° - 148.8° = 16.2°$ |
| $c = \dfrac{a \sin C}{\sin A} \approx 13.9$ | $c = \dfrac{a \sin C}{\sin A} \approx 5.39$ |

13. Given: $B = 150°$, $a = 10$, $c = 20$

$$b^2 = a^2 + c^2 - 2ac \cos B \approx 100 + 400 - 400(-0.8660) \approx 846.4 \implies b \approx 29.1$$

$$\sin C = \frac{c \sin B}{b} \approx \frac{20(0.5)}{29.09} \approx 0.3437 \implies C \approx 20.1°$$

$$A = 180° - B - C \approx 180° - 150° - 20.1° = 9.9°$$

**15.** Given: $B = 25°$, $a = 6.2$, $b = 4$

$$\sin A = \frac{a \sin B}{b} \approx 0.6551 \implies A \approx 40.9° \text{ or } 139.1°$$

Case 1: $A \approx 40.9°$

$C \approx 180° - 25° - 40.9° = 114.1°$

$c \approx 8.6$

Case 2: $A \approx 139.1°$

$C \approx 180° - 25° - 139.1° = 15.9°$

$c \approx 2.6$

**17.** $a = 4$, $b = 5$, $c = 7$

$$s = \frac{a + b + c}{2} = \frac{4 + 5 + 7}{2} = 8$$

$$\text{Area} = \sqrt{s(s - a)(s - b)(s - c)}$$
$$= \sqrt{8(4)(3)(1)} \approx 9.798$$

**19.** $A = 27°$, $b = 5$, $c = 8$

$$\text{Area} = \frac{1}{2}bc \sin A \approx \frac{1}{2}(5)(8)(0.4540) \approx 9.08$$

**21.** $\alpha = 180° - 31° = 149°$

$\phi = 180° - 149° - 17° = 14°$

$$x = \frac{50 \sin 17°}{\sin \phi} = \frac{50 \sin 17°}{\sin 14°} \approx 60.43$$

$h = x \sin 31°$

$\approx 60.43(0.5150) \approx 31.1$ meters

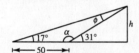

**23.** $\sin 28° = \dfrac{h}{75}$

$h = 75 \sin 28° \approx 35.21$ feet

$\cos 28° = \dfrac{x}{75}$

$x = 75 \cos 28° \approx 66.22$ feet

$\tan 45° = \dfrac{H}{x}$

$H = x \tan 45° \approx 66.22$ feet

Height of tree: $H - h \approx 31$ feet

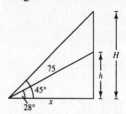

**25.** $d^2 = 850^2 + 1060^2 - 2(850)(1060) \cos 72°$

$\approx 1,289,251$

$d \approx 1135$ miles

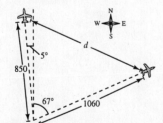

**27.** Initial point: $(1, 1)$

Terminal point: $(6, 4)$

$\mathbf{v} = \langle 6 - 1, 4 - 1 \rangle = \langle 5, 3 \rangle$

**29.** Initial point: $(3, 2)$

Terminal point: $(-1, 2)$

$\mathbf{v} = \langle -1 - 3, 2 - 2 \rangle = \langle -4, 0 \rangle$

**31.** Initial point: $(50, 0)$

Terminal point: $(-20, 50)$

$\mathbf{v} = \langle -20 - 50, 50 - 0 \rangle = \langle -70, 50 \rangle$

**33.** Initial point: $(0, 10)$

Terminal point: $(7, 3)$

$\mathbf{v} = \langle 7 - 0, 3 - 10 \rangle = \langle 7, -7 \rangle$

**35.** $\langle 8 \cos 120°, 8 \sin 120° \rangle = \left( -4, 4\sqrt{3} \right)$

**37.**  $\mathbf{v} = -10\mathbf{i} + 10\mathbf{j}$

$\|\mathbf{v}\| = \sqrt{(-10)^2 + (10)^2} = \sqrt{200} = 10\sqrt{2}$

$\tan \theta = \dfrac{10}{-10} = -1 \implies \theta = 135°$ since

$\mathbf{v}$ is in Quadrant II.

$\mathbf{v} = 10\sqrt{2}(\mathbf{i} \sin 135° + \mathbf{j} \cos 135°)$

**39.**  $\mathbf{u} = 6\mathbf{i} - 5\mathbf{j}$

$\dfrac{1}{\|\mathbf{u}\|}\mathbf{u} = \dfrac{1}{\sqrt{6^2 + 5^2}}(6\mathbf{i} - 5\mathbf{j}) = \dfrac{6}{\sqrt{61}}\mathbf{i} - \dfrac{5}{\sqrt{61}}\mathbf{j}$

$\qquad\qquad = \left\langle \dfrac{6}{\sqrt{61}}, -\dfrac{5}{\sqrt{61}} \right\rangle$

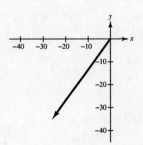

**41.**  $\mathbf{u} = 6\mathbf{i} - 5\mathbf{j}, \;\; \mathbf{v} = 10\mathbf{i} + 3\mathbf{j}$

$4\mathbf{u} - 5\mathbf{v} = (24\mathbf{i} - 20\mathbf{j}) - (50\mathbf{i} + 15\mathbf{j}) = -26\mathbf{i} - 35\mathbf{j}$

$\qquad\qquad\quad = \langle -26, -35 \rangle$

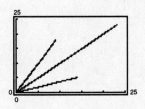

**43.**  $\mathbf{u} = 15[(\cos 20°)\mathbf{i} + (\sin 20°)\mathbf{j}]$

$\mathbf{v} = 20[(\cos 63°)\mathbf{i} + (\sin 63°)\mathbf{j}]$

$\mathbf{u} + \mathbf{v} \approx 23.1752\mathbf{i} + 22.9504\mathbf{j}$

$\|\mathbf{u} + \mathbf{v}\| \approx 32.6161$

$\tan \theta \approx \dfrac{22.9504}{23.1752} \implies \theta \approx 44.72°$

**45.**  $\tan \alpha = \frac{12}{5} \implies \sin \alpha = \frac{12}{13}$ and $\cos \alpha = \frac{5}{13}$

$\tan \beta = \frac{3}{4} \implies \sin(180° - \beta) = \frac{3}{5}$ and $\cos(180° - \beta) = -\frac{4}{5}$

$\mathbf{u} = 250\left(\frac{5}{13}\mathbf{i} + \frac{12}{13}\mathbf{j}\right)$

$\mathbf{v} = 100\left(-\frac{4}{5}\mathbf{i} + \frac{3}{5}\mathbf{j}\right)$

$\mathbf{w} = 200(0\mathbf{i} - \mathbf{j})$

$\mathbf{r} = \mathbf{u} + \mathbf{v} + \mathbf{w} = \left(\frac{1250}{13} - 80 + 0\right)\mathbf{i} + \left(\frac{3000}{13} + 60 - 200\right)\mathbf{j} = \frac{210}{13}\mathbf{i} + \frac{1180}{13}\mathbf{j}$

$\|\mathbf{r}\| = \sqrt{\left(\frac{210}{13}\right)^2 + \left(\frac{1180}{13}\right)^2} \approx 92.2 \text{ lb}$

$\tan \theta = \frac{1180}{210} \implies \theta \approx 79.9°$

**47.**  Rope One:  $\mathbf{u} = \|\mathbf{u}\|(\cos 30°\mathbf{i} - \sin 30°\mathbf{j}) = \|\mathbf{u}\|\left(\dfrac{\sqrt{3}}{2}\mathbf{i} - \dfrac{1}{2}\mathbf{j}\right)$

Rope Two:  $\mathbf{v} = \|\mathbf{u}\|(-\cos 30°\mathbf{i} - \sin 30°\mathbf{j}) = \|\mathbf{u}\|\left(-\dfrac{\sqrt{3}}{2}\mathbf{i} - \dfrac{1}{2}\mathbf{j}\right)$

Resultant:  $\mathbf{u} + \mathbf{v} = -\|\mathbf{u}\|\mathbf{j} = -180\mathbf{j}$

$\qquad\qquad\quad \|\mathbf{u}\| = 180$

Therefore, the tension on each rope is $\|\mathbf{u}\| = 180$ lb.

**49.**  $\mathbf{F} = -500\mathbf{j}$

$\mathbf{v} = (\cos 12°)\mathbf{i} + (\sin 12°)\mathbf{j}$

$\mathbf{w}_1 = \text{proj}_\mathbf{v}\mathbf{F} = \left(\dfrac{\mathbf{F} \cdot \mathbf{v}}{\|\mathbf{v}\|^2}\right)\mathbf{v} = (\mathbf{F} \cdot \mathbf{v})\mathbf{v} = -500(\sin 12°)\mathbf{v}$

The magnitude of $\mathbf{w}_1$ is $500 \sin 12°$ (since $\mathbf{v}$ is a unit vector). Therefore a force of $500 \sin 12° \approx 104$ pounds is required.

**51.**  $\mathbf{u} = \langle -2, 0 \rangle$

$\mathbf{v} = \langle 4, 8 \rangle$

$\mathbf{u} \cdot \mathbf{v} = -2(4) + 0(8) = -8$

**53.**  $\mathbf{u} = -3\mathbf{i} - 7\mathbf{j}$

$\mathbf{v} = 7\mathbf{i} - 3\mathbf{j}$

$\mathbf{u} \cdot \mathbf{v} = -3(7) + (-7)(-3)$

$= 0$

**55.**  $\cos \theta = \dfrac{\mathbf{u} \cdot \mathbf{v}}{\|\mathbf{u}\| \, \|\mathbf{v}\|}$

$\cos \dfrac{\pi}{3} = \dfrac{\mathbf{u} \cdot \mathbf{v}}{(5)(8)}$

$\left(\dfrac{1}{2}\right)(40) = \mathbf{u} \cdot \mathbf{v}$

$\mathbf{u} \cdot \mathbf{v} = 20$

**57.**  $P = (1, 2), Q = (4, 1), R = (5, 4)$

(a)  $\mathbf{u} = \overrightarrow{PQ} = \langle 4 - 1, 1 - 2 \rangle = \langle 3, -1 \rangle$

$\mathbf{v} = \overrightarrow{PR} = \langle 5 - 1, 4 - 2 \rangle = \langle 4, 2 \rangle$

(b)  $\|\mathbf{v}\| = \sqrt{4^2 + 2^2} = \sqrt{20} = 2\sqrt{5}$

(c)  $\mathbf{u} \cdot \mathbf{v} = 3(4) + (-1)(2) = 10$

(d)  $2\mathbf{u} + \mathbf{v} = \langle 6, -2 \rangle + \langle 4, 2 \rangle = \langle 10, 0 \rangle$

(e)  $\mathbf{w}_1 = \text{proj}_\mathbf{v}\mathbf{u} = \left(\dfrac{\mathbf{u} \cdot \mathbf{v}}{\|\mathbf{v}\|^2}\right)\mathbf{v} = \dfrac{10}{\left(2\sqrt{5}\right)^2}\mathbf{v}$

$= \dfrac{1}{2}\mathbf{v} = \langle 2, 1 \rangle$

(f)  $\mathbf{w}_2 = \mathbf{u} - \mathbf{w}_1 = \langle 3, -1 \rangle - \langle 2, 1 \rangle = \langle 1, -2 \rangle$

**59.**  $\mathbf{u} = \langle -1, 4 \rangle$

$\mathbf{v} = \langle 4, 4 \rangle$

$\|\mathbf{u}\| = \sqrt{17}, \|\mathbf{v}\| = 4\sqrt{2}$

$\cos \theta = \dfrac{\mathbf{u} \cdot \mathbf{v}}{\|\mathbf{u}\| \, \|\mathbf{v}\|} = \dfrac{12}{\left(\sqrt{17}\right)\left(4\sqrt{2}\right)}$

$\theta \approx 59°$

**61.**  $\mathbf{u} = 5\left[\cos\left(\dfrac{3\pi}{4}\right)\mathbf{i} + \sin\left(\dfrac{3\pi}{4}\right)\mathbf{j}\right]$

$= 5\left(-\dfrac{\sqrt{2}}{2}\mathbf{i} + \dfrac{\sqrt{2}}{2}\mathbf{j}\right)$

$\mathbf{v} = 2\left[\cos\left(\dfrac{2\pi}{3}\right)\mathbf{i} + \sin\left(\dfrac{2\pi}{3}\right)\mathbf{j}\right]$

$= 2\left(-\dfrac{1}{2}\mathbf{i} + \dfrac{\sqrt{3}}{2}\mathbf{j}\right)$

$\mathbf{u} \cdot \mathbf{v} = \left(-\dfrac{5\sqrt{2}}{2}\right)(-1) + \left(\dfrac{5\sqrt{2}}{2}\right)\left(\sqrt{3}\right)$

$= \dfrac{5\sqrt{2} + 5\sqrt{6}}{2}$

$\cos \theta = \dfrac{\mathbf{u} \cdot \mathbf{v}}{\|\mathbf{u}\| \, \|\mathbf{v}\|} = \dfrac{\dfrac{5\sqrt{2} + 5\sqrt{6}}{2}}{(5)(2)} = \dfrac{\sqrt{2} + \sqrt{6}}{4}$

$\theta = 15°$

**63.**  $\mathbf{u} = \langle 1, 0 \rangle, \mathbf{v} = \langle -2, 2 \rangle$

$\cos \theta = \dfrac{\mathbf{u} \cdot \mathbf{v}}{\|\mathbf{u}\| \, \|\mathbf{v}\|} = \dfrac{-2}{(1)\left(2\sqrt{2}\right)} = -\dfrac{1}{\sqrt{2}}$

$\theta = 135°$

**65.** $\mathbf{F}_1 = 60(\cos \alpha\, \mathbf{i} + \sin \alpha\, \mathbf{j})$

$\mathbf{F}_2 = 100(\cos \beta\, \mathbf{i} + \sin \beta\, \mathbf{j})$

$\mathbf{R} = \mathbf{F}_1 + \mathbf{F}_2 = (60\cos\alpha + 100\cos\beta)\mathbf{i} + (60\sin\alpha + 100\sin\beta)\mathbf{j}$

$\|\mathbf{R}\| = \sqrt{(60\cos\alpha + 100\cos\beta)^2 + (60\sin\alpha + 100\sin\beta)^2}$

$125 = \sqrt{3600\cos^2\alpha + 12{,}000\cos\alpha\cos\beta + 10{,}000\cos^2\beta + 3600\sin^2\alpha + 12{,}000\sin\alpha\sin\beta + 10{,}000\sin^2\beta}$

$125 = \sqrt{13{,}600 + 12{,}000(\cos\alpha\cos\beta + \sin\alpha\sin\beta)}$

$15{,}625 = 13{,}600 + 12{,}000\cos(\alpha - \beta)$

$\dfrac{2025}{12{,}000} = \cos(\alpha - \beta)$

$\alpha - \beta = 80.3°$

**67.** $\mathbf{u} = \langle 3, -2 \rangle$

$\mathbf{v} = \langle 4, 1 \rangle$

$\mathbf{w}_1 = \text{proj}_{\mathbf{v}}\mathbf{u} = \left(\dfrac{\mathbf{u}\cdot\mathbf{v}}{\|\mathbf{v}\|^2}\right)\mathbf{v} = \dfrac{10}{17}\mathbf{v} = \dfrac{10}{17}\langle 4, 1 \rangle$

$\mathbf{w}_2 = \mathbf{u} - \mathbf{w}_1 = \langle 3, -2 \rangle - \dfrac{10}{17}\langle 4, 1 \rangle = \dfrac{11}{17}\langle 1, -4 \rangle$

**69.** $\mathbf{u} = \langle 0, 6 \rangle$

$\mathbf{v} = \langle 2, 3 \rangle$

$\mathbf{w}_1 = \text{proj}_{\mathbf{v}}\mathbf{u} = \left(\dfrac{\mathbf{u}\cdot\mathbf{v}}{\|\mathbf{v}\|^2}\right)\mathbf{v} = \dfrac{18}{13}\langle 2, 3 \rangle$

$\mathbf{w}_2 = \mathbf{u} - \mathbf{w}_1 = \langle 0, 6 \rangle - \dfrac{18}{13}\langle 2, 3 \rangle = \dfrac{12}{13}\langle -3, 2 \rangle$

**71.** $\mathbf{u} = \langle 6, 4 \rangle$

$\mathbf{v} = \langle 12, 4 \rangle$

$\mathbf{w}_1 = \text{proj}_{\mathbf{v}}\mathbf{u} = \left(\dfrac{\mathbf{u}\cdot\mathbf{v}}{\|\mathbf{v}\|^2}\right)\mathbf{v} = \dfrac{88}{160}\langle 12, 4 \rangle = \dfrac{11}{5}\langle 3, 1 \rangle$

$\mathbf{w}_2 = \mathbf{u} - \mathbf{w}_1 = \langle 6, 4 \rangle - \dfrac{11}{5}\langle 3, 1 \rangle = \dfrac{3}{5}\langle -1, 3 \rangle$

**73.** $W = \|\text{proj}_{\overrightarrow{PQ}}\,\mathbf{F}\|\,\|\overrightarrow{PQ}\|$

$= (\cos 30°)\|\mathbf{F}\|\,\|\overrightarrow{PQ}\|$

$= \dfrac{\sqrt{3}}{2}(25)(10)$

$= 125\sqrt{3}$ foot-pounds

**75.** Position the triangle so that the vertices are located at $(0, 0)$, $(a, 0)$, and $(b, c)$.

Let $\mathbf{u} = \left\langle \dfrac{a+b}{2} - 0, \dfrac{c}{2} - 0 \right\rangle = \left\langle \dfrac{a+b}{2}, \dfrac{c}{2} \right\rangle$

$\mathbf{v} = \left\langle \dfrac{b}{2} - a, \dfrac{c}{2} - 0 \right\rangle = \left\langle \dfrac{b-2a}{2}, \dfrac{c}{2} \right\rangle$

$\mathbf{w} = \left\langle \dfrac{a}{2} - b, 0 - c \right\rangle = \left\langle \dfrac{a-2b}{2}, -c \right\rangle$

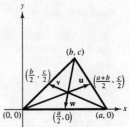

The point of intersection of these three vectors is $\left(\dfrac{a+b}{3}, \dfrac{c}{3}\right)$.

This can be found by solving the following system of linear equations:

Line through $(0, 0)$ and $\left(\dfrac{a+b}{2}, \dfrac{c}{2}\right)$: $y = \dfrac{c}{a+b}x$

Line through $(a, 0)$ and $\left(\dfrac{b}{2}, \dfrac{c}{2}\right)$: $y = \dfrac{-c}{2a-b}x + \dfrac{ac}{2a-b}$

Line through $(b, c)$ and $\left(\dfrac{a}{2}, 0\right)$: $y = \dfrac{2c}{2b-a}x - \dfrac{ac}{2b-a}$

The vector between $(0, 0)$ and $\left(\dfrac{a+b}{3}, \dfrac{c}{3}\right)$ is $\left\langle \dfrac{a+b}{3}, \dfrac{c}{3} \right\rangle = \dfrac{2}{3}\mathbf{u}$.

The vector between $(a, 0)$ and $\left(\dfrac{a+b}{3}, \dfrac{c}{3}\right)$ is $\left\langle \dfrac{b-2a}{3}, \dfrac{c}{3} \right\rangle = \dfrac{2}{3}\mathbf{v}$.

The vector between $(b, c)$ and $\left(\dfrac{a+b}{3}, \dfrac{c}{3}\right)$ is $\left\langle \dfrac{a-2b}{3}, -\dfrac{2c}{3} \right\rangle = \dfrac{2}{3}\mathbf{w}$.

**77.** $(\mathbf{w} \cdot \mathbf{u})\mathbf{u} + (\mathbf{w} \cdot \mathbf{v})\mathbf{v} = \left(\dfrac{\mathbf{w} \cdot \mathbf{v}}{\|\mathbf{v}\|^2}\right)\mathbf{u} + \left(\dfrac{\mathbf{w} \cdot \mathbf{v}}{\|\mathbf{v}\|^2}\right)\mathbf{v}$ Since $\mathbf{u}$ and $\mathbf{v}$ are unit vectors.

$\qquad\qquad\qquad\qquad\quad = \text{proj}_{\mathbf{u}}\mathbf{w} + \text{proj}_{\mathbf{v}}\mathbf{w}$

$\qquad\qquad\qquad\qquad\quad = \mathbf{w}$ Since these are orthogonal vector components of $\mathbf{w}$.

**79.** $\|\mathbf{u} + \mathbf{v}\|^2 = (\mathbf{u} + \mathbf{v}) \cdot (\mathbf{u} + \mathbf{v})$

$\qquad\qquad\quad = (\mathbf{u} + \mathbf{v}) \cdot \mathbf{u} + (\mathbf{u} + \mathbf{v}) \cdot \mathbf{v}$

$\qquad\qquad\quad = \mathbf{u} \cdot \mathbf{u} + 2\mathbf{u} \cdot \mathbf{v} + \mathbf{v} \cdot \mathbf{v}$

$\qquad\qquad\quad = \|\mathbf{u}\|^2 + 2\mathbf{u} \cdot \mathbf{v} + \|\mathbf{v}\|^2$

$\qquad\qquad\quad \leq \|\mathbf{u}\|^2 + 2\|\mathbf{u}\|\,\|\mathbf{v}\| + \|\mathbf{v}\|^2$

$\qquad\qquad\quad \leq (\|\mathbf{u}\| + \|\mathbf{v}\|)^2$

Therefore, $\|\mathbf{u} + \mathbf{v}\| \leq \|\mathbf{u}\| + \|\mathbf{v}\|$.

**81.** $\mathbf{u} = (\cos \alpha)\mathbf{i} + (\sin \alpha)\mathbf{j} \;\Rightarrow\; \|\mathbf{u}\| = 1$

$\quad\;\; \mathbf{v} = (\cos \beta)\mathbf{i} + (\sin \beta)\mathbf{j} \;\Rightarrow\; \|\mathbf{v}\| = 1$

Assume that $\alpha \geq \beta$. Then the angle between $\mathbf{u}$ and $\mathbf{v}$ is $\alpha - \beta$.

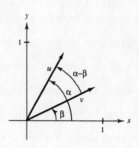

$\cos(\alpha - \beta) = \dfrac{\mathbf{u} \cdot \mathbf{v}}{\|\mathbf{u}\|\,\|\mathbf{v}\|}$

$\qquad\qquad\;\; = \dfrac{\cos \alpha \cos \beta + \sin \alpha \sin \beta}{(1)(1)}$

$\qquad\qquad\;\; = \cos \alpha \cos \beta + \sin \alpha \sin \beta$

**83.** $\mathbf{u} = \langle 1, 3 \rangle$

$\quad\;\; \mathbf{v} = \langle 1, 0 \rangle$

$\quad\;\; \mathbf{w} = \left(\dfrac{2\mathbf{u} \cdot \mathbf{v}}{\mathbf{v} \cdot \mathbf{v}}\right)\mathbf{v} - \mathbf{u}$

$\qquad\; = \dfrac{2}{1}\langle 1, 0 \rangle - \langle 1, 3 \rangle$

$\qquad\; = \langle 1, -3 \rangle$

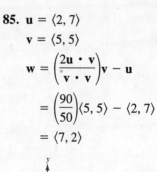

**85.** $\mathbf{u} = \langle 2, 7 \rangle$

$\quad\;\; \mathbf{v} = \langle 5, 5 \rangle$

$\quad\;\; \mathbf{w} = \left(\dfrac{2\mathbf{u} \cdot \mathbf{v}}{\mathbf{v} \cdot \mathbf{v}}\right)\mathbf{v} - \mathbf{u}$

$\qquad\; = \left(\dfrac{90}{50}\right)\langle 5, 5 \rangle - \langle 2, 7 \rangle$

$\qquad\; = \langle 7, 2 \rangle$

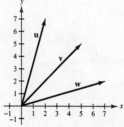

# ❑ Practice Test for Chapter 3

**For Exercises 1 and 2, use the Law of Sines to find the remaining sides and angles of the triangle.**

1. $A = 40°$, $B = 12°$, $b = 100$        2. $C = 150°$, $a = 5$, $c = 20$

3. Find the area of the triangle: $a = 3$, $b = 6$, $C = 130°$.

4. Determine the number of solutions to the triangle: $a = 10$, $b = 35$, $A = 22.5°$.

**For Exercises 5 and 6, use the Law of Cosines to find the remaining sides and angles of the triangle.**

5. $a = 49$, $b = 53$, $c = 38$        6. $C = 29°$, $a = 100$, $b = 300$

7. Use Heron's Formula to find the area of the triangle: $a = 4.1$, $b = 6.8$, $c = 5.5$.

8. A ship travels 40 miles due east, then adjusts its course 12° southward. After traveling 70 miles in that direction, how far is the ship from its point of departure?

9. $\mathbf{w} = 4\mathbf{u} - 7\mathbf{v}$ where $\mathbf{u} = 3\mathbf{i} + \mathbf{j}$ and $\mathbf{v} = -\mathbf{i} + 2\mathbf{j}$. Find $\mathbf{w}$.

10. Find a unit vector in the direction of $\mathbf{v} = 5\mathbf{i} - 3\mathbf{j}$.

11. Find the dot product and the angle between $\mathbf{u} = 6\mathbf{i} + 5\mathbf{j}$ and $\mathbf{v} = 2\mathbf{i} - 3\mathbf{j}$.

12. $\mathbf{v}$ is a vector of magnitude 4 making an angle of 30° with the positive $x$-axis. Find $\mathbf{v}$ in component form.

13. Find the projection of $\mathbf{u}$ onto $\mathbf{v}$ given $\mathbf{u} = \langle 3, -1 \rangle$ and $\mathbf{v} = \langle -2, 4 \rangle$.

14. Given $\|\mathbf{u}\| = 7$, $\theta\mathbf{u} = 35°$
$\|\mathbf{v}\| = 4$, $\theta\mathbf{v} = 123°$

Find the component form of $\mathbf{u} + \mathbf{v}$.

15. Find two vectors orthogonal to $\langle -3, 10 \rangle$.

# C H A P T E R   4
## Complex Numbers

# CHAPTER 4
## Complex Numbers

## Section 4.1    Complex Numbers

---

- You should know how to work with complex numbers.
- Operations on complex numbers
  - (a) Addition: $(a + bi) + (c + di) = (a + c) + (b + d)i$
  - (b) Subtraction: $(a + bi) - (c + di) = (a - c) + (b - d)i$
  - (c) Multiplication: $(a + bi)(c + di) = (ac - bd) + (ad + bc)i$
  - (d) Division: $\dfrac{a + bi}{c + di} = \dfrac{a + bi}{c + di} \cdot \dfrac{c - di}{c - di} = \dfrac{ac + bd}{c^2 + d^2} + \dfrac{bc - ad}{c^2 + d^2}i$
- The complex conjugate of $a + bi$ is $a - bi$:
  $$(a + bi)(a - bi) = a^2 + b^2$$
- The additive inverse of $a + bi$ is $-a - bi$.
- The multiplicative inverse of $a + bi$ is
  $$\frac{a - bi}{a^2 + b^2}.$$
- $\sqrt{-a} = \sqrt{a}\, i$ for $a > 0$.

---

**Solutions to Odd-Numbered Exercises**

**1.** $a + bi = -10 + 6i$

$\quad a = -10$

$\quad b = \phantom{-}6$

**3.** $(a - 1) + (b + 3)i = 5 + 8i$

$\quad a - 1 = 5 \quad \Rightarrow \quad a = 6$

$\quad b + 3 = 8 \quad \Rightarrow \quad b = 5$

**5.** $4 + \sqrt{-9} = 4 + 3i$

**7.** $2 - \sqrt{-27} = 2 - \sqrt{27}\,i = 2 - 3\sqrt{3}\,i$

**9.** $\sqrt{-75} = \sqrt{75}\,i = 5\sqrt{3}\,i$

**11.** $-6i + i^2 = -6i - 1 = -1 - 6i$

**13.** $8 = 8 + 0i = 8$

**15.** $\sqrt{-0.09} = \sqrt{0.09}\,i = 0.3i$

**17.** $(5 + i) + (6 - 2i) = 11 - i$

**19.** $(8 - i) - (4 - i) = 8 - i - 4 + i = 4$

**21.** $\left(-2 + \sqrt{-8}\right)\left(5 - \sqrt{50}\right) = -2 + 2\sqrt{2}\,i + 5 - 5\sqrt{2}\,i$

$\qquad\qquad\qquad\qquad\qquad = 3 - 3\sqrt{2}\,i$

**23.** $13i - (14 - 7i) = 13i - 14 + 7i = -14 + 20i$

**25.** $-\left(\frac{3}{2} + \frac{5}{2}i\right) + \left(\frac{5}{3} + \frac{11}{3}i\right) = -\frac{3}{2} - \frac{5}{2}i + \frac{5}{3} + \frac{11}{3}i$

$\qquad\qquad\qquad\qquad\qquad\qquad = -\frac{9}{6} - \frac{15}{6}i + \frac{10}{6} + \frac{22}{6}i$

$\qquad\qquad\qquad\qquad\qquad\qquad = \frac{1}{6} + \frac{7}{6}i$

**27.** $\sqrt{-6} \cdot \sqrt{-2} = \left(\sqrt{6}\,i\right)\left(\sqrt{2}\,i\right) = \sqrt{12}\,i^2 = \left(2\sqrt{3}\right)(-1) = -2\sqrt{3}\,i$

**29.** $\left(\sqrt{-10}\right)^2 = \left(\sqrt{10}\,i\right)^2 = 10i^2 = -10$

**31.** $(1 + i)(3 - 2i) = 3 - 2i + 3i - 2i^2$

$$= 3 + i + 2$$

$$= 5 + i$$

**33.** $6i(5 - 2i) = 30i - 12i^2 = 30i + 12 = 12 + 30i$

**35.** $\left(\sqrt{14} + \sqrt{10}\,i\right)\left(\sqrt{14} - \sqrt{10}\,i\right) = 14 - 10i^2 = 14 + 10 = 24$

**37.** $(4 + 5i)^2 = 16 + 40i + 25i^2 = 16 + 40i - 25$

$$= -9 + 40i$$

**39.** $(2 + 3i)^2 + (2 - 3i)^2 = 4 + 12i + 9i^2 + 4 - 12i + 9i^2$

$$= 4 + 12i - 9 + 4 - 12i - 9$$

$$= -10$$

**41.** $\sqrt{-6}\sqrt{-6} = \sqrt{6}\,i\sqrt{6}\,i = 6i^2 = -6$

**43.** The complex conjugate of $5 + 3i$ is $5 - 3i$.

$(5 + 3i)(5 - 3i) = 25 - 9i^2 = 25 + 9 = 34$

**45.** The complex conjugate of $-2 - \sqrt{5}\,i$ is $-2 + \sqrt{5}\,i$.

$\left(-2 - \sqrt{5}\,i\right)\left(-2 + \sqrt{5}\,i\right) = 4 - 5t^2 = 4 + 5 = 9$

**47.** The complex conjugate of $20i$ is $-20i$.

$(20i)(-20i) = -400i^2 = 400$

**49.** The complex conjugate of $\sqrt{8}$ is $\sqrt{8}$.

$\left(\sqrt{8}\right)\left(\sqrt{8}\right) = 8$

**51.** $\dfrac{6}{i} = \dfrac{6}{i} \cdot \dfrac{-i}{-i} = \dfrac{-6i}{-i^2} = \dfrac{-6i}{1} = -6i$

**53.** $\dfrac{4}{4 - 5i} = \dfrac{4}{4 - 5i} \cdot \dfrac{4 + 5i}{4 + 5i} = \dfrac{4(4 + 5i)}{16 + 25} = \dfrac{16 + 20i}{41} = \dfrac{16}{41} + \dfrac{20}{41}i$

**55.** $\dfrac{2 + i}{2 - i} = \dfrac{2 + i}{2 - i} \cdot \dfrac{2 + i}{2 + i} = \dfrac{4 + 4i + i^2}{4 + 1} = \dfrac{3 + 4i}{5} = \dfrac{3}{5} + \dfrac{4}{5}i$

**57.** $\dfrac{6 - 7i}{i} = \dfrac{6 - 7i}{i} \cdot \dfrac{-i}{-i} = \dfrac{-6i + 7i^2}{1} = -7 - 6i$

**59.** $\dfrac{1}{(4 - 5i)^2} = \dfrac{1}{16 - 40i + 25i^2} = \dfrac{1}{-9 - 40i} \cdot \dfrac{-9 + 40i}{-9 + 40i}$

$$= \dfrac{-9 + 40i}{81 + 1600} = \dfrac{-9 + 40i}{1681} = -\dfrac{9}{1681} + \dfrac{40}{1681}i$$

**61.** $\dfrac{2}{1 + i} - \dfrac{3}{1 - i} = \dfrac{2(1 - i) - 3(1 + i)}{(1 + i)(1 - i)}$

$$= \dfrac{2 - 2i - 3 - 3i}{1 + 1}$$

$$= \dfrac{-1 - 5i}{2}$$

$$= -\dfrac{1}{2} - \dfrac{5}{2}i$$

**63.** $\dfrac{i}{3 - 2i} + \dfrac{2i}{3 + 8i} = \dfrac{i(3 + 8i) + 2i(3 - 2i)}{(3 - 2i)(3 + 8i)}$

$$= \dfrac{3i + 8i^2 + 6i - 4i^2}{9 + 24i - 6i - 16i^2}$$

$$= \dfrac{4i^2 + 9i}{9 + 18i + 16}$$

$$= \dfrac{-4 + 9i}{25 + 18i} \cdot \dfrac{25 - 18i}{25 - 18i}$$

$$= \dfrac{-100 + 72i + 225i - 162i^2}{625 + 324}$$

$$= \dfrac{-100 + 297i + 162}{949}$$

$$= \dfrac{62 + 297i}{949}$$

$$= \dfrac{62}{949} + \dfrac{297}{949}i$$

**65.** $x^2 - 2x + 2 = 0; \; a = 1, \; b = -2, \; c = 2$

$$x = \dfrac{-(-2) \pm \sqrt{(-2)^2 - 4(1)(2)}}{2(1)} = \dfrac{2 \pm \sqrt{-4}}{2} = \dfrac{2 \pm 2i}{2} = 1 \pm i$$

**67.** $4x^2 + 16x + 17 = 0; \; a = 4, \; b = 16, \; c = 17$

$$x = \dfrac{-16 \pm \sqrt{(16)^2 - 4(4)(17)}}{2(4)}$$

$$= \dfrac{-16 \pm \sqrt{-16}}{8} = \dfrac{-16 + 4i}{8}$$

$$= -2 \pm \dfrac{1}{2}i$$

**69.** $4x^2 + 16x + 15 = 0; \; a = 4, \; b = 16, \; c = 15$

$$x = \dfrac{-16 \pm \sqrt{(16)^2 - 4(4)(15)}}{2(4)} = \dfrac{-16 \pm \sqrt{16}}{8} = \dfrac{-16 \pm 4}{8}$$

$$x = -\dfrac{12}{8} = -\dfrac{3}{2} \quad \text{or} \quad x = \dfrac{-20}{8} = -\dfrac{5}{2}$$

**71.** $16t^2 - 4t + 3 = 0; \; a = 16, \; b = -4, \; c = 3$

$$t = \dfrac{-(-4) \pm \sqrt{(-4)^2 - 4(16)(3)}}{2(16)}$$

$$= \dfrac{4 \pm \sqrt{-176}}{32} = \dfrac{4 \pm 4\sqrt{11}i}{32}$$

$$= \dfrac{1}{8} \pm \dfrac{\sqrt{11}}{8}i$$

**73.** $y = \frac{1}{4}(4x^2 - 20x + 25)$ 

$$0 = \frac{1}{4}(4x^2 - 20x + 25)$$
$$0 = 4x^2 - 20x + 25$$
$$0 = (2x - 5)^2$$
$$0 = 2x - 5$$
$$-2x = -5$$
$$x = \frac{5}{2}$$

x-intercept: $\left(\frac{5}{2}, 0\right)$

**75.** $y = -(x^2 - 4x + 5)$

$$0 = -(x^2 - 4x + 5)$$
$$0 = x^2 - 4x + 5$$
$$x = \frac{-(-4) \pm \sqrt{(-4)^2 - 4(1)(5)}}{2(1)}$$
$$= \frac{4 \pm 2i}{2} = 2 \pm i$$

No x-intercepts          No real solutions

**77.** The number of x-intercepts of the graph of $y = ax^2 + bx + c$ corresponds to the number of real solutions of the equation $0 = ax^2 + bx + c$. If there are no x-intercepts, the quadratic equation has two complex solutions.

**79.** $-6i^3 + i^2 = -6i^2i + i^2$

$$= -6(-1)i + (-1)$$
$$= 6i - 1$$
$$= -1 + 6i$$

**81.** $-5i^5 = -5i^2i^2i$

$$= -5(-1)(-1)i$$
$$= -5i$$

**83.** $\left(\sqrt{-75}\right)^3 \left(5\sqrt{3}i\right)^3 = 5^3\left(\sqrt{3}\right)^3 i^3$

$$125\left(3\sqrt{3}\right)(-1)$$
$$= -375\sqrt{3}i$$

**85.** $\dfrac{1}{i^3} = \dfrac{1}{-i} = \dfrac{1}{-i} \cdot \dfrac{i}{i} = \dfrac{i}{-i^2} = \dfrac{i}{1} = i$

**87.** $(2)^3 = 8$

$$(1 + \sqrt{3}i)^3 = (-1)^3 + 3(-1)^2(\sqrt{3}i) + 3(-1)(\sqrt{3}i)^2 + (\sqrt{3}i)^3$$
$$= -1 + 3\sqrt{3}i - 9i^2 + 3\sqrt{3}i^3$$
$$= -1 + 3\sqrt{3}i + 9 - 3\sqrt{3}i$$
$$= 8$$

$$\left(-1 - \sqrt{3}i\right)^3 = (-1)^3 + 3(-1)^2\left(-\sqrt{3}i\right) + 3(-1)\left(-\sqrt{3}i\right)^2 + \left(-\sqrt{3}i\right)^3$$
$$= -1 - 3\sqrt{3}i - 9i^2 - 3\sqrt{3}i^3$$
$$= -1 - 3\sqrt{3}i + 9 + 3\sqrt{3}i$$
$$= 8$$

**89.** $(a + bi) + (a - bi) = 2a$ which is a real number.

**91.** $(u + bi)(a - bi) = a^2 - (bi)^2 = a^2 - b^2i^2$

$$= a^2 + b^2 \text{ which is a real number.}$$

**93.** $(a_1 + b_1i) + (a_2 + b_2i) = (a_1 + a_2) + (b_1 + b_2)i$

The complex conjugate of the sum is $(a_1 + a_2) - (b_1 + b_2)i$, and the sum of the conjugates is

$$(a_1 - b_1i) + (a_2 - b_2i) = (a_1 + a_2) + (-b_1 - b_2)i$$
$$= (a_1 + a_2) - (b_1 + b_2)i.$$

Thus, the conjugate of the sum is the sum of the conjugates.

**95.** $(4 + 3x) + (8 - 6x - x^2) = -x^2 - 3x + 12$

**97.** $(2x - 5)^2 = (2x)^2 - 2(2x)(5) + (5)^2$
$$= 4x^2 - 20x + 25$$

**99.**
$$V = \frac{4}{3}\pi a^2 b$$
$$3V = 4\pi a^2 b$$
$$\frac{3V}{4\pi b} = a^2$$
$$\sqrt{\frac{3V}{4\pi b}} = a$$
$$a = \frac{1}{2}\sqrt{\frac{3V}{\pi b}}$$

**101.** Let $x$ = # liters withdrawn and replaced.
$$0.50(5 - x) + 1.00x = 0.60(5)$$
$$2.50 - 0.50x + 1.00x = 3.00$$
$$0.50x = 0.50$$
$$x = 1 \text{ liter}$$

# Section 4.2 Complex Solutions of Equations

---

- If $f$ is a polynomial with real coefficients of degree $n > 0$, then $f$ has exactly $n$ solutions (zeros, roots) in the complex number system.
- Given the quadratic equation, $ax^2 + bx + c = 0$, the discriminant can be used to determine the types of solutions.

  If $b^2 - 4ac < 0$, then both solutions are complex.

  If $b^2 - 4ac = 0$, then there is one repeating real solution.

  If $b^2 - 4ac > 0$, then both solutions are real.
- If $a + bi$ is a complex solution to a polynomial with real coefficients, then so is $a - bi$.

---

**Solutions to Odd-Numbered Exercises**

**1.** $x^3 - 4x + 5 = 0$ has degree three so there are three solutions in the complex number system.

**3.** $25 - x^4 = 0$ has degree four so there are four solutions in the complex number system.

**5.** $2x^2 - 5x + 5 = 0$

$b^2 - 4ac = (-5)^2 - 4(2)(5) = -15 < 0$

Both solutions are complex. There are no real solutions.

**7.** $\frac{1}{5}x^2 + \frac{6}{5}x - 8 = 0$

$b^2 - 4ac = \left(\frac{6}{5}\right)^2 - 4\left(\frac{1}{5}\right)(-8) = \frac{196}{25} > 0$

Both solutions are real.

**9.** $x^2 - 5 = 0$

$x^2 = 5$

$x = \pm\sqrt{5}$

**11.** $(x + 5)^2 - 6 = 0$

$(x + 5)^2 = 6$

$x + 5 = \pm\sqrt{6}$

$x = -5 \pm \sqrt{6}$

**13.** $x^2 - 8x + 16 = 0$

$(x - 4)^2 = 0$

$x = 4$

**15.** $x^2 + 2x + 5 = 0$

$x = \dfrac{-2 \pm \sqrt{2^2 - 4(1)(5)}}{2(1)}$

$= \dfrac{-2 \pm \sqrt{-16}}{2}$

$= \dfrac{-2 \pm 4i}{2}$

$= -1 \pm 2i$

**17.** $4x^2 - 4x + 5 = 0$

$x = \dfrac{-(-4) \pm \sqrt{(-4)^2 - 4(4)(5)}}{2(4)} = \dfrac{4 \pm \sqrt{-64}}{8} = \dfrac{4 \pm 8i}{8} = \dfrac{1}{2} \pm i$

**19.** $230 + 20x - 0.5x^2 = 0$

$x = \dfrac{-20 \pm \sqrt{(20)^2 - 4(-0.5)(230)}}{2(-0.5)} = \dfrac{-20 \pm \sqrt{860}}{-1} = 20 \pm 2\sqrt{215}$

**21.** $f(x) = x^3 - 4x^2 + x - 4 = x^2(x - 4) + 1(x - 4)$

$\qquad = (x - 4)(x^2 + 1)$

The only real zero of $f(x)$ is $x = 4$. This corresponds to the $x$-intercept of $(4, 0)$ on the graph.

**23.** $f(x) = x^4 + 4x^2 + 4 = (x^2 + 2)^2$

$f(x)$ has no real zeros and the graph of $f(x)$ has no $x$-intercepts.

**25.** $f(z) = z^2 - 2z + 2$

$f$ has no rational zeros.

By the Quadratic Formula, the zeros are

$z = \dfrac{2 \pm \sqrt{4 - 8}}{2} = 1 \pm i.$

$f(z) = [z - (1 + i)][z - (1 - i)]$

$\qquad = (z - 1 - i)(z - 1 + i)$

**27.** $f(x) = x^4 - 81$

$\qquad = (x^2 - 9)(x^2 + 9)$

$\qquad = (x + 3)(x - 3)(x + 3i)(x - 3i)$

The zeros of $f(x)$ are $x = \pm 3$ and $x = \pm 3i$.

**29.** $f(x) = 2x^3 + 3x^2 + 50x + 75$

Since $5i$ is a zero, so is $-5i$.

$$
\begin{array}{r|rrrr}
5i & 2 & 3 & 50 & 75 \\
   &   & 10i & -50 + 15i & -75 \\
\hline
   & 2 & 3 + 10i & 15i & 0
\end{array}
$$

$$
\begin{array}{r|rrr}
-5i & 2 & 3 + 10i & 15i \\
    &   & -10i & -15i \\
\hline
    & 2 & 3 & 0
\end{array}
$$

The zero of $2x + 3$ is $x = -\frac{3}{2}$. The zeros of $f(x)$ are $x = -\frac{3}{2}$ and $x = \pm 5i$.

<u>Alternate Solution</u>

Since $x = \pm 5i$ are zeros of $f(x)$, $(x + 5i)(x - 5i) = x^2 + 25$ is a factor of $f(x)$. By long division we have:

$$
\require{enclose}
\begin{array}{r}
2x + 3 \phantom{00000000} \\
x^2 + 0x + 25 \enclose{longdiv}{2x^3 + 3x^2 + 50x + 75} \\
\underline{2x^3 + 0x^2 + 50x \phantom{00000}} \\
3x^2 + 0x + 75 \\
\underline{3x^2 + 0x + 75} \\
0
\end{array}
$$

Thus, $f(x) = (x^2 + 25)(2x + 3)$ and the zeros of $f$ are $x = \pm 5i$ and $x = -\frac{3}{2}$.

**31.** $g(x) = 4x^3 + 23x^2 + 34x - 10$

Since $-3 + i$ is a zero, so is $-3 - i$.

| $-3 + i$ | 4 | 23 | 34 | $-10$ |
|---|---|---|---|---|
| | | $-12 + 4i$ | $-37 - i$ | 10 |
| | 4 | $11 + 4i$ | $-3 - i$ | 0 |

| $-3 - i$ | 4 | $11 + 4i$ | $-3 - i$ |
|---|---|---|---|
| | | $-12 - 4i$ | $3 + i$ |
| | 4 | $-1$ | 0 |

The zero of $4x - 1$ is $x = \frac{1}{4}$. The zeros of $g(x)$ are $x = -3 \pm i$ and $x = \frac{1}{4}$.

Alternate Solution

Since $-3 \pm i$ are zeros of $g(x)$,

$$[x - (-3 + i)][x - (-3 - i)] = [(x + 3) - i][(x + 3) + i]$$
$$= (x + 3)^2 - i^2$$
$$= x^2 + 6x + 10$$

is a factor of $g(x)$. By long division we have:

$$
\begin{array}{r}
4x - 1 \\
x^2 + 6x + 10\overline{)4x^3 + 23x^2 + 34x - 10} \\
\underline{4x^3 + 24x^2 + 40x} \\
-x^2 - 6x - 10 \\
\underline{-x^2 - 6x - 10} \\
0
\end{array}
$$

Thus, $g(x) = (x^2 + 6x + 10)(4x - 1)$ and the zeros of $g(x)$ are $x = -3 \pm i$ and $x = \frac{1}{4}$.

**33.** Since $-3 + \sqrt{2}i$ is a zero, so is $-3 - \sqrt{2}i$, and

$$\left[x - \left(-3 + \sqrt{2}i\right)\right]\left[x - \left(-3 - \sqrt{2}i\right)\right]$$
$$= \left[(x + 3) - \sqrt{2}i\right]\left[(x + 3) + \sqrt{2}i\right]$$
$$= (x + 3)^2 - \left(\sqrt{2}i\right)^2$$
$$= x^2 + 6x + 11$$

is a factor of $f(x)$. By long division, we have:

$$
\begin{array}{r}
x^2 - 3x + 2 \\
x^2 + 6x + 11\overline{)x^4 + 3x^3 - 5x^2 - 21x + 22} \\
\underline{x^4 + 6x^3 + 11x^2} \\
-3x^3 - 16x^2 - 21x \\
\underline{-3x^3 - 18x^2 - 33x} \\
2x^2 + 12x + 22 \\
\underline{2x^2 + 12x + 22} \\
0
\end{array}
$$

Thus,

$$f(x) = (x^2 + 6x + 11)(x^2 - 3x + 2)$$
$$= (x^2 + 6x + 11)(x - 1)(x - 2)$$

and the zeros of $f$ are $x = -3 \pm \sqrt{2}i$, $x = 1$, and $x = 2$.

**35.** Since $\frac{1}{2}\left(1 - \sqrt{5}i\right)$ is a zero, so is $\frac{1}{2}\left(1 + \sqrt{5}i\right)$, and

$$\left[x - \frac{1}{2}\left(1 - \sqrt{5}i\right)\right]\left[x - \frac{1}{2}\left(1 + \sqrt{5}i\right)\right]$$
$$= \left[\left(x - \frac{1}{2}\right) + \frac{1}{2}\sqrt{5}i\right]\left[\left(x - \frac{1}{2}\right) - \frac{1}{2}\sqrt{5}i\right]$$
$$= \left(x - \frac{1}{2}\right)^2 - \left(\frac{1}{2}\sqrt{5}i\right)^2$$
$$= x^2 - x + \frac{1}{4} + \frac{5}{4}$$
$$= x^2 - x + \frac{3}{2}$$

is a factor of $h(x)$. By long division, we have:

$$
\begin{array}{r}
8x - 6 \\
x^2 - x + \frac{3}{2}\overline{)8x^3 - 14x^2 + 18x - 9} \\
\underline{8x^3 - 8x^2 + 12x} \\
-6x^2 + 6x - 9 \\
\underline{-6x^2 + 6x - 9} \\
0
\end{array}
$$

Thus, $h(x) = \left(x^2 - x + \frac{3}{2}\right)(8x - 6)$ and the zeros of $h(x)$ are

$$x = \frac{3}{4} \text{ and } x = \frac{1}{2} \pm \frac{\sqrt{5}}{2}i.$$

**37.** $f(x) = (x - 1)(x - 5i)(x + 5i)$

$\quad = (x - 1)(x^2 + 25)$

$\quad = x^3 - x^2 + 25x - 25$

Note: $f(x) = a(x^3 - x^2 + 25x - 25)$ where $a$ is any nonzero real number, has the zeros 1 and $\pm 5i$.

**39.** $f(x) = (x - 2)[x - (4 + i)][x - (4 - i)]$

$\quad = (x - 2)[(x - 4) - i][(x - 4) + i]$

$\quad = (x - 2)[(x - 4)^2 - i^2]$

$\quad = (x - 2)(x^2 + 8x + 17)$

$\quad = x^3 - 10x^2 + 33x - 34$

**41.** Since $1 + \sqrt{3}i$ is a zero, so is $1 - \sqrt{3}i$.

$f(x) = (x + 5)(x + 5)\left[x - \left(1 + \sqrt{3}i\right)\right]\left[x - (1 - \sqrt{3}i)\right]$

$\quad = (x^2 + 10x + 25)\left[(x - 1) - \sqrt{3}i\right]\left[(x - 1) + \sqrt{3}i\right]$

$\quad = (x^2 + 10x + 25)\left[(x - 1)^2 - 3i^2\right]$

$\quad = (x^2 + 10x + 25)(x^2 - 2x + 4)$

$\quad = x^4 + 8x^3 + 9x^2 - 10x + 100$

**43.** If $-\frac{1}{2} + i$ is a zero, so is its conjugate, $-\frac{1}{2} - i$.

$f(x) = 4\left(x - \frac{3}{4}\right)(x + 2)\,2\left[x - \left(-\frac{1}{2} + i\right)\right]2\left[x - \left(-\frac{1}{2} - i\right)\right]$

$\quad = (4x - 3)(x + 2)[(2x + 1) - 2i][(2x + 1) + 2i]$

$\quad = (4x^2 + 5x - 6)[(2x + 1)^2 - (2i)^2]$

$\quad = (4x^2 + 5x - 6)(4x^2 + 4x + 1 + 4)$

$\quad = (4x^2 + 5x - 6)(4x^2 + 4x + 5)$

$\quad = 16x^4 + 36x^3 + 16x^2 + x - 30$

Note: $f(x) = a(16x^4 + 36x^3 + 16x^2 + x - 30)$, where $a$ is any nonzero real number, has the zeros $x = \frac{3}{4}$, $x = -2$, and $x = -\frac{1}{2} \pm i$. In fact, we used constant multiples of the linear factors in this equation to simplify the fractions. Without these constants, you would have

$f(x) = \left(x - \frac{3}{4}\right)(x + 2)\left(x + \frac{1}{2} - i\right)\left(x + \frac{1}{2} + i\right)$

$\quad = x^4 + \frac{9}{4}x^3 + x^2 + \frac{1}{16}x - \frac{15}{8}.$

**45.** $f(x) = x^4 + 6x^2 - 27$

$f(x) = (x + 3i)(x - 3i)(x + \sqrt{3})(x - \sqrt{3})$

$f(x) = x^4 - 6x^2 - 27$

$\quad = (x^2 - 3)(x^2 + 9)$

$\quad = (x + \sqrt{3})(x - \sqrt{3})(x + 3i)(x - 3i)$

**47.** $f(x) = x^3 + ix^2 + ix - 1$

(a)
$$
\begin{array}{r|rrrr}
i & 1 & i & i & -1 \\
  &   & i & -2 & -1 - 2i \\
\hline
  & 1 & 2i & -2 + i & -2 - 2i
\end{array}
$$

Since the remainder is not zero, $x = i$ is not a zero of $f$.

(b) The theorem that states that complex zeros occur in conjugate pairs has the condition that the coefficients of $f(x)$ must be real numbers. This polynomial has complex coefficients for $x^2$ and $x$.

**49.** $f(x) = x^4 - 4x^2 + k$

$$x^2 = \frac{-(-4) \pm \sqrt{(-4)^2 - 4(1)(k)}}{2(1)} = \frac{4 \pm 2\sqrt{4-k}}{2} = 2 \pm \sqrt{4-k}$$

$$x = \pm\sqrt{2 \pm \sqrt{4-k}}$$

(a) For there to be four distinct real roots, both $4 - k$ and $2 \pm \sqrt{4-k}$ must be positive. This occurs when $0 < k < 4$. Thus, some possible $k$-values are $k = 1$, $k = 2$, $k = 3$, $k = \frac{1}{2}$, $k = \sqrt{2}$, etc.

(b) For there to be two real roots, each of multiplicity 2, $4 - k$ must equal zero. Thus, $k = 4$.

(c) For there to be two real zeros and two complex zeros, $2 + \sqrt{4-k}$ must be positive and $2 - \sqrt{4-k}$ must be negative. This occurs when $k < 0$. Thus, some possible $k$-values are $k = -1$, $k = -2$, $k = -\frac{1}{2}$, etc.

(d) For there to be four complex zeros, $2 \pm \sqrt{4-k}$ must be complex. This occurs when $k > 4$. Some possible $k$-values are $k = 5$, $k = 6$, $k = 7.4$, etc.

**51.** (a) $f(x) = x^2(x + 2)(x - 3.5)$

This is not the correct function since $f(x)$ has a zero ($x$-intercept) at $x = 0$.

(b) $g(x) = (x + 2)(x - 3.5)$

This is not the correct function since it is second-degree and its graph would be a parabola.

(c) $h(x) = (x + 2)(x - 3.5)(x^2 + 1)$

This function matches the graph.

(d) $k(x) = (x + 1)(x + 2)(x - 3.5)$

This is not the correct function since $k(x)$ has a zero ($x$-intercept) at $x = -1$.

# Section 4.3   Trigonometric Form of a Complex Number

- You should be able to graphically represent complex numbers and know the following facts about them.
- The absolute value of the complex numbers $z = a + bi$ is $|z| = \sqrt{a^2 + b^2}$.
- The trigonometric form of the complex number $z = a + bi$ is $z = r(\cos \theta + i \sin \theta)$ where
    - (a) $a = r \cos \theta$
    - (b) $b = r \sin \theta$
    - (c) $r = \sqrt{a^2 + b^2}$; $r$ is called the modulus of $z$.
    - (d) $\tan \theta = \frac{b}{a}$; $\theta$ is called the argument of $z$.
- Given $z_1 = r_1(\cos \theta_1 + i \sin \theta_1)$ and $z_2 = r_2(\cos \theta_2 + i \sin \theta_2)$:
    - (a) $z_1 z_2 = r_1 r_2[\cos(\theta_1 + \theta_2) + i \sin(\theta_1 + \theta_2)]$
    - (b) $\dfrac{z_1}{z_2} = \dfrac{r_1}{r_2}[\cos(\theta_1 - \theta_2) + i \sin(\theta_1 - \theta_2)]$, $z_2 \neq 0$

## Solutions to Odd-Numbered Exercises

**1.** $|-5i| = \sqrt{0^2 + (-5)^2}$
$= \sqrt{25} = 5$

**3.** $|-4 + 4i| = \sqrt{(-4)^2 + (4)^2}$
$= \sqrt{32} = 4\sqrt{2}$

**5.** $|6 - 7i| = \sqrt{6^2 + (-7)^2}$
$= \sqrt{85}$

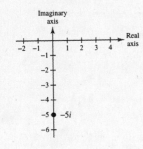

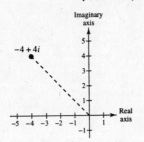

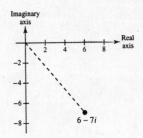

**7.** $z = 3i$

$r = \sqrt{0^2 + 3^2} = \sqrt{9} = 3$

$\tan \theta = \dfrac{3}{0}$, undefined $\Rightarrow \theta = \dfrac{\pi}{2}$

$z = 3\left(\cos \dfrac{\pi}{2} + i \sin \dfrac{\pi}{2}\right)$

**9.** $z = -2 - 2i$

$r = \sqrt{(-2)^2 + (-2)^2} = \sqrt{8} = 2\sqrt{2}$

$\tan \theta = \dfrac{-2}{-2} = 1$, $\theta$ is in Quadrant III.

$\theta = 225°$ or $\dfrac{5\pi}{4}$

$z = 2\sqrt{2}\left(\cos \dfrac{5\pi}{4} + i \sin \dfrac{5\pi}{4}\right)$

**11.** $z = 3 - 3i$

$r = \sqrt{3^2 + (-3)^2} = \sqrt{18} = 3\sqrt{2}$

$\tan \theta = \dfrac{-3}{3} = -1$, $\theta$ is in Quadrant IV $\Rightarrow \theta = \dfrac{7\pi}{4}$

$z = 3\sqrt{2}\left(\cos \dfrac{7\pi}{4} + i \sin \dfrac{7\pi}{4}\right)$

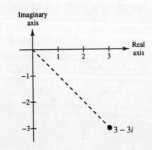

**13.** $z = \sqrt{3} + i$

$r = \sqrt{(\sqrt{3})^2 + 1^2} = \sqrt{4} = 2$

$\tan\theta = \dfrac{1}{\sqrt{3}} = \dfrac{\sqrt{3}}{3} \implies \theta = \dfrac{\pi}{6}$

$z = 2\left(\cos\dfrac{\pi}{6} + i\sin\dfrac{\pi}{6}\right)$

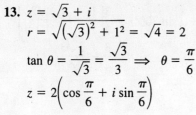

**15.** $z = -2(1 + \sqrt{3}i)$

$r = \sqrt{(-2)^2 + (-2\sqrt{3})^2} = \sqrt{16} = 4$

$\tan\theta = \dfrac{\sqrt{3}}{1} = \sqrt{3},\ \theta$ is in Quadrant III $\implies \theta = \dfrac{4\pi}{3}$

$z = 4\left(\cos\dfrac{4\pi}{3} + i\sin\dfrac{4\pi}{3}\right)$

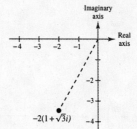

**17.** $z = 0 + 6i$

$r = \sqrt{0^2 + (6)^2} = \sqrt{36} = 6$

$\tan\theta = \dfrac{6}{0}$, undefined $\implies \theta = \dfrac{\pi}{2}$

$z = 6\left(\cos\dfrac{\pi}{2} + i\sin\dfrac{\pi}{2}\right)$

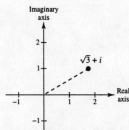

**19.** $z = -7 + 4i$

$r = \sqrt{(-7)^2 + (4)^2} = \sqrt{65}$

$\tan\theta = \dfrac{4}{-7},\ \theta$ is in Quadrant II $\implies \theta \approx 2.62$

$z \approx \sqrt{65}(\cos 2.62 + i\sin 2.62)$

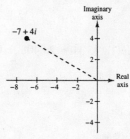

**21.** $z = 7 + 0i$

$r = \sqrt{(7)^2 + (0)^2} = \sqrt{49} = 7$

$\tan\theta = \dfrac{0}{7} = 0 \implies \theta = 0$

$z = 7(\cos 0 + i\sin 0)$

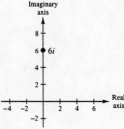

**23.** $z = 1 + 6i$

$r = \sqrt{1^2 + (6)^2} = \sqrt{37}$

$\tan\theta = \dfrac{6}{1} = 6 \implies \theta \approx 1.41$

$z \approx \sqrt{37}(\cos 1.41 + i\sin 1.41)$

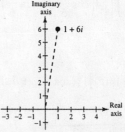

**25.** $z = 5 + 2i$

$r \approx 5.39$

$\theta \approx 0.38$

$z \approx 5.39(\cos 0.38 + i\sin 0.38)$

**27.** $z = 3\sqrt{2} - 7i,\ \theta$ in Quadrant IV

$r \approx 8.19$

$\theta \approx 5.26$

$z \approx 8.19(\cos 5.26 + i\sin 5.26)$

**29.** $2(\cos 150° + i \sin 150°) = 2\left[-\dfrac{\sqrt{3}}{2} + i\left(\dfrac{1}{2}\right)\right]$
$$= -\sqrt{3} + i$$

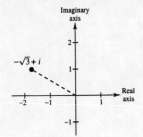

**31.** $\dfrac{3}{2}(\cos 300° + i \sin 300°) = \dfrac{3}{2}\left[\dfrac{1}{2} + i\left(-\dfrac{\sqrt{3}}{2}\right)\right]$
$$= \dfrac{3}{4} - \dfrac{3\sqrt{3}}{4}i$$

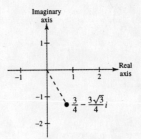

**33.** $3.75\left(\cos \dfrac{3\pi}{4} + i \sin \dfrac{3\pi}{4}\right) = -\dfrac{15\sqrt{2}}{8} + \dfrac{15\sqrt{2}}{8}i$

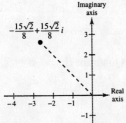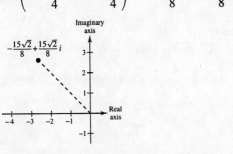

**35.** $4\left(\cos \dfrac{3\pi}{2} + i \sin \dfrac{3\pi}{2}\right) = 4(0 - i) = -4i$

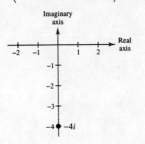

**37.** $3[\cos(18°45') + i \sin(18°45')] \approx 2.8408 + 0.9643i$

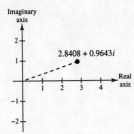

**39.** $5\left(\cos \dfrac{\pi}{9} + i \sin \dfrac{\pi}{9}\right) \approx 4.70 + 1.71i$

**41.** $4(\cos 216.5° + i \sin 216.5°) \approx -3.22 - 2.38i$

**43.** $\left[3\left(\cos \dfrac{\pi}{3} + i \sin \dfrac{\pi}{3}\right)\right]\left[4\left(\cos \dfrac{\pi}{6} + i \sin \dfrac{\pi}{6}\right)\right] = (3)(4)\left[\cos\left(\dfrac{\pi}{3} + \dfrac{\pi}{6}\right) + i \sin\left(\dfrac{\pi}{6} + \dfrac{\pi}{3}\right)\right]$
$$= 12\left(\cos \dfrac{\pi}{2} + i \sin \dfrac{\pi}{2}\right)$$

**45.** $\left[\dfrac{5}{3}(\cos 140° + i \sin 140°)\right]\left[\dfrac{2}{3}(\cos 60° + i \sin 60°)\right] = \left(\dfrac{5}{3}\right)\left(\dfrac{2}{3}\right)[\cos(140° + 60°) + i \sin(140° + 60°)]$
$$= \dfrac{10}{9}(\cos 200° + i \sin 200°)$$

**47.** $\dfrac{2(\cos 120° + i \sin 120°)}{4(\cos 40° + i \sin 40°)} = \dfrac{2}{4}[\cos(120° - 40°) + i \sin(120° - 40°)]$
$$= \dfrac{1}{2}(\cos 80° + i \sin 80°)$$

**49.** $\dfrac{\cos(5\pi/3) + i\sin(5\pi/3)}{\cos \pi + i\sin \pi} = \cos\left(\dfrac{5\pi}{3} - \pi\right) + i\sin\left(\dfrac{5\pi}{3} - \pi\right) = \cos\left(\dfrac{2\pi}{3}\right) + i\sin\left(\dfrac{2\pi}{3}\right)$

**51.** $\dfrac{12(\cos 52° + i\sin 52°)}{3(\cos 110° + i\sin 110°)} = 4[\cos(52° - 110°) + i\sin(52° - 110°)]$

$= 4[\cos(-58°) + i\sin(-58°)]$

**53.** (a) $2 + 2i = 2\sqrt{2}(\cos 45° + i\sin 45°)$

$1 - i = \sqrt{2}[\cos(-45°) + i\sin(-45°)]$

(b) $(2 + 2i)(1 - i) = [2\sqrt{2}(\cos 45° + i\sin 45°)][\sqrt{2}(\cos(-45°) + i\sin(-45°))] = 4(\cos 0° + i\sin 0°) = 4$

(c) $(2 + 2i)(1 - i) = 2 - 2i + 2i - 2i^2 = 2 + 2 = 4$

**55.** (a) $-2i = 2[\cos(-90°) + i\sin(-90°)]$

$1 + i = \sqrt{2}(\cos 45° + i\sin 45°)$

(b) $-2i(1 + i) = 2[\cos(-90°) + i\sin(-90°)][\sqrt{2}(\cos 45° + i\sin 45°)]$

$= 2\sqrt{2}[\cos(-45°) + i\sin(-45°)]$

$= 2\sqrt{2}\left[\dfrac{1}{\sqrt{2}} - \dfrac{1}{\sqrt{2}}i\right] = 2 - 2i$

(c) $-2i(1 + i) = -2i - 2i^2 = -2i + 2 = 2 - 2i$

**57.** (a) $5 = 5(\cos 0° + i\sin 0°)$

$2 + 3i \approx \sqrt{13}(\cos 56.31° + i\sin 56.31°)$

(b) $\dfrac{5}{2 + 3i} \approx \dfrac{5(\cos 0° + i\sin 0°)}{\sqrt{13}[\cos(-56.31°) + i\sin(-56.31°)]} = \dfrac{5\sqrt{13}}{13}[\cos(-56.31°) + i\sin(-56.31°)] \approx 0.7692 - 1.154i$

(c) $\dfrac{5}{2 + 3i} = \dfrac{5}{2 + 3i} \cdot \dfrac{2 - 3i}{2 - 3i} = \dfrac{10 - 15i}{13} = \dfrac{10}{13} - \dfrac{15}{13}i \approx 0.7692 - 1.154i$

**59.** $\dfrac{z_1}{z_2} = \dfrac{r_1(\cos\theta_1 + i\sin\theta_1)}{r_2(\cos\theta_2 + i\sin\theta_2)} \cdot \dfrac{\cos\theta_2 + i\sin\theta_2}{\cos\theta_2 + i\sin\theta_2}$

$= \dfrac{r_1}{r_2(\cos^2\theta_2 + \sin^2\theta_2)}[\cos\theta_1\cos\theta_2 + \sin\theta_1\sin\theta_2 + i(\sin\theta_1\cos\theta_2 - \sin\theta_2\cos\theta_1)]$

$= \dfrac{r_1}{r_2}[\cos(\theta_1 - \theta_2) + i\sin(\theta_1 - \theta_2)]$

**61.** (a) $z\bar{z} = [r(\cos\theta + i\sin\theta)][r(\cos(-\theta) + i\sin(-\theta))]$

$= r^2[\cos(\theta - \theta) + i\sin(\theta - \theta)]$

$= r^2[\cos 0 + i\sin 0]$

$= r^2$

(b) $\dfrac{z}{\bar{z}} = \dfrac{r(\cos\theta + i\sin\theta)}{r[\cos(-\theta) + i\sin(-\theta)]}$

$= \dfrac{r}{r}[\cos(\theta - (-\theta)) + i\sin(\theta - (-\theta))]$

$= \cos(2\theta) + i\sin(2\theta)$

**63.** Let $z = x + iy$ such that:

$|z| = 2 \implies 2 = \sqrt{x^2 + y^2}$

$\implies 4 = x^2 + y^2$: circle with radius of 2

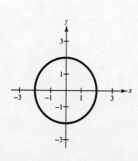

# Section 4.4    DeMoivre's Theorem

- You should know DeMoivre's Theorem: If $z = r(\cos\theta + i\sin\theta)$, then for any positive integer $n$,
  $$z^n = r^n(\cos n\theta + i\sin n\theta).$$
- You should know that for any positive integer $n$, $z = r(\cos\theta + i\sin\theta)$ has $n$ distinct $n$th roots given by
  $$\sqrt[n]{r}\left[\cos\left(\frac{\theta + 2\pi k}{n}\right) + i\sin\left(\frac{\theta + 2\pi k}{n}\right)\right]$$
  where $k = 0, 1, 2, \ldots, n - 1$.

## Solutions to Odd-Numbered Exercises

**1.** $(1 + i)^5 = \left[\sqrt{2}\left(\cos\dfrac{\pi}{4} + i\sin\dfrac{\pi}{4}\right)\right]^5$

$= (\sqrt{2})^5\left(\cos\dfrac{5\pi}{4} + i\sin\dfrac{5\pi}{4}\right)$

$= 4\sqrt{2}\left(-\dfrac{\sqrt{2}}{2} - \dfrac{\sqrt{2}}{2}i\right)$

$= -4 - 4i$

**3.** $(-1 + i)^{10} = \left[\sqrt{2}\left(\cos\dfrac{3\pi}{4} + i\sin\dfrac{3\pi}{4}\right)\right]^{10}$

$= (\sqrt{2})^{10}\left(\cos\dfrac{30\pi}{4} + i\sin\dfrac{30\pi}{4}\right)$

$= 32\left[\cos\left(\dfrac{3\pi}{2} + 6\pi\right) + i\sin\left(\dfrac{3\pi}{2} + 6\pi\right)\right]$

$= 32\left(\cos\dfrac{3\pi}{2} + i\sin\dfrac{3\pi}{2}\right)$

$= 32[0 + i(-1)]$

$= -32i$

**5.** $2\left(\sqrt{3} + i\right)^7 = 2\left[2\left(\cos\dfrac{\pi}{6} + i\sin\dfrac{\pi}{6}\right)\right]^7$

$= 2\left[2^7\left(\cos\dfrac{7\pi}{6} + i\sin\dfrac{7\pi}{6}\right)\right]$

$= 256\left(-\dfrac{\sqrt{3}}{2} - \dfrac{1}{2}i\right)$

$= -128\sqrt{3} - 128i$

**7.** $[5(\cos 20° + i\sin 20°)]^3 = 5^3(\cos 60° + i\sin 60°) = \dfrac{125}{2} + \dfrac{125\sqrt{3}}{2}i$

**9.** $\left(\cos\dfrac{5\pi}{4} + i\sin\dfrac{5\pi}{4}\right)^{10} = \cos\dfrac{25\pi}{2} + i\sin\dfrac{25\pi}{2}$

$= \cos\left(12\pi + \dfrac{\pi}{2}\right) + i\sin\left(12\pi + \dfrac{\pi}{2}\right) = \cos\dfrac{\pi}{2} + i\sin\dfrac{\pi}{2} = i$

**11.** $[5(\cos 3.2 + i\sin 3.2)]^4 = 5^4(\cos 12.8 + i\sin 12.8)$

$\approx 608.02 + 144.69i$

**13.** $(3 - 2i)^5 = -597 - 122i$

**15.** $[3(\cos 15° + i\sin 15°)]^4 = 81(\cos 60° + i\sin 60°)$

$= \dfrac{81}{2} + \dfrac{8\sqrt{3}}{2}i$

**17.** $\left[-\dfrac{1}{2}(1 + \sqrt{3}i)\right]^6 = \left[\cos\dfrac{4\pi}{3} + i\sin\dfrac{4\pi}{3}\right]^6$

$= \cos 8\pi + i\sin 8\pi$

$= 1$

**19.** (a) In trigonometric form we have:

$2(\cos 30° + i \sin 30°)$

$2(\cos 150° + i \sin 150°)$

$2(\cos 270° + i \sin 270°)$

(b) There are three roots evenly spaced around a circle of radius 2. Therefore, they represent the cube roots of some number of modulus 8. Cubing them shows that they are all cube roots of $8i$.

(c) $[2(\cos 30° + i \sin 30°)]^3 = 8i$

$[2(\cos 150° + i \sin 150°)]^3 = 8i$

$[2(\cos 270° + i \sin 270°)]^3 = 8i$

**21.** (a) Square roots of $5(\cos 120° + i \sin 120°)$:

$$\sqrt{5}\left[\cos\left(\frac{120° + 360°k}{2}\right) + i \sin\left(\frac{120° + 360°k}{2}\right)\right], \; k = 0, 1$$

$k = 0: \; \sqrt{5}(\cos 60° + i \sin 60°)$

$k = 1: \; \sqrt{5}(\cos 240° + i \sin 240°)$

(c) $\dfrac{\sqrt{5}}{2} + \dfrac{\sqrt{15}}{2}i, \; -\dfrac{\sqrt{5}}{2} - \dfrac{\sqrt{15}}{2}i$

(b)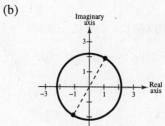

**23.** (a) Fourth roots of $16\left(\cos \dfrac{4\pi}{3} + i \sin \dfrac{4\pi}{3}\right)$:

$$\sqrt[4]{16}\left[\cos\left(\frac{(4\pi/3) + 2k\pi}{4}\right) + i \sin\left(\frac{(4\pi/3) + 2k\pi}{4}\right)\right], \; k = 0, 1, 2, 3$$

$k = 0: 2\left(\cos \dfrac{\pi}{3} + i \sin \dfrac{\pi}{3}\right)$

$k = 1: 2\left(\cos \dfrac{5\pi}{6} + i \sin \dfrac{5\pi}{6}\right)$

$k = 2: 2\left(\cos \dfrac{4\pi}{3} + i \sin \dfrac{4\pi}{3}\right)$

$k = 3: 2\left(\cos \dfrac{11\pi}{6} + i \sin \dfrac{11\pi}{6}\right)$

(c) $1 + \sqrt{3}i, \; -\sqrt{3} + i, \; -1 - \sqrt{3}i, \; \sqrt{3} - i$

(b)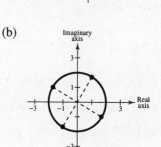

**25.** (a) Square roots of $-25i = 25\left(\cos \dfrac{3\pi}{2} + i \sin \dfrac{3\pi}{2}\right)$:

$$\sqrt{25}\left[\cos\left(\frac{(3\pi/2) + 2k\pi}{2}\right) + i \sin\left(\frac{(3\pi/2) + 2k\pi}{2}\right)\right], \; k = 0, 1$$

$k = 0: 5\left(\cos \dfrac{3\pi}{4} + i \sin \dfrac{3\pi}{4}\right)$

$k = 1: 5\left(\cos \dfrac{7\pi}{4} + i \sin \dfrac{7\pi}{4}\right)$

(c) $-\dfrac{5\sqrt{2}}{2} + \dfrac{5\sqrt{2}}{2}i, \; \dfrac{5\sqrt{2}}{2} - \dfrac{5\sqrt{2}}{2}i$

(b)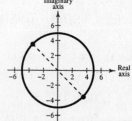

**27.** (a) Cube roots of $-\dfrac{125}{3}(1 + \sqrt{3}i) = 125\left(\cos\dfrac{4\pi}{3} + i\sin\dfrac{4\pi}{3}\right)$:

(b)

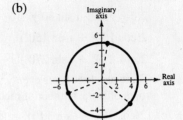

$$\sqrt[3]{125}\left[\cos\left(\frac{(4\pi/3) + 2k\pi}{3}\right) + i\sin\left(\frac{(4\pi/3) + 2k\pi}{3}\right)\right],\ k = 0, 1, 2$$

$$k = 0:\ 5\left(\cos\frac{4\pi}{9} + i\sin\frac{4\pi}{9}\right)$$

$$k = 1:\ 5\left(\cos\frac{10\pi}{9} + i\sin\frac{10\pi}{9}\right)$$

$$k = 2:\ 5\left(\cos\frac{16\pi}{9} + i\sin\frac{16\pi}{9}\right)$$

(c) $0.8682 + 4.924i,\ -4.698 - 1.710i,\ 3.830 - 3.214i$

**29.** (a) Cube roots of $8 = 8(\cos 0 + i\sin 0)$:

(b)

$$\sqrt[3]{8}\left[\cos\left(\frac{2k\pi}{3}\right) + i\sin\left(\frac{2k\pi}{3}\right)\right],\ k = 0, 1, 2$$

$$k = 0:\ 2(\cos 0 + i\sin 0)$$

$$k = 1:\ 2\left(\cos\frac{2\pi}{3} + i\sin\frac{2\pi}{3}\right)$$

$$k = 2:\ 2\left(\cos\frac{4\pi}{3} + i\sin\frac{4\pi}{3}\right)$$

(c) $2,\ -1 + \sqrt{3}i,\ -1 - \sqrt{3}i$

**31.** (a) Fifth roots of $1 = \cos 0 + i\sin 0$:

(b)

$$\cos\left(\frac{2k\pi}{5}\right) + i\sin\left(\frac{2k\pi}{5}\right),\ k = 0, 1, 2, 3, 4$$

$$k = 0:\ \cos 0 + i\sin 0$$

$$k = 1:\ \cos\frac{2\pi}{5} + i\sin\frac{2\pi}{5}$$

$$k = 2:\ \cos\frac{4\pi}{5} + i\sin\frac{4\pi}{5}$$

$$k = 3:\ \cos\frac{6\pi}{5} + i\sin\frac{6\pi}{5}$$

$$k = 4:\ \cos\frac{8\pi}{5} + i\sin\frac{8\pi}{5}$$

(c) $1, 0.3090 + 0.9511i, -0.8090 + 0.5878i, -0.8090 - 0.5878i, 0.3090 - 0.9511i$

**33.** (a) The cube roots of $-125 = 125(\cos 180° + i\sin 180°)$ are:

$5(\cos 60° + i\sin 60°)$

$5(\cos 180° + i\sin 180°)$

$5(\cos 300° + i\sin 300°)$

(c) $\dfrac{5}{2} + \dfrac{5\sqrt{3}}{2}i,\ -5,\ \dfrac{5}{2} - \dfrac{5\sqrt{3}}{2}i$

(b)

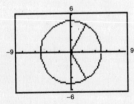

**35. (a)** The fifth roots of $128(-1 + i) = 128\sqrt{2}(\cos 135° + i \sin 135°)$ are:

$2\sqrt[5]{4\sqrt{2}}(\cos 27° + i \sin 27°)$

$2\sqrt[5]{4\sqrt{2}}(\cos 99° + i \sin 99°)$

$2\sqrt[5]{4\sqrt{2}}(\cos 171° + i \sin 171°)$

$2\sqrt[5]{4\sqrt{2}}(\cos 243° + i \sin 243°)$

$2\sqrt[5]{4\sqrt{2}}(\cos 315° + i \sin 315°)$

**(b)**

**(c)** $2.52 + 1.28i, -0.44 + 2.79i, -2.79 + 0.44i,$
$-1.28 - 2.52i, 2 - 2i$

**37.** $x^4 - i = 0$

$x^4 = i$

The solutions are the fourth roots of $i = \cos \dfrac{\pi}{2} + i \sin \dfrac{\pi}{2}$:

$$\sqrt[4]{1}\left[\cos\left(\frac{(\pi/2) + 2k\pi}{4}\right) + i \sin\left(\frac{(\pi/2) + 2k\pi}{4}\right)\right], \ k = 0, 1, 2, 3$$

$k = 0: \cos \dfrac{\pi}{8} + i \sin \dfrac{\pi}{8}$

$k = 1: \cos \dfrac{5\pi}{8} + i \sin \dfrac{5\pi}{8}$

$k = 2: \cos \dfrac{9\pi}{8} + i \sin \dfrac{9\pi}{8}$

$k = 3: \cos \dfrac{13\pi}{8} + i \sin \dfrac{13\pi}{8}$

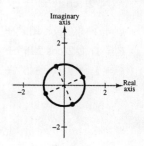

**39.** $x^5 + 243 = 0$

$x^5 = -243$

The solutions are the fifth roots of $-243 = 243(\cos \pi + i \sin \pi)$:

$$\sqrt[5]{243}\left[\cos\left(\frac{\pi + 2k\pi}{5}\right) + i \sin\left(\frac{\pi + 2k\pi}{5}\right)\right], \ k = 0, 1, 2, 3, 4$$

$k = 0: 3\left(\cos \dfrac{\pi}{5} + i \sin \dfrac{\pi}{5}\right)$

$k = 1: 3\left(\cos \dfrac{3\pi}{5} + i \sin \dfrac{3\pi}{5}\right)$

$k = 2: 3(\cos \pi + i \sin \pi) = -3$

$k = 3: 3\left(\cos \dfrac{7\pi}{5} + i \sin \dfrac{7\pi}{5}\right)$

$k = 4: 3\left(\cos \dfrac{9\pi}{5} + i \sin \dfrac{9\pi}{5}\right)$

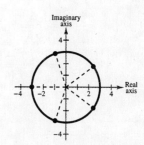

**41.** $x^3 + 64i = 0$

$\qquad x^3 = -64i$

The solutions are the cube roots of $-64i = 64\left(\cos\dfrac{3\pi}{2} + i\sin\dfrac{3\pi}{2}\right)$:

$$\sqrt[3]{64}\left[\cos\left(\frac{(3\pi/2) + 2k\pi}{3}\right) + i\sin\left(\frac{(3\pi/2) + 2k\pi}{3}\right)\right],\ k = 0, 1, 2$$

$k = 0:\ 4\left(\cos\dfrac{\pi}{2} + i\sin\dfrac{\pi}{2}\right) = 4i$

$k = 1:\ 4\left(\cos\dfrac{7\pi}{6} + i\sin\dfrac{7\pi}{6}\right) = -2\sqrt{3} - 2i$

$k = 2:\ 4\left(\cos\dfrac{11\pi}{6} + i\sin\dfrac{11\pi}{6}\right) = 2\sqrt{3} - 2i$

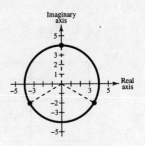

**43.** $x^3 - (1 - i) = 0$

$\qquad x^3 = 1 - i = \sqrt{2}(\cos 315° + i\sin 315°)$

The solutions are the cube roots of $1 - i$:

$$\sqrt[3]{\sqrt{2}}\left[\cos\left(\frac{315° + 360°k}{3}\right) + i\sin\left(\frac{315° + 360°k}{3}\right)\right],\ k = 0, 1, 2$$

$k = 0:\ \sqrt[6]{2}(\cos 105° + i\sin 105°)$

$k = 1:\ \sqrt[6]{2}(\cos 225° + i\sin 225°)$

$k = 2:\ \sqrt[6]{2}(\cos 345° + i\sin 345°)$

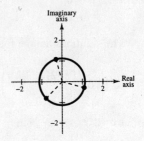

**For 45 and 47, use the following figure.**

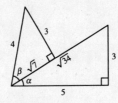

**45.** $\cos(\alpha + \beta) = \cos\alpha\cos\beta - \sin\alpha\sin\beta$

$\qquad = \dfrac{5}{\sqrt{34}} \cdot \dfrac{\sqrt{7}}{4} - \dfrac{3}{\sqrt{34}} \cdot \dfrac{3}{4}$

$\qquad = \dfrac{5\sqrt{7} - 9}{4\sqrt{34}}$

**47.** $\tan 2\alpha = \dfrac{2\tan\alpha}{1 - \tan^2\alpha}$

$\qquad = \dfrac{2(3/5)}{1 - (3/5)^2}$

$\qquad = \dfrac{6}{5} \cdot \dfrac{25}{16} = \dfrac{15}{8}$

# ❑ Review Exercises for Chapter 4

**Solutions to Odd-Numbered Exercises**

**1.** $3 + \sqrt{-25} = 3 + 5i$

**3.** $3i + 6i^3 = 3i - 6i = -3i$

**5.** $(7 + 5i) + (-4 + 2i) = 3 + 7i$

**7.** $\left(\dfrac{\sqrt{2}}{2} - \dfrac{\sqrt{2}}{2}i\right) - \left(\dfrac{\sqrt{2}}{2} + \dfrac{\sqrt{2}}{2}i\right) = -\sqrt{2}i$

**9.** $5i(13 - 8i) = 65i - 40i^2 = 40 + 65i$

**11.** $(10 - 8i)(2 - 3i) = 20 - 30i - 16i + 24i^2 = -4 - 46i$

**13.** $\dfrac{6 + i}{i} = \dfrac{6 + i}{i} \cdot \dfrac{-i}{-i} = \dfrac{-6i - i^2}{1} = 1 - 6i$

**15.** $\dfrac{4}{-3i} = \dfrac{4}{-3i} \cdot \dfrac{3i}{3i} = \dfrac{12i}{9} = \dfrac{4}{3}i$

**17.** $6x^2 + x - 2 = 0$

$b^2 - 4ac = 1^2 - 4(6)(-2) = 49 > 0$

Two real solutions

**19.** $9x^2 - 12x + 4 = 0$

$b^2 - 4ac = (-12)^2 - 4(9)(4) = 0$

One real solution

**21.** $0.13x^2 - 0.45x + 0.65 = 0$

$b^2 - 4ac = (-0.45)^2 - 4(0.13)(0.65) = -0.1355 < 0$

No real solutions

**23.** $-x^2 + 2x + 15 = 0$

$b^2 - 4ac = 2^2 - 4(-1)(15) = 64 > 0$

Two real solutions

**25.** $x^2 - 2x = 0$

$x(x - 2) = 0$

$x = 0, x = 2$

**27.** $x^2 + 8x + 10 = 0$

$x^2 + 8x = -10$

$x^2 + 8x + 16 = -10 + 16$

$(x + 4)^2 = 6$

$x + 4 = \pm\sqrt{6}$

$x = -4 \pm \sqrt{6}$

**29.** $2x^2 + 2x + 3 = 0$

$x = \dfrac{-2 \pm \sqrt{2^2 - 4(2)(3)}}{2(2)}$

$= \dfrac{-2 \pm \sqrt{-20}}{4}$

$= \dfrac{-2 \pm 2\sqrt{5}i}{4}$

$= -\dfrac{1}{2} \pm \dfrac{\sqrt{5}}{2}i$

**31.** $f(x) = x^3 + 4x^2 - 7x - 10$

From the graph we see that $f(x)$ has three real zeros which correspond with the $x$-intercepts. The zeros are $-5$, $-1$, and $2$.

**33.** $f(x) = x^4 - 8x^3 + 21x^2 - 20x$

From the graph we see that $f(x)$ has two real zeros which correspond with the $x$-intercepts at 0 and 4. That means that the other two zeros are complex. To find them, we divide $f(x)$ by $x(x - 4) = x^2 - 4x$.

$$
\begin{array}{r}
x^2 - 4x + 5 \\
x^2 - 4x \overline{\smash{)}\, x^4 - 8x^3 + 21x^2 - 20x + 0} \\
\underline{x^4 - 4x^3} \\
-4x^3 + 21x^2 \\
\underline{-4x^3 + 16x^2} \\
5x^2 - 20x \\
\underline{5x^2 - 20x} \\
0
\end{array}
$$

The other factor of $f(x)$ is $x^2 - 4x + 5$ and by the Quadratic Formula, its zeros are $2 \pm i$. The zeros of $f(x)$ are $0, 4, 2 \pm i$.

**35.** $f(x) = 4x^3 - 11x^2 + 10x - 3$

Zero: $1 \implies x - 1$ is a factor of $f(x)$.

$$
\begin{array}{r}
4x^2 - 7x + 3 \\
x - 1 \overline{)\, 4x^3 - 11x^2 + 10x - 3} \\
\underline{4x^3 - 4x^2} \\
-7x^2 + 10x \\
\underline{-7x^2 + 7x} \\
3x - 3 \\
\underline{3x - 3} \\
0
\end{array}
$$

Thus, $f(x) = (x - 1)(4x^2 - 7x + 3)$

$\qquad = (x - 1)(x - 1)(4x - 3)$

$\qquad = (4x - 3)(x - 1)^2.$

The zeros of $f$ are $\frac{3}{4}$ and 1.

**37.** $f(x) = x^3 + 3x^2 - 5x + 25$

Zero: $-5 \implies x + 5$ is a factor of $f(x)$.

$$
\begin{array}{r}
x^2 - 2x + 5 \\
x + 5 \overline{)\, x^3 + 3x^2 - 5x + 25} \\
\underline{x^3 + 5x^2} \\
-2x^2 - 5x \\
\underline{-2x^2 - 10x} \\
5x + 25 \\
\underline{5x + 25} \\
0
\end{array}
$$

Thus, $f(x) = (x + 5)(x^2 - 2x + 5)$ and by the Quadratic Formula, the zeros of $x^2 - 2x + 5$ are $1 \pm 2i$. The zeros of $f$ are $-5$ and $1 \pm 2i$.

$f(x) = (x + 5)[x - (1 + 2i)][x - (1 - 2i)]$

$\qquad = (x + 5)(x - 1 - 2i)(x - 1 + 2i)$

**39.** $h(x) = x^3 - 18x^2 + 106x - 200$

Zero: $7 + i \implies 7 - i$ is also a zero and $(x - 7 - i)(x - 7 + i) = x^2 - 14x + 50$ is a factor of $h(x)$.

$$
\begin{array}{r}
x - 4 \\
x^2 - 14x + 50 \overline{)\, x^3 - 18x^2 + 106x - 200} \\
\underline{x^3 - 14x^2 + 50x} \\
-4x^2 + 56x - 200 \\
\underline{-4x^2 + 56x - 200} \\
0
\end{array}
$$

Thus, $h(x) = (x - 4)(x^2 - 14x + 50)$

$\qquad = (x - 4)(x - 7 - i)(x - 7 + i).$

The zeros of $h$ are $4, 7 \pm i$.

**41.** $f(x) = x^4 + 5x^3 + 2x^2 - 50x - 84$

Zero: $-3 + \sqrt{5}i \implies -3 - \sqrt{5}i$ is also a zero and $\left(x + 3 - \sqrt{5}i\right)\left(x + 3 + \sqrt{5}i\right) = x^2 + 6x + 14$ is a factor of $f(x)$.

$$
\begin{array}{r}
x^2 - x - 6 \\
x^2 + 6x + 14 \overline{)\, x^4 + 5x^3 + 2x^2 - 50x - 84} \\
\underline{x^4 + 6x^3 + 14x^2} \\
-x^3 - 12x^2 - 50x \\
\underline{-x^3 - 6x^2 - 14x} \\
-6x^2 - 36x - 84 \\
\underline{-6x^2 - 36x - 84} \\
0
\end{array}
$$

Thus,

$f(x) = (x^2 - x - 6)(x^2 + 6x + 14)$

$\qquad = (x + 2)(x - 3)\left(x + 3 - \sqrt{5}i\right)\left(x + 3 + \sqrt{5}i\right)$

The zeros of $f$ are $-2, 3, -3 \pm \sqrt{5}i$.

**43.** $f(x) = (x - 0)(x - 0)(x - 3i)(x + 3i)$

$\qquad = x^2(x^2 + 9)$

$\qquad = x^4 + 9x^2$

**45.** $f(x) = (x + 1)(x + 1)(3x - 1)(2x + 1)$

$\qquad = (x^2 + 2x + 1)(6x^2 + x - 1)$

$\qquad = 6x^4 + 13x^3 + 7x^2 - x - 1$

**47.** $f(x) = (3x - 2)(x - 4)\left(x - \sqrt{3}i\right)\left(x + \sqrt{3}i\right)$

$\qquad = (3x^2 - 14x + 8)(x^2 + 3)$

$\qquad = 3x^4 - 14x^3 + 17x^2 - 42x + 24$

**49.** $f(x) = (2x - 1)^2(x + 1)^3$

$\qquad = (4x^2 - 4x + 1)(x^3 + 3x^2 + 3x + 1)$

$\qquad = 4x^5 + 8x^4 + x^3 - 5x^2 - x + 1$

**51.** $P = xp - C$

$$9,000,000 = x(140 - 0.0001x) - (80x - 150,000)$$

$$9,000,000 = -0.0001x^2 + 60x + 150,000$$

$$0.0001x^2 - 60x + 8,850,000 = 0$$

$$b^2 - 4ac = (-60)^2 - 4(0.0001)(8,850,000) > 0$$

$x \approx 338,730$ units or $x \approx 261,270$ units

$p \approx \$106.13 \qquad p \approx \$113.87$

There are two possible values for $p$ here. Either $p \approx \$106.13$ or $p \approx \$113.87$ will yield a profit of $\approx 9$ million dollars.

**53.** $|4| = 4$

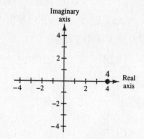

**55.** $|3 + 4i| = \sqrt{3^2 + 4^2} = 5$

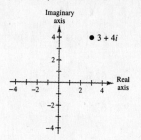

**57.** $z = -5i = 5(\cos 270° + i \sin 270°)$

**59.** $z = 2 + 2i$

$|z| = \sqrt{2^2 + 2^2} = 2\sqrt{2}$

$\tan \theta = \frac{2}{2} = 1 \implies \theta = 45°$ since the complex number is in Quadrant I.

$z = 2\sqrt{2}(\cos 45° + i \sin 45°)$

**61.** $5 - 5i$

$r = \sqrt{5^2 + (-5)^2} = \sqrt{50} = 5\sqrt{2}$

$\tan \theta = -\frac{5}{5} = -1 \implies \theta = 315°$ since the complex number is in Quadrant IV.

$5 - 5i = 5\sqrt{2}(\cos 315° + i \sin 315°)$

**63.** $5 + 12i$

$r = \sqrt{5^2 + 12^2} = \sqrt{169} = 13$

$\tan \theta = \frac{12}{15} \implies \theta \approx 67.38°$ since the complex number is in Quadrant I.

$5 + 12i \approx 13(\cos 67.38° + i \sin 67.38°)$

**65.** $z = \left(\cos \dfrac{\pi}{3} + i \sin \dfrac{\pi}{3}\right)$

$= 6\left(\dfrac{1}{2} + \dfrac{\sqrt{3}}{2}i\right)$

$= 3 + 3\sqrt{3}i$

**67.** $z = 10[\cos(-30°) + i \sin(-30°)]$

$= 10\left(\dfrac{\sqrt{3}}{2} - \dfrac{1}{2}i\right) = 5\sqrt{3} - 5i$

**69.** $100(\cos 240° + i \sin 240°) = 100\left(-\dfrac{1}{2} - \dfrac{\sqrt{3}}{2}i\right)$

$= -50 - 50\sqrt{3}i$

**71.** $13(\cos 0 + i \sin 0) = 13(1 + 0i) = 13$

**73.** (a) $z_1 = -5 = 5(\cos 180° + i \sin 180°)$

$z_2 = 5i = 5(\cos 90° + i \sin 90°)$

(b) $z_1 z_2 = (5)(5)[\cos(180° + 90°) + i \sin(180° + 90°)]$

$= 25(\cos 270° + i \sin 270°)$

$\dfrac{z_1}{z_2} = \dfrac{5}{5}[\cos(180° - 90°) + i \sin(180° + 90°)]$

$= \cos 90° + i \sin 90°$

**75.** (a) $z_1 = -3(1 + i) = 3\sqrt{2}(\cos 225° + i \sin 225°)$

$= 3\sqrt{2}\left(\cos \dfrac{5\pi}{4} + i \sin \dfrac{5\pi}{4}\right)$

$z_2 = 2(\sqrt{3} + i) = 4(\cos 30° + i \sin 30°)$

$= 4\left(\cos \dfrac{\pi}{6} + i \sin \dfrac{\pi}{6}\right)$

(b) $z_1 z_2 = (3\sqrt{2})(4)\left[\cos\left(\dfrac{5\pi}{4} + \dfrac{\pi}{6}\right) + i \sin\left(\dfrac{5\pi}{4} + \dfrac{\pi}{6}\right)\right]$

$= 12\sqrt{2}\left(\cos \dfrac{17\pi}{12} + i \sin \dfrac{17\pi}{12}\right)$

$\dfrac{z_1}{z_2} = \dfrac{3\sqrt{2}}{4}\left[\cos\left(\dfrac{5\pi}{4} - \dfrac{\pi}{6}\right) + i \sin\left(\dfrac{5\pi}{4} - \dfrac{\pi}{6}\right)\right]$

$= \dfrac{3\sqrt{2}}{4}\left(\cos \dfrac{13\pi}{12} + i \sin \dfrac{13\pi}{12}\right)$

**77.** $\left[5\left(\cos\dfrac{\pi}{12} + i\sin\dfrac{\pi}{12}\right)\right]^4 = 5^4\left(\cos\dfrac{4\pi}{12} + i\sin\dfrac{4\pi}{12}\right)$

$$= 625\left(\cos\dfrac{\pi}{3} + i\sin\dfrac{\pi}{3}\right)$$

$$= 625\left(\dfrac{1}{2} + \dfrac{\sqrt{3}}{2}i\right)$$

$$= \dfrac{625}{2} + \dfrac{625\sqrt{3}}{2}i$$

**79.** $(2 + 3i)^6 \approx \left[\sqrt{13}(\cos 56.3° + i\sin 56.3°)\right]^6$

$$= 13^3(\cos 337.9° + i\sin 337.9°)$$

$$\approx 13^3(0.9263 - 0.3769i)$$

$$\approx 2035 - 828i$$

**81.** Sixth roots of $-729i = 729\left(\cos\dfrac{3\pi}{2} + i\sin\dfrac{3\pi}{2}\right)$:

$$\sqrt[6]{729}\left(\cos\dfrac{(3\pi/2) + 2k\pi}{6} + i\sin\dfrac{(3\pi/2) + 2k\pi}{6}\right),$$

$k = 0, 1, 2, 3, 4, 5$

$k = 0:\ 3\left(\cos\dfrac{\pi}{4} + i\sin\dfrac{\pi}{4}\right)$

$k = 1:\ 3\left(\cos\dfrac{7\pi}{12} + i\sin\dfrac{7\pi}{12}\right)$

$k = 2:\ 3\left(\cos\dfrac{11\pi}{12} + i\sin\dfrac{11\pi}{12}\right)$

$k = 3:\ 3\left(\cos\dfrac{5\pi}{4} + i\sin\dfrac{5\pi}{4}\right)$

$k = 4:\ 3\left(\cos\dfrac{19\pi}{12} + i\sin\dfrac{19\pi}{12}\right)$

$k = 5:\ 3\left(\cos\dfrac{23\pi}{12} + i\sin\dfrac{23\pi}{12}\right)$

**83.** Cube roots of $-1 = 1(\cos \pi + i\sin \pi)$

$$\sqrt[3]{1}\left[\cos\left(\dfrac{\pi + 2\pi k}{3}\right) + i\sin\left(\dfrac{\pi + 2\pi k}{3}\right)\right],$$

$k = 0, 1, 2$

$k = 0:\ \cos\dfrac{\pi}{3} + i\sin\dfrac{\pi}{3} = \dfrac{1}{2} + \dfrac{\sqrt{3}}{2}i$

$k = 1:\ \cos \pi + i\sin \pi = -1$

$k = 2:\ \cos\dfrac{5\pi}{3} + i\sin\dfrac{5\pi}{3} = \dfrac{1}{2} - \dfrac{\sqrt{3}}{2}i$

**85.** $x^4 + 81 = 0$

$\quad x^4 = -81$   Solve by finding the fourth roots of $-81$.

$\quad -81 = 81(\cos \pi + i\sin \pi)$

$$\sqrt[4]{-81} = \sqrt[4]{81}\left[\cos\left(\dfrac{\pi + 2\pi k}{4}\right) + i\sin\left(\dfrac{\pi + 2\pi k}{4}\right)\right], k = 0, 1, 2, 3$$

$k = 0:\ 3\left(\cos\dfrac{\pi}{4} + i\sin\dfrac{\pi}{4}\right) = \dfrac{3\sqrt{2}}{2} + \dfrac{3\sqrt{2}}{2}i$

$k = 1:\ 3\left(\cos\dfrac{3\pi}{4} + i\sin\dfrac{3\pi}{4}\right) = -\dfrac{3\sqrt{2}}{2} + \dfrac{3\sqrt{2}}{2}i$

$k = 2:\ 3\left(\cos\dfrac{5\pi}{4} + i\sin\dfrac{5\pi}{4}\right) = -\dfrac{3\sqrt{2}}{2} - \dfrac{3\sqrt{2}}{2}i$

$k = 3:\ 3\left(\cos\dfrac{7\pi}{4} + i\sin\dfrac{7\pi}{4}\right) = \dfrac{3\sqrt{2}}{2} - \dfrac{3\sqrt{2}}{2}i$

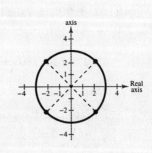

**87.**
$$(x^3 - 1)(x^2 + 1) = 0$$
$$(x - 1)(x^2 + x + 1)(x + i)(x - i) = 0$$

By the Quadratic Formula, the zeros of $x^2 + x + 1$ are $-\dfrac{1}{2} \pm \dfrac{\sqrt{3}}{2}i$.

The zeros of the equation are $1, -\dfrac{1}{2} \pm \dfrac{\sqrt{3}}{2}i, \pm i$.

In trigonometric form we have the cube roots of 1:

$$\cos 0 + i \sin 0$$

$$\cos \frac{2\pi}{3} + i \sin \frac{2\pi}{3}$$

$$\cos \frac{4\pi}{3} + i \sin \frac{4\pi}{3}$$

and the square root of 1:

$$\cos \frac{\pi}{2} + i \sin \frac{\pi}{2}$$

$$\cos \frac{3\pi}{2} + i \sin \frac{3\pi}{2}$$

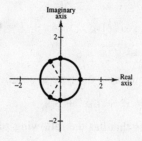

**89.** $f(x) = (x - 5)(x + 5) = x^2 - 25$

**91.** $f(x) = (x - 3)(x - 3i)(x + 3i)$
$$= (x - 3)(x^2 + 9)$$
$$= x^3 - 3x^2 + 9x - 27$$

**93.** $f(x) = (x - 1 - i)(x - 1 + i)(x + 1 - 2i)(x + 1 + 2i)$
$$= (x^2 - 2x + 2)(x^2 + 2x + 5)$$
$$= x^4 + 3x^2 - 6x + 10$$

**95.** The complex conjugates are reflections in the real axis.

# ❑ **Practice Test for Chapter 4**

1. Write $4 + \sqrt{-81} - 3i^2$ in standard form.

2. Write the result in standard form: $\dfrac{3 + i}{5 - 4i}$

3. Use the Quadratic Formula to solve $x^2 - 4x + 7 = 0$.

4. True or false: $\sqrt{-6}\sqrt{-6} = \sqrt{36} = 6$

5. Use the discriminant to determine the type of solutions of $3x^2 - 8x + 7 = 0$.

6. Find all the zeros of $f(x) = x^4 + 13x^2 + 36$.

7. Find a polynomial function that has the following zeros: $3, -1 \pm 4i$

8. Use the zero $x = 4 + i$ to find all the zeros of $f(x) = x^3 - 10x^2 + 33x - 34$.

9. Give the trigonometric form of $z = 5 - 5i$.

10. Give the standard form of $z = 6(\cos 225° + i \sin 225°)$.

11. Multiply $[7(\cos 23° + i \sin 23°)][4(\cos 7° + i \sin 7°)]$.

12. Divide $\dfrac{9\left(\cos \dfrac{5\pi}{4} + i \sin \dfrac{5\pi}{4}\right)}{3(\cos \pi + i \sin \pi)}$

13. Find $(2 + 2i)^8$.

14. Find the cube roots of $8\left(\cos \dfrac{\pi}{3} + i \sin \dfrac{\pi}{3}\right)$.

15. Find all the solutions to $x^4 + i = 0$.

# C H A P T E R   5
# Exponential and Logarithmic Functions

# CHAPTER 5
## Exponential and Logarithmic Functions

## Section 5.1    Exponential Functions and Their Graphs

- You should know that a function of the form $y = a^x$, where $a > 0$, $a \neq 1$, is called an exponential function with base $a$.
- You should be able to graph exponential functions.
- You should know formulas for compound interest.

    (a) For $n$ compoundings per year: $A = P\left(1 + \dfrac{r}{n}\right)^{nt}$.

    (b) For continuous compoundings: $A = Pe^{rt}$.

### Solutions to Odd-Numbered Exercises

**1.** $(3.4)^{5.6} \approx 946.852$

**3.** $(1.005)^{400} \approx 7.352$

**5.** $5^{-\pi} \approx 0.006$

**7.** $100^{\sqrt{2}} \approx 673.639$

**9.** $e^{-3/4} \approx 0.472$

**11.** $f(x) = 3^{x-2}$

$= 3^x 3^{-2}$

$= 3^x\left(\dfrac{1}{3^2}\right)$

$= \dfrac{1}{9}(3^x)$

$= h(x)$

Thus, $f(x) \neq g(x)$, but $f(x) = h(x)$.

**13.** $f(x) = 16(4^{-x})$    and    $f(x) = 16(4^{-x})$

$\quad = 4^2(4^{-x}) \qquad\qquad = 16(2^2)^{-x}$

$\quad = 4^{2-x} \qquad\qquad\quad = 16(2^{-2x})$

$\quad = \left(\dfrac{1}{4}\right)^{-(2-x)} \qquad = h(x)$

$\quad = \left(\dfrac{1}{4}\right)^{x-2}$

$\quad = g(x)$

Thus, $f(x) = g(x) = h(x)$.

**15.** $f(x) = 2^x$ rises to the right.

Asymptote: $y = 0$

Intercept: $(0, 1)$

Matches graph (d).

**17.** $f(x) = 2^{-x}$ falls to the right.

Asymptote: $y = 0$

Intercept: $(0, 1)$

Matches graph (a).

**19.** $g(x) = 5^x$

| $x$ | $-2$ | $-1$ | 0 | 1 | 2 |
|-----|------|------|---|---|---|
| $g(x)$ | $\frac{1}{25}$ | $\frac{1}{5}$ | 1 | 5 | 25 |

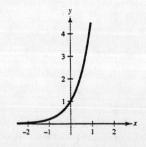

**21.** $f(x) = \left(\dfrac{1}{5}\right)^x = 5^{-x}$

| x | −2 | −1 | 0 | 1 | 2 |
|---|---|---|---|---|---|
| y | 25 | 5 | 1 | $\frac{1}{5}$ | $\frac{1}{25}$ |

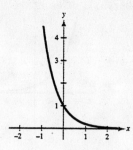

**23.** $h(x) = 5^{x-2}$

| x | −1 | 0 | 1 | 2 | 3 |
|---|---|---|---|---|---|
| y | $\frac{1}{125}$ | $\frac{1}{25}$ | $\frac{1}{5}$ | 1 | 5 |

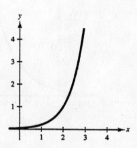

**25.** $g(x) = 5^{-x} - 3$

| x | −1 | 0 | 1 | 2 |
|---|---|---|---|---|
| y | 2 | −2 | $-2\frac{4}{5}$ | $-2\frac{24}{25}$ |

Asymptote: $y = -3$

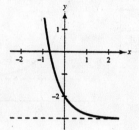

**27.** $y = 2^{-x^2}$

| x | ±2 | ±1 | 0 |
|---|---|---|---|
| y | $\frac{1}{16}$ | $\frac{1}{2}$ | 1 |

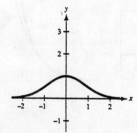

**29.** $f(x) = 3^{x-2} + 1$

| x | −1 | 0 | 1 | 2 | 3 | 4 |
|---|---|---|---|---|---|---|
| y | $1\frac{1}{27}$ | $1\frac{1}{9}$ | $1\frac{1}{3}$ | 2 | 4 | 10 |

Asymptote: $y = 1$

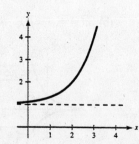

**31.** $y = 1.08^{-5x}$

| $x$ | $-1$ | 0 | 1 | 2 |
|---|---|---|---|---|
| $y$ | 1.47 | 1 | 0.68 | 0.46 |

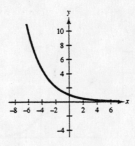

**33.** $s(t) = 2e^{0.12t}$

| $t$ | $-4$ | 0 | 4 | 8 |
|---|---|---|---|---|
| $s(t)$ | 1.24 | 2 | 3.23 | 5.22 |

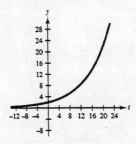

**35.** $g(x) = 1 + e^{-x}$

| $x$ | $-2$ | $-1$ | 0 | 1 | 2 |
|---|---|---|---|---|---|
| $y$ | 8.39 | 3.72 | 2 | 1.37 | 1.14 |

Asymptote: $y = 1$

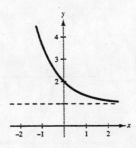

**37.** $y = 3^x$ and $y = 4^x$

| $x$ | $-2$ | $-1$ | 0 | 1 | 2 |
|---|---|---|---|---|---|
| $3^x$ | $\frac{1}{9}$ | $\frac{1}{3}$ | 1 | 3 | 9 |
| $4^x$ | $\frac{1}{16}$ | $\frac{1}{4}$ | 1 | 4 | 16 |

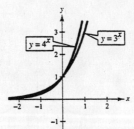

(a) $4^x < 3^x$ when $x < 0$.

(b) $4^x > 3^x$ when $x > 0$.

**39.** $f(x) = 3^x$

(a) $g(x) = f(x - 2) = 3^{x-2}$

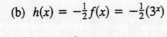

(b) $h(x) = -\frac{1}{2}f(x) = -\frac{1}{2}(3^x)$

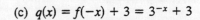

(c) $q(x) = f(-x) + 3 = 3^{-x} + 3$

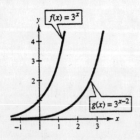

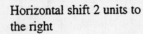

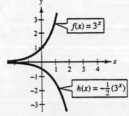

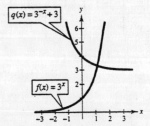

Horizontal shift 2 units to the right

Vertical shrink and a reflection about the $x$-axis

Reflection about the $y$-axis and a vertical translation 3 units upward

**41.** (a) $f(x) = x^2 e^{-x}$                      (b) $g(x) = x2^{3-x}$

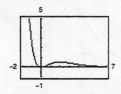
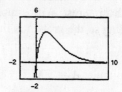

Decreasing: $(-\infty, 0),\ (2, \infty)$                 Decreasing: $(1.44, \infty)$
Increasing: $(0, 2)$                               Increasing: $(-\infty, 1.44)$
Relative maximum: $(2, 4e^{-2})$             Relative maximum: $(1.44, 4.25)$
Relative minimum: $(0, 0)$

**43.** The exponential function, $y = e^x$, increases at a faster rate than the polynomial function, $y = x^n$.

**45.** $f(x) = \left(1 + \dfrac{0.5}{x}\right)^x$ and $g(x) = e^{0.5}$ (Horizontal line)

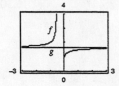

As $x \to \infty, f(x) \to g(x)$.

**47.** $P = \$2500$, $r = 12\%$, $t = 10$ years

Compounded $n$ times per year: $A = 2500\left(1 + \dfrac{0.12}{n}\right)^{10n}$

Compounded continuously: $A = 2500e^{0.12(10)}$

| $n$ | 1 | 2 | 4 | 12 | 365 | Continuous Compounding |
|---|---|---|---|---|---|---|
| $A$ | \$7,764.62 | \$8,017.84 | \$8,155.09 | \$8250.97 | \$8298.66 | \$8,300.29 |

**49.** $P = \$2500$, $r = 12\%$, $t = 20$ years

Compounded $n$ times per year: $A = 2500\left(1 + \dfrac{0.12}{n}\right)^{20n}$

Compounded continuously: $A = 2500e^{0.12(20)}$

| $n$ | 1 | 2 | 4 | 12 | 365 | Continuous Compounding |
|---|---|---|---|---|---|---|
| $A$ | \$24,115.73 | \$25,714.29 | \$26,602.23 | \$27,231.38 | \$27,547.07 | \$27,557.94 |

**51.** $A = Pe^{rt}$

$100{,}000 = Pe^{0.09t}$

$\dfrac{100{,}000}{e^{0.09t}} = P$

$P = 100{,}000e^{-0.09t}$

| $t$ | 1 | 10 | 20 | 30 | 40 | 50 |
|---|---|---|---|---|---|---|
| $P$ | \$91,393.12 | \$40,656.97 | \$16,529.89 | \$6,720.55 | \$2,732.37 | \$1,110.90 |

**53.** $A = 25{,}000e^{(0.0875)(25)} \approx \$222{,}822.57$

**55.** (a) The graph that is increasing faster represents 7% compounded annually. When interest is compounded, you earn interest on that interest. With simple interest there is no compounding so the growth is linear.

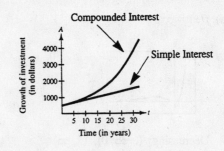

Compounded Interest

Simple Interest

Growth of investment (in dollars)

Time (in years)

(b) Compound interest formula: $A = 500\left(1 + \dfrac{0.07}{1}\right)^{(1)t}$

$$= 500(1.07)^t$$

Simple interest formula: $A = Prt + P$

$$= 500(0.07)t + 500$$

**57.** $C(10) = 23.95(1.04)^{10} \approx \$35.45$

**59.** $P(t) = 100e^{0.2197t}$

(a) $P(0) = 100$

(b) $P(5) \approx 300$

(c) $P(10) \approx 900$

**61.** $Q = 25\left(\frac{1}{2}\right)^{t/1620}$

(a) When $t = 0$, $Q = 25\left(\frac{1}{2}\right)^{0/1620} = 25(1) = 25$ units.

(b) When $t = 1000$, $Q = 25\left(\frac{1}{2}\right)^{1000/1620} \approx 16.30$ units.

(c)

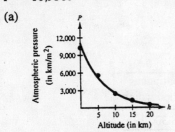

**63.** $P = 10{,}958e^{-0.15h}$

(a)

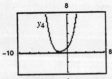

Atmospheric pressure (in km/m²)

Altitude (in km)

(b)

| $h$ | 0 | 5 | 10 | 15 | 20 |
|---|---|---|---|---|---|
| $P$ | 10,958 | 5176 | 2445 | 1155 | 546 |

The model is a "good fit."

(c) $P(8) \approx 3300 \text{ kg/m}^2$

(d) $2000 = 10{,}958e^{-0.15h}$ when $x \approx 11.3$ km.

**65.** False, $e \neq \dfrac{271{,}801}{99{,}990}$.

Since $e$ is an irrational number it cannot equal a rational number.

**67.** Since $\sqrt{2} \approx 1.414$ we know that $1 < \sqrt{2} < 2$.

Thus, $2^1 < 2^{\sqrt{2}} < 2^2$
$2 < 2^{\sqrt{2}} < 4$.

**69.** $y_4 = 1 + \dfrac{x}{1!} + \dfrac{x^2}{2!} + \dfrac{x^3}{3!} + \dfrac{x^4}{4!}$

$e^x = 1 + \dfrac{x}{1!} + \dfrac{x^2}{2!} + \dfrac{x^3}{3!} + \dfrac{x^4}{4!} + \dfrac{x^5}{5!} + \cdots$

**71.** $2x - 7y + 14 = 0$

$2x + 14 = 7y$

$\frac{1}{7}(2x + 14) = y$

**73.** $x^2 + y^2 = 25$

$y^2 = 25 - x^2$

$y = \pm\sqrt{25 - x^2}$

# Section 5.2    Logarithmic Functions and Their Graphs

- You should know that a function of the form $y = \log_a x$, where $a > 0$, $a \neq 1$, and $x > 0$, is called a logarithm of $x$ to base $a$.

- You should be able to convert from logarithmic form to exponential form and vice versa.

  $$y = \log_a x \iff a^y = x$$

- You should know the following properties of logarithms.

  (a) $\log_a 1 = 0$ since $a^0 = 1$.

  (b) $\log_a a = 1$ since $a^1 = a$.

  (c) $\log_a a^x = x$ since $a^x = a^x$.

  (d) If $\log_a x = \log_a y$, then $x = y$.

- You should know the definition of the natural logarithmic function.

  $$\log_e x = \ln x, \, x > 0$$

- You should know the properties of the natural logarithmic function.

  (a) $\ln 1 = 0$ since $e^0 = 1$.

  (b) $\ln e = 1$ since $e^1 = e$.

  (c) $\ln e^x = x$ since $e^x = e^x$.

  (d) If $\ln x = \ln y$, then $x = y$.

- You should be able to graph logarithmic functions.

## Solutions to Odd-Numbered Exercises

**1.** $\log_4 64 = 3 \implies 4^3 = 64$

**3.** $\log_7 \frac{1}{49} = -2 \implies 7^{-2} = \frac{1}{49}$

**5.** $\log_{32} 4 = \frac{2}{5} \implies 32^{2/5} = 4$

**7.** $\ln 1 = 0 \implies e^0 = 1$

**9.** $5^3 = 125 \implies \log_5 125 = 3$

**11.** $81^{1/4} = 3 \implies \log_{81} 3 = \frac{1}{4}$

**13.** $6^{-2} = \frac{1}{36} \implies \log_6 \frac{1}{36} = -2$

**15.** $e^3 = 20.0855 \ldots \implies \ln 20.0855 \ldots = 3$

**17.** $e^x = 4 \implies \ln 4 = x$

**19.** $\log_2 16 = \log_2 2^4 = 4$

**21.** $\log_{16} 4 = \log_{16} 16^{1/2} = \frac{1}{2}$

**23.** $\log_7 1 = \log_7 7^0 = 0$

**25.** $\log_{10} 0.01 = \log_{10} 10^{-2} = -2$

**27.** $\log_8 32 = \log_8 8^{5/3} = \frac{5}{3}$

**29.** $\ln e^3 = 3$

**31.** $\log_{10} 345 \approx 2.538$

**33.** $\log_{10} 145 \approx 2.161$

**35.** $\ln 18.42 \approx 2.913$

**37.** $\ln(1 + \sqrt{3}) \approx 1.005$

**39.** $\ln 0.32 \approx -1.139$

**41.** $f(x) = 3^x$, $g(x) = \log_3 x$

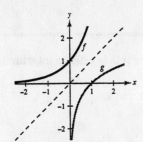

$f$ and $g$ are inverses. Their graphs are reflected about the line $y = x$.

**43.** $f(x) = e^x$, $g(x) = \ln x$

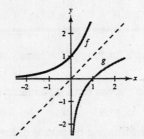

$f$ and $g$ are inverses. Their graphs are reflected about the line $y = x$.

**45.** $f(x) = \log_3 x + 2$
Asymptote: $x = 0$
Point on graph: $(1, 2)$
Matches graph (c).

**47.** $f(x) = -\log_3(x + 2)$
Asymptote: $x = -2$
Point on graph: $(-1, 0)$
Matches graph (d).

**49.** $f(x) = \log_3(1 - x)$
Asymptote: $x = 1$
Point on graph: $(0, 0)$
Matches graph (b).

**51.** $f(x) = \log_4 x$

Domain: $x > 0$ $\implies$ The domain is $(0, \infty)$.

Vertical asymptote: $x = 0$

$x$-intercept: $(1, 0)$

$y = \log_4 x \implies 4^y = x$

| $x$ | $\frac{1}{4}$ | 1 | 4 | 2 |
|---|---|---|---|---|
| $y$ | $-1$ | 0 | 1 | $\frac{1}{2}$ |

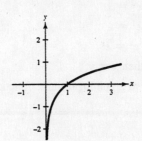

**53.** $y = -\log_3 x + 2$

Domain: $(0, \infty)$

Vertical asymptote: $x = 0$

$x$-intercept: $-\log_3 x + 2 = 0$

$$2 = \log_3 x$$
$$3^2 = x$$
$$9 = x$$

The $x$-intercept is $(9, 0)$.

$y = -\log_2 x + 2$

$\log_3 x = 2 - y \implies 3^{2-y} = x$

| $x$ | 27 | 9 | 3 | 1 | $\frac{1}{3}$ |
|---|---|---|---|---|---|
| $y$ | $-1$ | 0 | 1 | 2 | 3 |

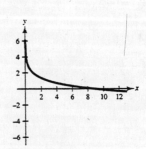

**55.** $f(x) = -\log_6(x + 2)$

Domain: $x + 2 > 0 \implies x > -2$

The domain is $(-2, \infty)$.

Vertical asymptote: $x + 2 = 0 \implies x = -2$

$x$-intercept:    $0 = -\log_6(x + 2)$

$$0 = \log_6(x + 2)$$

$$6^0 = x + 2$$

$$1 = x + 2$$

$$-1 = x$$

The $x$-intercept is $(-1, 0)$.

$$y = -\log_6(x + 2)$$

$$-y = \log_6(x + 2)$$

$$6^{-y} - 2 = x$$

| $x$ | 4 | $-1$ | $-1\frac{5}{6}$ | $-1\frac{35}{36}$ |
|---|---|---|---|---|
| $y$ | $-1$ | 0 | 1 | 2 |

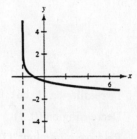

**57.** $y = \log_{10}\left(\dfrac{x}{5}\right)$

Domain: $\dfrac{x}{5} > 0 \implies x > 0$

The domain is $(0, \infty)$.

Vertical asymptote: $\dfrac{x}{5} = 0 \implies x = 0$

The vertical asymptote is the $y$-axis.

$x$-intercept:   $\log_{10}\left(\dfrac{x}{5}\right) = 0$

$$\frac{x}{5} = 10^0$$

$$\frac{x}{5} = 1 \implies x = 5$$

The $x$-intercept is $(5, 0)$.

| $x$ | 1 | 2 | 3 | 4 | 5 | 6 | 7 |
|---|---|---|---|---|---|---|---|
| $y$ | $-0.70$ | $-0.40$ | $-0.22$ | $-0.10$ | 0 | 0.08 | 0.15 |

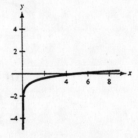

**59.** $f(x) = \ln(x - 2)$

Domain: $x - 2 > 0 \implies x > 2$

The domain is $(2, \infty)$.

Vertical asymptote: $x - 2 = 0 \implies x = 2$

$x$-intercept:   $0 = \ln(x - 2)$

$$e^0 = x - 2$$

$$3 = x$$

The $x$-intercept is $(3, 0)$.

| $x$ | 2.5 | 3 | 4 | 5 |
|---|---|---|---|---|
| $y$ | $-0.69$ | 0 | 0.69 | 1.10 |

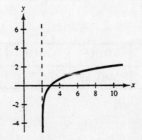

**61.** $g(x) = \ln(-x)$

Domain: $-x > 0 \implies x < 0$

The domain is $(-\infty, 0)$.

Vertical asymptote: $-x = 0 \implies x = 0$

$x$-intercept: $\quad 0 = \ln(-x)$

$\qquad\qquad\quad e^0 = -x$

$\qquad\qquad\quad -1 = x$

The $x$-intercept is $(-1, 0)$.

| $x$ | $-0.5$ | $-1$ | $-2$ | $-3$ |
|-----|--------|------|------|------|
| $y$ | $-0.69$ | $0$ | $0.69$ | $1.10$ |

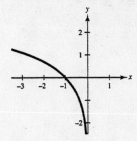

**63.** $f(x) = |\ln x|$

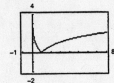

Increasing on $(1, \infty)$

Decreasing on $(0, 1)$

Relative minimum: $(1, 0)$

**65.** $f(x) = \dfrac{x}{2} - \ln\dfrac{x}{4}$

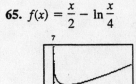

Increasing on $(2, \infty)$

Decreasing on $(0, 2)$

Relative minimum: $\left(2, 1 - \ln\frac{1}{2}\right)$

**67.** (a) $f(x) = \ln x$

$\qquad g(x) = \sqrt{x}$

The natural log function grows at a slower rate than the square root function.

(b) $f(x) = \ln x$

$\qquad g(x) = \sqrt[4]{x}$

The natural log function grows at a slower rate than the fourth root function.

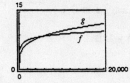

**69.** $y_1 = \ln x$

$y_2 = x - 1$

$y_3 = (x - 1) - \frac{1}{2}(x - 1)^2$

$y_4 = (x - 1) - \frac{1}{2}(x - 1)^2 + \frac{1}{3}(x - 1)^3$

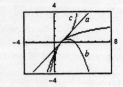

**71.** $f(t) = 80 - 17 \log_{10}(t + 1), \ 0 \le t \le 12$

(a) $f(0) = 80 - 17 \log_{10} 1 = 80.0$

(b) $f(4) = 80 - 17 \log_{10} 5 \approx 68.1$

(c) $f(10) = 80 - 17 \log_{10} 11 \approx 62.3$

**73.** $t = \dfrac{\ln 2}{r}$

| $r$ | 0.005 | 0.01 | 0.015 | 0.02 | 0.025 | 0.03 |
|-----|-------|------|-------|------|-------|------|
| $t$ | 138.6 yr | 69.3 yr | 46.2 yr | 34.7 yr | 27.7 yr | 23.1 yr |

**75.** $y = 80.4 - 11 \ln x$

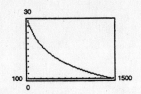

$y(300) = 80.4 - 11 \ln 300 \approx 17.66 \text{ ft}^3/\text{min}$

**77.** $W = 19,440(\ln 9 - \ln 3) \approx 21,357$ ft-lb

**79.** $t = 10.042 \ln\left(\dfrac{1316.35}{1316.35 - 1250}\right) \approx 30$ years

**81.** Total amount $= (1316.35)(30)(12) = \$473,886$

Interest $= 473,886 - 150,000 = \$323,886$

**83.** $f(x) = \dfrac{\ln x}{x}$

(a)

| $x$ | 1 | 5 | 10 | $10^2$ | $10^4$ | $10^6$ |
|-----|---|---|----|--------|--------|--------|
| $f(x)$ | 0 | 0.322 | 0.230 | 0.046 | 0.00092 | 0.0000138 |

(b) As $x \to \infty$, $f(x) \to 0$.

(c)
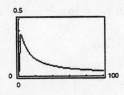

**85.** $8n - 3$

**87.** $83.95 + 37.50t$     Parts and labor

# Section 5.3    Properties of Logarithms

■ You should know the following properties of logarithms.

(a) $\log_a x = \dfrac{\log_b x}{\log_b a}$

(b) $\log_a(uv) = \log_a u + \log_a v$      $\ln(uv) = \ln u + \ln v$

(c) $\log_a(u/v) = \log_a u - \log_a v$      $\ln(u/v) = \ln u - \ln v$

(d) $\log_a u^n = n \log_a u$      $\ln u^n = n \ln u$

■ You should be able to rewrite logarithmic expressions using these properties.

**Solutions to Odd-Numbered Exercises**

**1.** $f(x) = \log_{10} x$

$g(x) = \dfrac{\ln x}{\ln 10}$

$f(x) = g(x)$

**3.** $\log_3 5 = \dfrac{\log_{10} 5}{\log_{10} 3}$

**5.** $\log_2 x = \dfrac{\log_{10} x}{\log_{10} 2}$

**7.** $\log_3 5 = \dfrac{\ln 5}{\ln 3}$

**9.** $\log_2 x = \dfrac{\ln x}{\ln 2}$

**11.** $\log_3 7 = \dfrac{\log_{10} 7}{\log_{10} 3} = \dfrac{\ln 7}{\ln 3} \approx 1.771$

**13.** $\log_{1/2} 4 = \dfrac{\log_{10} 4}{\log_{10} (1/2)} = \dfrac{\ln 4}{\ln (1/2)} = -2.000$

**15.** $\log_9 (0.4) = \dfrac{\log_{10} 0.4}{\log_{10} 9} = \dfrac{\ln 0.4}{\ln 9} \approx -0.417$

**17.** $\log_{15} 1250 = \dfrac{\log_{10} 1250}{\log_{10} 15} = \dfrac{\ln 1250}{\ln 15} \approx 2.633$

**19.** $\log_{10} 5x = \log_{10} 5 + \log_{10} x$

**21.** $\log_{10} \dfrac{5}{x} = \log_{10} 5 - \log_{10} x$

**23.** $\log_8 x^4 = 4 \log_8 x$

**25.** $\ln \sqrt{z} = \ln z^{1/2} = \tfrac{1}{2} \ln z$

**27.** $\ln xyz = \ln x + \ln y + \ln z$

**29.** $\ln \sqrt{a-1} = \tfrac{1}{2} \ln(a-1)$

**31.** $\ln z(z-1)^2 = \ln z + \ln(z-1)^2$

$\qquad = \ln z + 2 \ln(z-1)$

**33.** $\ln \sqrt[3]{\dfrac{x}{y}} = \dfrac{1}{3} \ln \dfrac{x}{y}$

$\qquad = \dfrac{1}{3} [\ln x - \ln y]$

$\qquad = \dfrac{1}{3} \ln x - \dfrac{1}{3} \ln y$

**35.** $\ln \left( \dfrac{x^4 \sqrt{y}}{z^5} \right) = \ln x^4 \sqrt{y} - \ln z^5$

$\qquad = \ln x^4 + \ln \sqrt{y} - \ln z^5$

$\qquad = 4 \ln x + \dfrac{1}{2} \ln y - 5 \ln z$

**37.** $\log_b \left( \dfrac{x^2}{y^2 z^3} \right) = \log_b x^2 - \log_b y^2 z^3$

$\qquad = \log_b x^2 - [\log_b y^2 + \log_b z^3]$

$\qquad = 2 \log_b x - 2 \log_b y - 3 \log_b z$

**39.** $y_1 = \ln[x^3(x+4)]$

$y_2 = 3 \ln x + \ln(x+4)$

$y_1 = y_2$

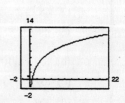

**41.** $\ln x + \ln 2 = \ln 2x$

**43.** $\log_4 z - \log_4 y = \log_4 \dfrac{z}{y}$

**45.** $2 \log_2 (x+4) = \log_2 (x+4)^2$

**47.** $\tfrac{1}{3} \log_3 5x = \log_3 (5x)^{1/3} = \log_3 \sqrt[3]{5x}$

**49.** $\ln x - 3 \ln(x+1) = \ln x - \ln(x+1)^3$

$\qquad = \ln \dfrac{x}{(x+1)^3}$

**51.** $\ln(x-2) - \ln(x+2) = \ln \left( \dfrac{x-2}{x+2} \right)$

**53.** $\ln x - 2[\ln(x + 2) + \ln(x - 2)] = \ln x - 2\ln(x + 2)(x - 2)$

$$= \ln x - 2\ln(x^2 - 4)$$

$$= \ln x - \ln(x^2 - 4)^2$$

$$= \ln \frac{x}{(x^2 - 4)^2}$$

**55.** $\frac{1}{3}[2\ln(x + 3) + \ln x - \ln(x^2 - 1)] = \frac{1}{3}[\ln(x + 3)^2 + \ln x - \ln(x^2 - 1)]$

$$= \frac{1}{3}[\ln x(x + 3)^2 - \ln(x^2 - 1)]$$

$$= \frac{1}{3}\ln \frac{x(x + 3)^2}{x^2 - 1}$$

$$= \ln \sqrt[3]{\frac{x(x + 3)^2}{x^2 - 1}}$$

**57.** $\frac{1}{3}[\ln y + 2\ln(y + 4)] - \ln(y - 1) = \frac{1}{3}[\ln y + \ln(y + 4)^2] - \ln(y - 1)$

$$= \frac{1}{3}\ln y(y + 4)^2 - \ln(y - 1)$$

$$= \ln \sqrt[3]{y(y + 4)^2} - \ln(y - 1)$$

$$= \ln \frac{\sqrt[3]{y(y + 4)^2}}{y - 1}$$

**59.** $2\ln 3 - \frac{1}{2}\ln(x^2 + 1) = \ln 3^2 - \ln \sqrt{x^2 + 1}$

$$= \ln \frac{9}{\sqrt{x^2 + 1}}$$

**61.** $y_1 = 2[\ln 8 - \ln(x^2 + 1)]$

$y_2 = \ln\left[\dfrac{64}{(x^2 + 1)^2}\right]$

$y_1 = y_2$

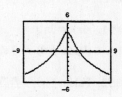

**63.** $y_1 = \ln x^2$

$y_2 = 2\ln x$

$y_1 = y_2$ for $x > 0$

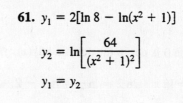

They are not equivalent. The domain of $f(x)$ is all real numbers except 0. The domain of $g(x)$ is $x > 0$.

**65.** $\log_2\left(\frac{32}{4}\right) = \log_2 32 - \log_2 4$ by Property 2.

**67.** $f(x) = \ln \dfrac{x}{2}$,  $g(x) = \dfrac{\ln x}{\ln 2}$,  $h(x) = \ln x - \ln 2$

$f(x) = h(x)$ by Property 2.

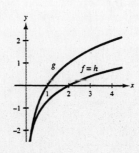

**69.** $\log_3 9 = 2\log_3 3 = 2$

**71.** $\log_4 16^{1.2} = 1.2(\log_4 16) = 1.2(2) = 2.4$

**73.** $\log_3(-9)$ is undefined. $-9$ is not in the domain of $\log_3 x$.

**75.** $\log_5 75 - \log_5 3 = \log_5 \frac{75}{3} = \log_5 25 = \log_5 5^2 = 2\log_5 5 = 2$

**77.** $\ln e^2 - \ln e^5 = 2 - 5 = -3$

**79.** $\log_{10} 0$ is undefined. $0$ is not in the domain of $\log_{10} x$.

**81.** $\ln e^{4.5} = 4.5$

**83.** $\log_4 8 = \log_4 2^3 = 3\log_4 2 = 3\log_4 \sqrt{4} = 3\log_4 4^{1/2} = 3(\frac{1}{2})\log_4 4 = \frac{3}{2}$

**85.** $\log_5 \frac{1}{250} = \log_5 1 - \log_5 250 = 0 - \log_5(125 \cdot 2)$

$$= -\log_5(5^3 \cdot 2) = -[\log_5 5^3 + \log_5 2]$$

$$= -[3\log_5 5 + \log_5 2] = -3 - \log_5 2$$

**87.** $\ln(5e^6) = \ln 5 + \ln e^6 = \ln 5 + 6 = 6 + \ln 5$

**89.** $f(t) = 90 - 15\log_{10}(t+1), \ 0 \le t \le 12$

(a) $f(0) = 90$  (b) $f(6) \approx 77$  (c) $f(12) \approx 73$

(d) $\quad 75 = 90 - 15\log_{10}(t+1)$  (e) $f(t) = 90 - \log_{10}(t+1)^{15}$  (f)

$\quad -15 = -15\log_{10}(t+1)$

$\quad\quad 1 = \log_{10}(t+1)$

$\quad\quad 10^1 = t+1$

$\quad\quad t = 9$ months

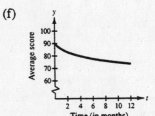

**91.** $f(x) = \ln x$

False, $f(0) \ne 0$ since $0$ is not in the domain of $f(x)$. $f(1) = \ln 1 = 0$

**93.** False. $f(x) - f(2) = \ln x - \ln 2 = \ln\frac{x}{2} \ne \ln(x-2)$

**95.** False. $f(u) = 2f(v) \implies \ln u = 2\ln v \implies \ln u = \ln v^2 \implies u = v^2$

**97.** Let $x = \log_b u$ and $y = \log_b v$, then $b^x = u$ and $b^y = v$.

$$\frac{u}{v} = \frac{b^x}{b^y} = b^{x-y}$$

$$\log_b\left(\frac{u}{v}\right) = \log_b(b^{x-y}) = x - y = \log_b u - \log_b v$$

**99.** $\dfrac{24xy^{-2}}{16x^{-3}y} = \dfrac{24xx^3}{16yy^2} = \dfrac{3x^4}{2y^3}, x \ne 0$

**101.** $(18x^3y^4)^{-3}(18x^3y^4)^3 = \dfrac{(18x^3y^4)^3}{(18x^3y^4)^3} = 1$ if $x \ne 0, y \ne 0$.

# Section 5.4    Exponential and Logarithmic Equations

■ To solve an exponential equation, isolate the exponential expression, then take the logarithm of both sides. Then solve for the variable.

1. $\log_a a^x = x$

2. $\ln e^x = x$

■ To solve a logarithmic equation, rewrite it in exponential form. Then solve for the variable.

1. $a^{\log_a x} = x$

2. $e^{\ln x} = x$

■ If $a > 0$ and $a \neq 1$ we have the following:

1. $\log_a x = \log_a y \implies x = y$

2. $a^x = a^y \implies x = y$

**Solutions to Odd-Numbered Exercises**

**1.** $4^{2x-7} = 64$

(a) $x = 5$

$4^{2(5)-7} = 4^3 = 64$

Yes, $x = 5$ is a solution.

(b)　　　$x = 2$

$4^{2(2)-7} = 4^{-3} = \frac{1}{64} \neq 64$

No, $x = 2$ is not a solution.

**3.** $3e^{x+2} = 75$

(a) $x = -2 + e^{25}$

$3e^{(-2+e^{25})+2} = 3e^{e^{25}} \neq 75$

No, $x = -2 + e^{25}$ is not a solution.

(b) $x = -2 + \ln 25$

$3e^{(-2+\ln 25)+2} = 3e^{\ln 25} = 3(25) = 75$

Yes, $x = -2 + \ln 25$ is a solution.

(c) $x \approx 1.2189$

$3e^{1.2189+2} = 3e^{3.2189} \approx 75$

Yes, $x \approx 1.2189$ is a solution.

**5.** $\log_4(3x) = 3 \implies 3x = 4^3 \implies 3x = 64$

(a) $x \approx 20.3560$

$3(20.3560) = 61.0680 \neq 64$

No, $x \approx 20.3560$ is not a solution.

(b) $x = -4$

$3(-4) = -12 \neq 64$

No, $x = -4$ is not a solution.

(c) $x = \frac{64}{3}$

$3\left(\frac{64}{3}\right) = 64$

Yes, $x = \frac{64}{3}$ is a solution.

**7.** $f(x) = g(x)$

$2^x = 8$

$2^x = 2^3$

$x = 3$

Point of intersection: $(3, 8)$

**9.** $f(x) = g(x)$

$\log_3 x = 2$

$x = 3^2$

$x = 9$

Point of intersection: $(9, 2)$

**11.** $4^x = 16$

$4^x = 4^2$

$x = 2$

**13.** $7^x = \frac{1}{49}$

$7^x = 7^{-2}$

$x = -2$

**15.** $\left(\frac{3}{4}\right)^x = \frac{27}{64}$

$\left(\frac{3}{4}\right)^x = \left(\frac{3}{4}\right)^3$

$x = 3$

**17.** $\log_4 x = 3$

$x = 4^3$

$x = 64$

**19.** $\log_{10} x = -1$

$\quad x = 10^{-1}$

$\quad x = \frac{1}{10}$

**21.** $\log_{10} 10^{x^2} = x^2$

**23.** $e^{\ln(5x+2)} = 5x + 2$

**25.** $e^{\ln x^2} = x^2$

**27.** $e^x = 10$

$\quad x = \ln 10 \approx 2.303$

**29.** $7 - 2e^x = 5$

$\quad -2e^x = -2$

$\quad\quad e^x = 1$

$\quad\quad\ x = \ln 1 = 0$

**31.** $e^{3x} = 12$

$\quad 3x = \ln 12$

$\quad x = \dfrac{\ln 12}{3} \approx 0.828$

**33.** $500e^{-x} = 300$

$\quad e^{-x} = \dfrac{3}{5}$

$\quad -x = \ln\dfrac{3}{5}$

$\quad x = -\ln\dfrac{3}{5} = \ln\dfrac{5}{3} \approx 0.511$

**35.** $\quad e^{2x} - 4e^x - 5 = 0$

$\quad (e^x - 5)(e^x + 1) = 0$

$\quad e^x = 5 \ $ or $\ e^x = -1 \quad$ (No solution)

$\quad x = \ln 5 \approx 1.609$

**37.** $20(100 - e^{x/2}) = 500$

$\quad 100 - e^{x/2} = 25$

$\quad -e^{x/2} = -75$

$\quad e^{x/2} = 75$

$\quad \dfrac{x}{2} = \ln 75$

$\quad x = 2\ln 75 \approx 8.635$

**39.** $10^x = 42$

$\quad x = \log_{10} 42 \approx 1.623$

**41.** $\quad 3^{2x} = 80$

$\quad \ln 3^{2x} = \ln 80$

$\quad 2x\ln 3 = \ln 80$

$\quad x = \dfrac{\ln 80}{2\ln 3} \approx 1.994$

**43.** $5^{-t/2} = 0.20$

$\quad 5^{-t/2} = \dfrac{1}{5}$

$\quad 5^{-t/2} = 5^{-1}$

$\quad -\dfrac{t}{2} = -1$

$\quad t = 2$

**45.** $\quad 2^{3-x} = 565$

$\quad \ln 2^{3-x} = \ln 565$

$\quad (3 - x)\ln 2 = \ln 565$

$\quad 3\ln 2 - x\ln 2 = \ln 565$

$\quad -x\ln 2 = \ln 565 - \ln 2^3$

$\quad x\ln 2 = \ln 8 - \ln 565$

$\quad x = \dfrac{\ln 8 - \ln 565}{\ln 2} \approx -6.142$

**47.** $g(x) = 6e^{1-x} - 25$

The zero is $x \approx -0.427$.

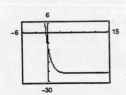

**49.** $g(t) = e^{0.09t} - 3$

The zero is $x \approx 12.207$.

**51.** $8(10^{3x}) = 12$

$$10^{3x} = \frac{12}{8}$$

$$3x = \log_{10}\left(\frac{3}{2}\right)$$

$$x = \tfrac{1}{3}\log_{10}\left(\frac{3}{2}\right) \approx 0.059$$

**53.** $\left(1 + \dfrac{0.065}{365}\right)^{365t} = 4$

$$\ln\left(1 + \frac{0.065}{365}\right)^{365t} = \ln 4$$

$$365t \ln\left(1 + \frac{0.065}{365}\right) = \ln 4$$

$$t = \frac{\ln 4}{365 \ln\left(1 + \frac{0.065}{365}\right)} \approx 21.330$$

**55.** $\ln x = -3$

$$x = e^{-3} \approx 0.050$$

**57.** $\ln 2x = 2.4$

$$2x = e^{2.4}$$

$$x = \frac{e^{2.4}}{2} \approx 5.512$$

**59.** $\ln \sqrt{x + 2} = 1$

$$\sqrt{x + 2} = e^1$$

$$x + 2 = e^2$$

$$x = e^2 - 2 \approx 5.389$$

**61.** $\log_{10}(z - 3) = 2$

$$z - 3 = 10^2$$

$$z = 10^2 + 3 = 103$$

**63.** $\ln x + \ln(x - 2) = 1$

$$\ln[x(x - 2)] = 1$$

$$x(x - 2) = e^1$$

$$x^2 - 2x - e = 0$$

$$x = \frac{2 \pm \sqrt{4 + 4e}}{2}$$

$$= \frac{2 \pm 2\sqrt{1 + e}}{2}$$

$$= 1 \pm \sqrt{1 + e}$$

Using the positive value for $x$, we have
$x = 1 + \sqrt{1 + e} \approx 2.928$.

**65.** $\log_{10}(x + 4) - \log_{10} x = \log_{10}(x + 2)$

$$\log_{10}\left(\frac{x + 4}{x}\right) = \log_{10}(x + 2)$$

$$\frac{x + 4}{x} = x + 2$$

$$x + 4 = x^2 + 2x \quad \text{Quadratic}$$

$$0 = x^2 + x - 4 \quad \text{Formula}$$

$$x = \frac{-1 \pm \sqrt{17}}{2}$$

Choosing the positive value of $x$ (the negative value is extraneous), we have
$$x = \frac{-1 + \sqrt{17}}{2} \approx 1.562.$$

**67.** $\log_3 x + \log_3(x^2 - 8) = \log_3 8x$

$$\log_3 x(x^2 - 8) = \log_3 8x$$

$$x(x^2 - 8) = 8x$$

$$x^3 - 8x = 8x$$

$$x^3 - 16x = 0$$

$$x(x + 4)(x - 4) = 0$$

$$x = 0, \; x = -4, \; \text{or} \; x = 4$$

The only solution that is in the domain is $x = 4$. Both $x = 0$ and $x = -4$ are extraneous.

**69.** $\ln(x + 5) = \ln(x - 1) - \ln(x + 1)$

$$\ln(x + 5) = \ln\left(\frac{x - 1}{x + 1}\right)$$

$$x + 5 = \frac{x - 1}{x + 1}$$

$$(x + 5)(x + 1) = x - 1$$

$$x^2 + 6x + 5 = x - 1$$

$$x^2 + 5x + 6 = 0$$

$$(x + 2)(x + 3) = 0$$

$$x = -2 \; \text{or} \; x = -3$$

Both of these solutions are extraneous, so the equation has no solution.

**71.** $6 \log_3(0.5x) = 11$

$\log_3(0.5x) = \frac{11}{6}$

$0.5x = 3^{11/6}$

$x = 2(3^{11/6}) \approx 14.988$

**73.** $2 \ln x = 7$

$\ln x = \frac{7}{2}$

$x = e^{7/2} \approx 33.115$

**75.** $\ln x + \ln(x^2 + 1) = 8$

$\ln x(x^2 + 1) = 8$

$x(x^2 + 1) = e^8$

$x^3 + x - e^8 = 0$

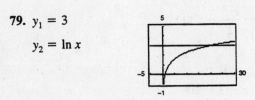

From the graph we have $x \approx 14.369$.

**77.** $y_1 = 7$

$y_2 = 2^x$

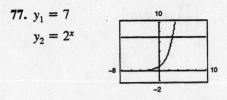

From the graph we have $x \approx 2.807$ when $y = 7$.

**79.** $y_1 = 3$

$y_2 = \ln x$

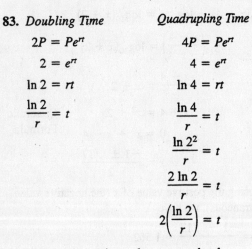

From the graph we have $x \approx 20.806$ when $y = 3$.

**81.** $A = Pe^{rt}$

$2000 = 1000e^{0.085t}$

$2 = e^{0.085t}$

$\ln 2 = 0.085t$

$\frac{\ln 2}{0.085} = t$

$t \approx 8.2$ years

**83.** *Doubling Time*

$2P = Pe^{rt}$

$2 = e^{rt}$

$\ln 2 = rt$

$\frac{\ln 2}{r} = t$

*Quadrupling Time*

$4P = Pe^{rt}$

$4 = e^{rt}$

$\ln 4 = rt$

$\frac{\ln 4}{r} = t$

$\frac{\ln 2^2}{r} = t$

$\frac{2 \ln 2}{r} = t$

$2\left(\frac{\ln 2}{r}\right) = t$

Yes, it takes twice as long to quadruple.

**85.** $A = Pe^{rt}$

$3000 = 1000e^{0.085t}$

$3 = e^{0.085t}$

$\ln 3 = 0.085t$

$\frac{\ln 3}{0.085} = t$

$t \approx 12.9$ years

**87.** $p = 500 - 0.5(e^{0.004x})$

(a)    $p = 350$

$350 = 500 - 0.5(e^{0.004x})$

$300 = e^{0.004x}$

$0.004x = \ln 300$

$x \approx 1426$ units

(b)    $p = 300$

$300 = 500 - 0.5(e^{0.004x})$

$400 = e^{0.004x}$

$0.004x = \ln 400$

$x \approx 1498$ units

**89.** $V = 6.7e^{-48.1/t}$, $t \geq 0$

(a)

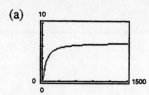

(b) As $x \to \infty$, $V \to 6.7$.

Horizontal asymptote: $y = 6.7$

The yield will approach
6.7 million cubic feet per acre.

(c) $1.3 = 6.7e^{-48.1/t}$

$\dfrac{1.3}{6.7} = e^{-48.1/t}$

$\ln\left(\dfrac{13}{67}\right) = \dfrac{-48.1}{t}$

$t = \dfrac{-48.1}{\ln(13/67)} \approx 29.3$ years

**91.** (a) From the graph shown in the textbook, we see horizontal asymptotes at $y = 0$ and $y = 100$. These represent the lower and upper percent bounds.

(b) Males

$$50 = \frac{100}{1 + e^{-0.6114(x - 69.71)}}$$

$$1 + e^{-0.6114(x - 69.71)} = 2$$

$$e^{-0.6114(x - 69.71)} = 1$$

$$-0.6114(x - 69.71) = \ln 1$$

$$-0.6114(x - 69.71) = 0$$

$$x = 69.71 \text{ inches}$$

Females

$$50 = \frac{100}{1 + e^{-0.66607(x - 64.51)}}$$

$$1 + e^{-0.66607(x - 64.51)} = 2$$

$$e^{-0.66607(x - 64.51)} = 1$$

$$-0.66607(x - 64.51) = \ln 1$$

$$-0.66607(x - 64.51) = 0$$

$$x = 64.51 \text{ inches}$$

**93.** $T = 20[1 + 7(2^{-h})]$

(a) From the graph in the textbook we see a horizontal asymptote at $y = 20$. This represents the room temperature.

(b) $100 = 20[1 + 7(2^{-h})]$

$5 = 1 + 7(2^{-h})$

$4 = 7(2^{-h})$

$\dfrac{4}{7} = 2^{-h}$

$\ln\left(\dfrac{4}{7}\right) = \ln 2^{-h}$

$\ln\left(\dfrac{4}{7}\right) = -h \ln 2$

$\dfrac{\ln\left(\frac{4}{7}\right)}{-\ln 2} = h$

$h \approx 0.81$ hour

**95.** $\sqrt{48x^2y^5} = \sqrt{16x^2y^4 3y} = 4|x|y^2\sqrt{3y}$

**97.** $\sqrt[3]{25}\sqrt[3]{15} = \sqrt[3]{375} = \sqrt[3]{125 \cdot 3} = 5\sqrt[3]{3}$

# Section 5.5    Exponential and Logarithmic Models

---

■ You should be able to solve compound interest problems.

    (a) Compound interest formulas:

        1. $A = P\left(1 + \dfrac{r}{n}\right)^{nt}$

        2. $A = Pe^{rt}$

    (b) Doubling time:

        1. $t = \dfrac{\ln 2}{n \ln[1 + (r/n)]}$, $n$ compoundings per year

        2. $t = \dfrac{\ln 2}{r}$, continuous compounding

■ You should be able to solve growth and decay problems.

    (a) Exponential growth if $b > 0$ and $y = ae^{bx}$.

    (b) Exponential decay if $b > 0$ and $y = ae^{-bx}$.

■ You should be able to use the Gaussian model

    $y = ae^{-(x-b)^2/c}$.

■ You should be able to use the logistics growth model

    $y = \dfrac{a}{1 + be^{-(x-c)/d}}$.

■ You should be able to use the logarithmic models

    $y = \ln(ax + b)$ and $y = \log_{10}(ax + b)$.

---

## Solutions to Odd-Numbered Exercises

**1.** $y = 2e^{x/4}$

This is an exponential growth model. Matches graph (c)

**3.** $y = \frac{1}{16}(x^2 + 8x + 32)$

This is a quadratic function. Its graph is a parabola. Matches graph (a)

**5.** $y = \ln(x + 1)$

This is a logarithmic model. Matches graph (d)

**7.** Since $A = 1000e^{0.12t}$, the time to double is given by $2000 = 1000e^{0.12t}$ and we have

$t = \dfrac{\ln 2}{0.12} \approx 5.78$ years.

Amount after 10 years: $A = 1000e^{1.2} \approx \$3320.12$

**9.** Since $A = 750e^{rt}$ and $A = 1500$ when $t = 7.75$, we have the following.

$1500 = 750e^{7.75r}$

$r = \dfrac{\ln 2}{7.75} \approx 0.0894 = 8.94\%$

Amount after 10 years: $A = 750e^{0.0894(10)} \approx \$1833.67$

**11.** Since $A = 500e^{rt}$ and $A = 1292.85$ when $t = 10$, we have the following.

$$1292.85 = 500e^{10r}$$

$$r = \frac{\ln(1292.85/500)}{10} \approx 0.9095 = 9.5\%$$

The time to double is given by

$$1000 = 500e^{0.095t}$$

$$t = \frac{\ln 2}{0.095} \approx 7.30 \text{ years.}$$

**13.** Since $A = Pe^{0.045t}$ and $A = 10,000.00$ when $t = 10$, we have the following.

$$10,000.00 = Pe^{0.045(10)}$$

$$\frac{10,000.00}{e^{0.045(10)}} = P \approx \$6376.28$$

The time to double is given by

$$t = \frac{\ln 2}{0.045} \approx 15.40 \text{ years.}$$

**15.** $$500,000 = P\left(1 + \frac{0.075}{12}\right)12(20)$$

$$P = \frac{500,000}{\left(1 + \frac{0.075}{12}\right)}12(20) = \$112,087.09$$

**17.** $P = 1000, r = 11\%$

(a) $n = 1$

$$t = \frac{\ln 2}{\ln(1 + 0.11)} \approx 6.642 \text{ years}$$

(b) $n = 12$

$$t = \frac{\ln 2}{12 \ln\left(1 + \frac{0.11}{12}\right)} \approx 6.330 \text{ years}$$

(c) $n = 365$

$$t = \frac{\ln 2}{365 \ln\left(1 + \frac{0.11}{365}\right)} \approx 6.302 \text{ years}$$

(d) Continuously

$$t = \frac{\ln 2}{0.11} \approx 6.301 \text{ years}$$

**19.** $3P = Pe^{rt}$

$3 = e^{rt}$

$\ln 3 = rt$

$\dfrac{\ln 3}{r} = t$

| $r$ | 2% | 4% | 6% | 8% | 10% | 12% |
|---|---|---|---|---|---|---|
| $t = \dfrac{\ln 3}{r}$ | 54.93 | 27.47 | 18.31 | 13.73 | 10.99 | 9.16 |

**21.** $3P = P(1 + r)^t$

$3 = (1 + r)^t$

$\ln 3 = \ln(1 + r)^t$

$\ln 3 = t \ln(1 + r)$

$\dfrac{\ln 3}{\ln(1 + r)} = t$

| $r$ | 2% | 4% | 6% | 8% | 10% | 12% |
|---|---|---|---|---|---|---|
| $t = \dfrac{\ln 3}{\ln(1 + r)}$ | 55.48 | 28.01 | 18.85 | 14.27 | 11.53 | 9.69 |

**23.** Continuous compounding results in faster growth.

$A = 1 + 0.075[\![t]\!]$ and $A = e^{0.07t}$

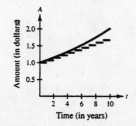

**25.** $\frac{1}{2}C = Ce^{k(1620)}$

$k = \dfrac{\ln 0.5}{1620}$

Given $C = 10$ grams, after 1000 years we have

$y = 10e^{[(\ln 0.5)/1620](1000)}$

$\approx 6.52$ grams.

**27.** $\frac{1}{2}C = Ce^{k(5730)}$

$k = \dfrac{\ln 0.5}{5730}$

Given $y = 2$ grams after 1000 years, we have

$2 = Ce^{[(\ln 0.5)/5730](1000)}$

$C \approx 2.26$ grams.

**29.** $\frac{1}{2}C = Ce^{k(24,360)}$

$k = \dfrac{\ln 0.5}{24,360}$

Given $y = 2.1$ grams after 1000 years, we have

$2.1 = Ce^{[(\ln 0.5)/24,360](1000)}$

$C \approx 2.16$ grams.

**31.** $\quad y = ae^{bx}$

$1 = ae^{b(0)} \implies 1 = a$

$10 = e^{b(3)}$

$\ln 10 = 3b$

$\dfrac{\ln 10}{3} = b \qquad \implies \quad b \approx 0.7675$

Thus, $y = e^{0.7675x}$.

**33.** $\quad y = ae^{bx}$

$1 = ae^{b(0)} \implies 1 = a$

$\dfrac{1}{4} = e^{b(3)}$

$\ln\left(\dfrac{1}{4}\right) = 3b$

$\dfrac{\ln\left(\frac{1}{4}\right)}{3} = b \qquad \implies \quad b \approx -0.4621$

Thus, $y = e^{-0.4621x}$.

**35.** $\quad P = 105,300e^{0.015t}$

$150,000 = 105,300e^{0.015t}$

$\ln \frac{1500}{1053} = 0.015t$

$t \approx 23.59$

The population will reach 150,000 during 2013.
[Note: 1990 + 23.59]

**37.** For 1945, use $t = -45$.

$1350 = 2500d^{k(-45)}$

$\ln\left(\frac{1350}{2500}\right) = -45k \implies k \approx 0.0137$

For 2010, use $t = 20$.

$P = 2500e^{0.0137(20)} \approx 3288$ people

**39.** $\quad y = ae^{bt}$

$4.22 = ae^{b(0)} \implies a = 4.22$

$6.49 = 4.22e^{b(10)}$

$\dfrac{6.49}{4.22} = e^{10b}$

$\ln\left(\dfrac{6.49}{4.22}\right) = 10b \quad \implies \quad b \approx 0.0430$

$y = 4.22e^{0.0430t}$

When $t = 20$,
$y = 4.22e^{0.0430(20)} \approx 9.97$ million.

**41.** $\quad y = ae^{bt}$

$3.00 = ae^{b(0)} \implies a = 3$

$2.74 = 3e^{b(10)}$

$\dfrac{2.74}{3} = e^{10b}$

$\ln\left(\dfrac{2.74}{3}\right) = 10b \quad \implies \quad b \approx -0.0091$

$y = 3e^{-0.0091t}$

When $t = 20$,
$y = 3e^{-0.0091(20)} \approx 2.50$ million.

**43.** $b$ is determined by the growth rate. The greater the rate of growth, the greater the value of $b$.

**45.** $\quad N = 100e^{kt}$

$300 = 100e^{5k}$

$k = \dfrac{\ln 3}{5} \approx 0.2197$

$N = 100e^{0.2197t}$

$200 = 100e^{0.2197t}$

$t = \dfrac{\ln 2}{0.2197} \approx 3.15$ hours

**47.**   $y = Ce^{kt}$

$$\frac{1}{2}C = Ce^{(1620)k}$$

$$\ln\frac{1}{2} = 1620k$$

$$k = \frac{\ln(1/2)}{1620}$$

When $t = 100$, we have

$$y = Ce^{[\ln(1/2)/1620](100)} \approx 0.958C = 95.8\%C.$$

After 100 years, approximately 95.8% of the radioactive radium will remain.

**49.** $(0, 22{,}000),\ (2, 13{,}000)$

(a) $m = \dfrac{13{,}000 - 22{,}000}{2 - 0} = -4500$

$b = 22{,}000$

Thus, $V = -4500t + 22{,}000.$

(b)     $a = 22{,}000$

$13{,}000 = 22{,}000e^{k(2)}$

$$\frac{13}{22} = e^{2k}$$

$$\ln\left(\frac{13}{22}\right) = 2k \Longrightarrow k \approx -0.263$$

Thus, $V = 22{,}000e^{-0.263t}.$

(c) The exponential model depreciates faster in the first two years.

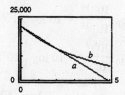

(d)

| $t$ | 1 | 3 |
|---|---|---|
| $V = -4500t + 22{,}000$ | \$17,500 | \$8500 |
| $V = 22{,}000e^{-0.263t}$ | \$16,912 | \$9995 |

(e) The slope of the linear model means that the car depreciates \$4500 per year.

**51.** $S(t) = 100(1 - e^{kt})$

(a)   $15 = 100(1 - e^{k(1)})$

$-85 = -100e^k$

$k = \ln 0.85$

$k \approx -0.1625$

$S(t) = 100(1 - e^{-0.1625t})$

(c) $S(5) = 100(1 - e^{-0.1625(5)})$

$\approx 55.625 = 55{,}625$ units

(b)

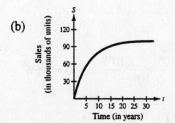

**53.** $S = 10(1 - e^{kx})$

$x = 5$ (in hundreds), $S = 2.5$ (in thousands)

(a) $2.5 = 10(1 - e^{k(5)})$

$0.25 = 1 - e^{5k}$

$e^{5k} = 0.75$

$5k = \ln 0.75$

$k \approx -0.0575$

$S = 10(1 - e^{-0.0575x})$

(b) When $x = 7$,
$S = 10(1 - e^{-0.0575(7)}) \approx 3.314$
which corresponds to 3314 units.

**55.** $N = 30(1 - e^{kt})$

(a)  $N = 19$, $t = 20$

$19 = 30(1 - e^{20k})$

$20k = \ln \frac{11}{30}$

$k \approx -0.050$

$N = 30(1 - e^{-0.050t})$

(b) $N = 25$

$25 = 30(1 - e^{-0.05t})$

$\frac{5}{30} = e^{-0.05t}$

$t = -\frac{1}{0.05} \ln \frac{5}{30} \approx 36$ days

(c) No, this is not a linear function.

**57.** $R = \log_{10} \dfrac{I}{I_0} = \log_{10} I$ since $I_0 = 1$.

(a)  $R = \log_{10} 80{,}500{,}000 \approx 7.91$

(b)  $R = \log_{10} 48{,}275{,}000 \approx 7.68$

**59.** $\beta(I) = 10 \log_{10} \dfrac{I}{I_0}$ where $I_0 = 10^{-16}$ watt/cm$^2$.

(a)  $\beta(10^{-14}) = 10 \log_{10} \dfrac{10^{-14}}{10^{-16}} = 10 \log_{10} 10^2 = 20$ decibels

(b)  $\beta(10^{-9}) = 10 \log_{10} \dfrac{10^{-9}}{10^{-16}} = 10 \log_{10} 10^7 = 70$ decibels

(c)  $\beta(10^{-6.5}) = 10 \log_{10} \dfrac{10^{-6.5}}{10^{-16}} = 10 \log_{10} 10^{9.5} = 95$ decibels

(d)  $\beta(10^{-4}) = 10 \log_{10} \dfrac{10^{-4}}{10^{-16}} = 10 \log_{10} 10^{12} = 120$ decibels

**61.** $\beta = 10 \log_{10} \dfrac{I}{I_0}$

$10^{\beta/10} = \dfrac{I}{I_0}$

$I = I_0 10^{\beta/10}$

% decrease $= \dfrac{I_0 10^{9.3} - I_0 10^{8.0}}{I_0 10^{9.3}} \times 100 \approx 95\%$

**63.** pH $= -\log_{10}[\text{H}^+] = -\log_{10}[2.3 \times 10^{-5}] \approx 4.64$

**65.** $5.8 = -\log_{10}[\text{H}^+]$

$10^{-5.8} = \text{H}^+$

$\text{H}^+ \approx 1.58 \times 10^{-6}$ moles per liter

**67.** $2.5 = -\log_{10}[\text{H}^+]$

$10^{-2.5} = \text{H}^+$ for the fruit.

$9.5 = -\log_{10}[\text{H}^+]$

$10^{-9.5} = \text{H}^+$ for the antacid tablet.

$\dfrac{10^{-2.5}}{10^{-9.5}} = 10^7$

**69.** Interest: $u = M - \left(M - \dfrac{Pr}{12}\right)\left(1 + \dfrac{r}{12}\right)^{12t}$

Principle: $v = \left(M - \dfrac{Pr}{12}\right)\left(1 + \dfrac{r}{12}\right)^{12t}$

(a) $P = 120,000$, $t = 35$, $r = 0.095$, $M = 985.93$

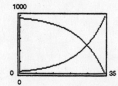

(b) In the early years of the mortgage, the majority of the monthly payment goes toward interest. The principle and interest are nearly equal when $t \approx 27.676 \approx 28$ years.

(c) $P = 120,000$, $t = 20$, $r = 0.095$, $M = 1118.56$

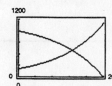

The interest is still the majority of the monthly payment in the early years. Now the principle and interest are nearly equal when $t \approx 12.675 \approx 12.7$ years.

**71.** $t_1 = 40.757 + 0.556s - 15.817 \ln s$

$t_2 = 1.2259 + 0.0023s^2$

(a) Linear model: $t_3 \approx 0.2729s - 6.0143$

Exponential model: $t_4 \approx 1.5385e^{1.0291s}$

(b)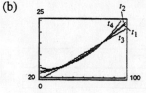

(c)

| s | 30 | 40 | 50 | 60 | 70 | 80 | 90 |
|---|---|---|---|---|---|---|---|
| $t_1$ | 3.6 | 4.6 | 6.7 | 9.4 | 12.5 | 15.9 | 19.6 |
| $t_2$ | 3.3 | 4.9 | 7.0 | 9.5 | 12.5 | 15.9 | 19.9 |
| $t_3$ | 2.2 | 4.9 | 7.6 | 10.4 | 13.1 | 15.8 | 18.5 |
| $t_4$ | 3.7 | 4.9 | 6.6 | 8.8 | 11.8 | 15.8 | 21.2 |

(d) Model $t_1$: $S_1 = |3.4 - 3.6| + |5 - 4.6| + |7 - 6.7| + |9.3 - 9.4| + |12 - 12.5| +$
$|15.8 - 15.9| + |20 - 19.6| = 2.0$

Model $t_2$: $S_2 = |3.4 - 3.3| + |5 - 4.9| + |7 - 7| + |9.3 - 9.5| + |12 - 12.5| +$
$|15.8 - 15.9| + |20 - 19.9| = 1.1$

Model $t_3$: $S_3 = |3.4 - 2.2| + |5 - 4.9| + |7 - 7.6| + |9.3 - 10.4| + |12 - 13.1| +$
$|15.8 - 15.8| + |20 - 18.5| = 5.6$

Model $t_4$: $S_4 = |3.4 - 3.7| + |5 - 4.9| + |7 - 6.6| + |9.3 - 8.8| + |12 - 11.8| +$
$|15.8 - 15.8| + |20 - 21.2| = 2.7$

$t_2$, the Quadratic model, is the best fit with the data.

**73.** Answers will vary.

**75.**

$$
-4 \begin{array}{|rrrr} 4 & 4 & -39 & 36 \\ & -16 & 48 & -36 \\ \hline 4 & -12 & 9 & 0 \end{array}
$$

Thus, $\dfrac{4x^3 + 4x^2 - 39x + 36}{x + 4} = 4x^2 - 12x + 9.$

**77.**

$$
4 \begin{array}{|rrrr} 2 & -8 & 3 & -9 \\ & 8 & 0 & 12 \\ \hline 2 & 0 & 3 & 3 \end{array}
$$

Thus, $\dfrac{2x^3 - 8x^2 + 3x - 9}{x - 4} = 2x^2 + 3 + \dfrac{3}{x - 4}.$

# ☐ Review Exercises for Chapter 5

### Solutions to Odd-Numbered Exercises

**1.** $f(x) = 4^x$

Intercept: $(0, 1)$

Horizontal asymptote: $x$-axis

Increasing on: $(-\infty, \infty)$

Matches graph (e)

**3.** $f(x) = -4^x$

Intercept: $(0, -1)$

Horizontal asymptote: $x$-axis

Decreasing on: $(-\infty, \infty)$

Matches graph (a)

**5.** $f(x) = \log_4 x$

Intercept: $(1, 0)$

Vertical asymptote: $y$-axis

Increasing on: $(0, \infty)$

Matches graph (d)

**7.** $f(x) = 0.3^x$

| $x$ | $-2$ | $-1$ | $0$ | $1$ | $2$ |
|---|---|---|---|---|---|
| $y$ | 11.11 | 3.33 | 1 | 0.3 | 0.09 |

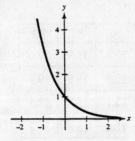

**9.** $h(x) = e^{-x/2}$

| $x$ | $-2$ | $-1$ | $0$ | $1$ | $2$ |
|---|---|---|---|---|---|
| $y$ | 2.72 | 1.65 | 1 | 0.61 | 0.37 |

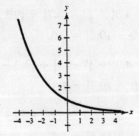

**11.** $f(x) = e^{x+2}$

| $x$ | $-3$ | $-2$ | $-1$ | $0$ | $1$ |
|---|---|---|---|---|---|
| $y$ | 0.37 | 1 | 2.72 | 7.39 | 20.09 |

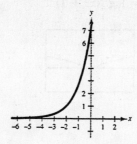

**13.** $g(x) = 200e^{4/x}$

As $x \to \infty$, $g(x) \to 200$ so we have a horizontal asymptote at $y = 200$.

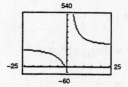

**15.** $A = 3500\left(1 + \dfrac{0.105}{n}\right)^{10n}$ or $A = 3500e^{(0.105)(10)}$

| $n$ | 1 | 2 | 4 | 12 | 365 | Continuous Compounding |
|---|---|---|---|---|---|---|
| $A$ | \$9,499.28 | \$9,738.91 | \$9,867.22 | \$9,956.20 | \$10,000.27 | \$10,001.78 |

**17.** $200{,}000 = Pe^{0.08t}$

$$P = \frac{200{,}000}{e^{0.08t}}$$

| $t$ | 1 | 10 | 20 | 30 | 40 | 50 |
|---|---|---|---|---|---|---|
| $P$ | \$184,623.27 | \$89,865.79 | \$40,379.30 | \$18,143.59 | \$8,152.44 | \$3,663.13 |

**19.** $F(t) = 1 - e^{-t/3}$

(a) $F\left(\frac{1}{2}\right) \approx 0.154$

(b) $F(2) \approx 0.487$

(c) $F(5) \approx 0.811$

**21.** (a) $A = 50{,}000e^{(0.0875)(35)} \approx \$1{,}069{,}047.14$

(b) The doubling time is

$$\frac{\ln 2}{0.0875} \approx 7.9 \text{ years.}$$

**23.** $g(x) = \log_2 x \implies 2^y = x$

Domain: $(0, \infty)$

Vertical asymptote: $x = 0$

| $x$ | $\frac{1}{4}$ | $\frac{1}{2}$ | 1 | 2 | 4 |
|---|---|---|---|---|---|
| $y$ | $-2$ | $-1$ | 0 | 1 | 2 |

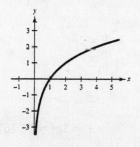

**25.** $f(x) = \ln x + 3$

Domain: $(0, \infty)$.

Vertical asymptote: $x = 0$

| $x$ | 1 | 2 | 3 | $\frac{1}{2}$ | $\frac{1}{4}$ |
|---|---|---|---|---|---|
| $f(x)$ | 3 | 3.69 | 4.10 | 2.31 | 1.61 |

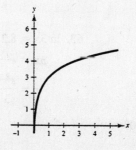

**27.** $h(x) = \ln(e^{x-1})$
$= (x - 1) \ln e$
$= x - 1$

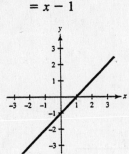

**29.** $y = \log_{10}(x^2 + 1)$

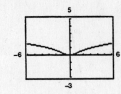

**31.**    $4^3 = 64$
$\log_4 64 = 3$

**33.** $\log_{10} 1000 = \log_{10} 10^3 = 3$

**35.** $\ln e^7 = 7$

**37.** $\log_4 9 = \dfrac{\log_{10} 9}{\log_{10} 4} \approx 1.585$

$\log_4 9 = \dfrac{\ln 9}{\ln 4} \approx 1.585$

**39.** $\log_{12} 200 = \dfrac{\log_{10} 200}{\log_{10} 12} \approx 2.132$

$\log_{12} 200 = \dfrac{\ln 200}{\ln 12} \approx 2.132$

**41.** $\log_5 5x^2 = \log_5 5 + \log_5 x^2$
$= 1 + 2\log_5|x|$

**43.** $\log_{10} \dfrac{5\sqrt{y}}{x^2} = \log_{10} 5\sqrt{y} - \log_{10} x^2$

$= \log_{10} 5 + \log_{10}\sqrt{y} - \log_{10} x^2$

$= \log_{10} 5 + \dfrac{1}{2}\log_{10} y - 2\log_{10}|x|$

**45.** $\log_2 5 + \log_2 x = \log_2 5x$

**47.** $\dfrac{1}{2}\ln|2x - 1| - 2\ln|x + 1| = \ln\sqrt{|2x - 1|} - \ln|x + 1|^2$

$= \ln \dfrac{\sqrt{|2x - 1|}}{(x + 1)^2}$

**49.** True; by the inverse properties, $\log_b b^{2x} = 2x$.

**51.** False; $\ln x + \ln y = \ln(xy) \neq \ln(x + y)$

**53.** True, $\log\left(\dfrac{10}{x}\right) = \log 10 - \log x = 1 - \log x$.

**55.** $S = 25 - \dfrac{13\ln(10/12)}{\ln 3} \approx 27.16$ miles

**57.** $e^x = 12$
$x = \ln 12 \approx 2.485$

**59.** $3e^{-5x} = 132$
$e^{-5x} = 44$
$-5x = \ln 44$
$x = \dfrac{\ln 44}{-5} \approx -0.757$

**61.**    $e^{2x} - 7e^x + 10 = 0$
$(e^x - 2)(e^x - 5) = 0$
$e^x = 2$  or  $e^x = 5$
$x = \ln 2$      $x = \ln 5$
$x \approx 0.693$      $x \approx 1.609$

**63.** $\ln 3x = 8.2$
$3x = e^{8.2}$
$x = \dfrac{e^{8.2}}{3} \approx 1213.650$

**65.** $\ln x - \ln 3 = 2$
$\ln \dfrac{x}{3} = 2$
$\dfrac{x}{3} = e^2$
$x = 3e^2 \approx 22.167$

**67.**
$$\log(x - 1) = \log(x - 2) - \log(x + 2)$$

$$\log(x - 1) = \log\left(\frac{x - 2}{x + 2}\right)$$

$$x - 1 = \frac{x - 2}{x + 2}$$

$$(x - 1)(x + 2) = x - 2$$

$$x^2 + x - 2 = x - 2$$

$$x^2 = 0$$

$$x = 0$$

Since $x = 0$ is not in the domain of $\ln(x - 1)$ or of $\ln(x - 2)$, it is an extraneous solution. The equation has no solution.

**71.** $2 \ln(x + 3) + 3x = 8$

Graph $y_1 = 2 \ln(x + 3) + 3x - 8$

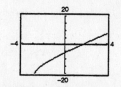

The $x$-intercept is at $x \approx 1.64$.

**75.** $p = 500 - 0.5e^{0.004x}$

(a) $p = 450$

$$450 = 500 - 0.5e^{0.004x}$$

$$0.5e^{0.004x} = 50$$

$$e^{0.004x} = 100$$

$$0.004x = \ln 100$$

$$x \approx 1151 \text{ units}$$

**77.** (a) $\dfrac{\ln 2}{r} = 5$

$$\ln 2 = 5r$$

$$r = \frac{\ln 2}{5} \approx 0.1386 = 13.86\%$$

(b) $A = 10,000e^{0.1386(1)} \approx \$11,486.65$

**69.** $2^{0.6x} - 3x = 0$

Graph $y_1 = 2^{0.6x} - 3x$

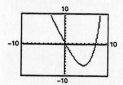

The $x$-intercepts are at $x \approx 0.39$ and at $x \approx 7.48$.

**73.**
$$y = ae^{bx}$$

$$2 = ae^{b(0)} \implies a = 2$$

$$3 = 2e^{b(4)}$$

$$1.5 = e^{4b}$$

$$\ln 1.5 = 4b \implies b \approx 0.1014$$

Thus, $y \approx 2e^{0.1014x}$

(b) $p = 400$

$$400 = 500 - 0.5e^{0.004x}$$

$$0.5e^{0.004x} = 100$$

$$e^{0.004x} = 200$$

$$0.004x = \ln 200$$

$$x \approx 1325 \text{ units}$$

**79.** $R = \log_{10} I$ since $I_0 = 1$.

(a) $\log_{10} I = 8.4$

$$I = 10^{8.4}$$

(b) $\log_{10} I = 6.85$

$$I = 10^{6.85}$$

(c) $\log_{10} I = 9.1$

$$I = 10^{9.1}$$

# ❑ **Practice Test for Chapter 5**

1. Solve for $x$: $x^{3/5} = 8$.

2. Solve for $x$: $3^{x-1} = \frac{1}{81}$.

3. Graph $f(x) = 2^{-x}$.

4. Graph $g(x) = e^x + 1$.

5. If \$5000 is invested at 9% interest, find the amount after three years if the interest is compounded

   (a) monthly        (b) quarterly        (c) continuously.

6. Write the equation in logarithmic form: $7^{-2} = \frac{1}{49}$.

7. Solve for $x$: $x - 4 = \log_2 \frac{1}{64}$.

8. Given $\log_b 2 = 0.3562$ and $\log_b 5 = 0.8271$, evaluate $\log_b \sqrt[4]{8/25}$.

9. Write $5 \ln x - \frac{1}{2} \ln y + 6 \ln z$ as a single logarithm.

10. Using your calculator and the change of base formula, evaluate $\log_9 28$.

11. Use your calculator to solve for $N$: $\log_{10} N = 0.6646$

12. Graph $y = \log_4 x$.

13. Determine the domain of $f(x) = \log_3(x^2 - 9)$.

14. Graph $y = \ln(x - 2)$.

15. True or false: $\dfrac{\ln x}{\ln y} = \ln(x - y)$

16. Solve for $x$: $5^x = 41$

17. Solve for $x$: $x - x^2 = \log_5 \frac{1}{25}$

18. Solve for $x$: $\log_2 x + \log_2(x - 3) = 2$

19. Solve for $x$: $\dfrac{e^x + e^{-x}}{3} = 4$

20. Six thousand dollars is deposited into a fund at an annual interest rate of 13%. Find the time required for the investment to double if the interest is compounded continuously.

# C H A P T E R   6
# Topics in Analytic Geometry

# C H A P T E R   6
## Topics in Analytic Geometry

## Section 6.1   Lines

---

- The **inclination** of a non-horizontal line is the positive angle $\theta$ $(\theta < 180°)$ measured counterclockwise from the $x$-axis to the line. A horizontal line has an inclination of zero.
- If a nonvertical line has inclination of $\theta$ and slope $m$, then $m = \tan \theta$.
- If two non-perpendicular lines have slopes $m_1$ and $m_2$, then the angle between the lines is given by
$$\tan \theta = \left| \frac{m_2 - m_1}{1 + m_1 m_2} \right|.$$
- The distance between a point $(x_1, y_1)$ and a line $Ax + By + C = 0$ is given by
$$d = \frac{|Ax_1 + By_1 + C|}{\sqrt{A^2 + B^2}}.$$

---

**Solutions to Odd-Numbered Exercises**

**1.** $m = \tan 30° = \dfrac{\sqrt{3}}{3}$

**3.** $m = \tan 135° = -1$

**5.** $m = \tan 38.2° \approx 0.7869$

**7.** $m = \tan 110° \approx -2.7475$

**9.** $m = -1$

$-1 = \tan \theta$

$\theta = 180° + \arctan(-1) = 135°$

**11.** $m = \frac{3}{4}$

$\frac{3}{4} = \tan \theta$

$\theta = \arctan\left(\frac{3}{4}\right) \approx 36.9°$

**13.** $(6, 1), (10, 8)$

$m = \dfrac{8 - 1}{10 - 6} = \dfrac{7}{4}$

$\dfrac{7}{4} = \tan \theta$

$\theta = \arctan\left(\dfrac{7}{4}\right) \approx 60.3°$

**15.** $(-2, 20), (10, 0)$

$m = \dfrac{0 - 20}{10 - (-2)} = -\dfrac{20}{12} = -\dfrac{5}{3}$

$-\dfrac{5}{3} = \tan \theta$

$\theta = 180° + \arctan\left(-\dfrac{5}{3}\right) \approx 121.0°$

**17.** $5x - y + 3 = 0$

$y = 5x + 3 \Rightarrow m = 5$

$5 = \tan \theta$

$\theta = \arctan 5 \approx 78.7°$

**19.** $5x + 3y = 0$

$y = -\frac{5}{3}x \Rightarrow m = -\frac{5}{3}$

$-\frac{5}{3} = \tan \theta$

$\theta = 180° + \arctan\left(-\frac{5}{3}\right) \approx 121.0°$

**21.** Slope:  $m = \tan 6.5° \approx 0.1139$

Change in elevation:  $\dfrac{x}{2(5280)} = \sin 6.5°$

$$x = 10{,}560 \sin 6.5°$$

$$x \approx 1195 \text{ feet}$$

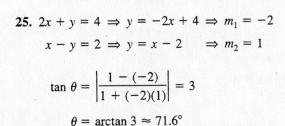

**23.** (a)  $\tan \theta = \dfrac{1}{3}$

$$\theta = \arctan\left(\dfrac{1}{3}\right) \approx 18.4°$$

(b)  $\dfrac{5}{x} = \sin 18.4°$

$$x = \dfrac{5}{\sin 18.4°}$$

$$x \approx 15.8 \text{ meters}$$

**25.**  $2x + y = 4 \Rightarrow y = -2x + 4 \Rightarrow m_1 = -2$

$x - y = 2 \Rightarrow y = x - 2 \quad \Rightarrow m_2 = 1$

$$\tan \theta = \left| \dfrac{1 - (-2)}{1 + (-2)(1)} \right| = 3$$

$$\theta = \arctan 3 \approx 71.6°$$

**27.**  $x - y = 0 \quad \Rightarrow y = x \qquad \Rightarrow m_1 = 1$

$3x - 2y = -1 \Rightarrow y = \dfrac{3}{2}x + \dfrac{1}{2} \Rightarrow m_2 = \dfrac{3}{2}$

$$\tan \theta = \left| \dfrac{(3/2) - 1}{1 + (1)(3/2)} \right| = \dfrac{1}{5}$$

$$\theta = \arctan\left(\dfrac{1}{5}\right) \approx 11.3°$$

**29.**  $x - 2y = 5 \Rightarrow y = \dfrac{1}{2}x - \dfrac{5}{2} \quad \Rightarrow m_2 = \dfrac{1}{2}$

$6x + 2y = 7 \Rightarrow y = -3x + \dfrac{7}{2} \Rightarrow m_2 = -3$

$$\tan \theta = \left| \dfrac{-3 - (1/2)}{1 + (1/2)(-3)} \right| = 7$$

$$\theta = \arctan 7 \approx 81.9°$$

**31.**  $x + 2y = 4 \Rightarrow y = -\dfrac{1}{2}x + 2 \Rightarrow m_1 = -\dfrac{1}{2}$

$x - 2y = 1 \Rightarrow y = \dfrac{1}{2}x - \dfrac{1}{2} \quad \Rightarrow m_2 = \dfrac{1}{2}$

$$\tan \theta = \left| \dfrac{(1/2) - (-1/2)}{1 + (-1/2)(1/2)} \right| = \dfrac{4}{3}$$

$$\theta = \arctan\left(\dfrac{4}{3}\right) \approx 53.1°$$

**33.**  $0.05x - 0.03y = 0.21 \Rightarrow y = \dfrac{5}{3}x - 7 \quad \Rightarrow m_1 = \dfrac{5}{3}$

$0.07x - 0.02y = 0.16 \Rightarrow y = -\dfrac{7}{2}x + 8 \Rightarrow m_2 = -\dfrac{7}{2}$

$$\tan \theta = \left| \dfrac{(-7/2) - (5/3)}{1 + (5/3)(-7/2)} \right| = \dfrac{31}{29}$$

$$\theta = \arctan\left(\dfrac{31}{29}\right) \approx 46.9°$$

**35.** Let $A = (2, 1)$, $B = (4, 4)$, and $C = (6, 2)$.

Slope of $AB$:  $m_1 = \dfrac{1 - 4}{2 - 4} = \dfrac{3}{2}$

Slope of $BC$:  $m_2 = \dfrac{4 - 2}{4 - 6} = -1$

Slope of $AC$:  $m_3 = \dfrac{1 - 2}{2 - 6} = \dfrac{1}{4}$

$\tan A = \left| \dfrac{(1/4) - (3/2)}{1 + (3/2)(1/4)} \right| = \dfrac{5/4}{11/8} = \dfrac{10}{11}$

$A = \arctan\left(\dfrac{10}{11}\right) \approx 42.3°$

$\tan B = \left| \dfrac{(3/2) - (-1)}{1 + (-1)(3/2)} \right| = \dfrac{5/2}{1/2} = 5$

$B = \arctan 5 \approx 78.7°$

$\tan C = \left| \dfrac{-1 - (1/4)}{1 + (1/4)(-1)} \right| = \dfrac{5/4}{3/4} = \dfrac{5}{3}$

$C = \arctan\left(\dfrac{5}{3}\right) \approx 59.0°$

**37.** Let $A = (-4, -1)$, $B = (3, 2)$, and $C = (1, 0)$.

Slope of $AB$:  $m_1 = \dfrac{-1 - 2}{-4 - 3} = \dfrac{3}{7}$

Slope of $BC$:  $m_2 = \dfrac{2 - 0}{3 - 1} = 1$

Slope of $AC$:  $m_3 = \dfrac{-1 - 0}{-4 - 1} = \dfrac{1}{5}$

$\tan A = \left| \dfrac{(1/5) - (3/7)}{1 + (3/7)(1/5)} \right| = \dfrac{8/35}{38/35} = \dfrac{4}{19}$

$A = \arctan\left(\dfrac{4}{19}\right) \approx 11.9°$

$\tan B = \left| \dfrac{1 - (3/7)}{1 + (3/7)(1)} \right| = \dfrac{4/7}{10/7} = \dfrac{2}{5}$

$B = \arctan\left(\dfrac{2}{5}\right) \approx 21.8°$

$C = 180° - A - B$

$\approx 180° - 11.9° - 21.8° = 146.3°$

**39.** (a)  $m_1 = \tan 48°$

$m_2 = \tan(180° - 63°) = \tan 117°$

$\tan \theta = \left| \dfrac{\tan 117° - \tan 48°}{1 + \tan 48° \tan 117°} \right|$

$\theta = 69°$

(b)  $\dfrac{x}{6700} = \sin 63°$

$x = 6700 \sin 63°$

$\approx 5970 \text{ feet}$

$\dfrac{x}{3250} = \sin 48°$

$x = 3250 \sin 48°$

$\approx 2415 \text{ feet}$

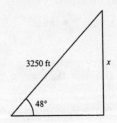

**41.** $(0, 0) \Rightarrow x_1 = 0$ and $y_1 = 0$

$4x + 3y = 0 \Rightarrow A = 4, B = 3,$ and $C = 0$

$d = \dfrac{|4(0) + 3(0) + 0|}{\sqrt{4^2 + 3^2}} = \dfrac{0}{5} = 0$

Note: The point is *on* the line.

**43.** $(2, 3) \Rightarrow x_1 = 2$ and $y_1 = 3$

$4x + 3y - 10 = 0 \Rightarrow A = 4, B = 3,$ and $C = -10$

$d = \dfrac{|4(2) + 3(3) + (-10)|}{\sqrt{4^2 + 3^2}} = \dfrac{7}{5} = 1.4$

**45.** $(6, 2) \Rightarrow x_1 = 6$ and $y_1 = 2$

$x + 1 = 0 \Rightarrow A = 1, B = 0,$ and $C = 1$

$$d = \frac{|1(6) + 0(2) + 1|}{\sqrt{1^2 + 0^2}} = 7$$

**47.** $(0, 8) \Rightarrow x_1 = 0$ and $y_1 = 8$

$6x - y = 0 \Rightarrow A = 6, B = -1,$ and $C = 0$

$$d = \frac{|6(0) + (-1)(8) + 0|}{\sqrt{6^2 + (-1)^2}}$$

$$= \frac{8}{\sqrt{37}} = \frac{8\sqrt{37}}{37} \approx 1.3152$$

**49.** (a) The slope of the line through $AC$ is $m = \dfrac{0 - 1}{0 - 3} = \dfrac{1}{3}$.

The equation of the line is $y - 0 = \dfrac{1}{3}(x - 0) \Rightarrow \dfrac{1}{3}x - y = 0$.

The distance between the line and $B = (1, 5)$ is $d = \dfrac{|(1/3)(1) + (-1)(5) + 0|}{\sqrt{(1/3)^2 + (-1)^2}} = \dfrac{14/3}{\sqrt{10}/3} = \dfrac{7\sqrt{10}}{5}$.

(b) The distance between $A$ and $C$ is $d = \sqrt{(0 - 3)^2 + (0 - 1)^2} = \sqrt{10}$.

$$A = \frac{1}{2}\left(\sqrt{10}\right)\left(\frac{7\sqrt{10}}{5}\right) = 7 \text{ square units}$$

**51.** (a) The slope of the line through $AC$ is $m = \dfrac{(1/2) - 0}{(-1/2) - (5/2)} = -\dfrac{1}{6}$.

The equation of the line through $AC$ is $y - 0 = -\dfrac{1}{6}\left(x - \dfrac{5}{2}\right) \Rightarrow 2x + 12y - 5 = 0$.

The distance between the line and $B = (2, 3)$ is $d = \dfrac{|2(2) + 12(3) + (-5)|}{\sqrt{2^2 + 12^2}} = \dfrac{35}{\sqrt{148}} = \dfrac{35\sqrt{37}}{74}$.

(b) The distance between $A$ and $C$ is $d = \sqrt{\left[\left(-\dfrac{1}{2}\right) - \left(\dfrac{5}{2}\right)\right]^2 + \left[\left(\dfrac{1}{2}\right) - 0\right]^2} = \dfrac{\sqrt{37}}{2}$.

$$A = \frac{1}{2}\left(\frac{\sqrt{37}}{2}\right)\left(\frac{35\sqrt{37}}{74}\right) = \frac{35}{8} \text{ square units}$$

**53.** $x + y = 1 \Rightarrow (0, 1) \Rightarrow x_1 = 0$ and $y_1 = 1$

$x + y = 5 \Rightarrow A = 1, B = 1,$ and $C = -5$

$$d = \frac{|1(0) + 1(1) + (-5)|}{\sqrt{1^2 + 1^2}} = \frac{4}{\sqrt{2}} = 2\sqrt{2} \approx 2.8284$$

**55.** (a) $(0, 0) \Rightarrow x_1 = 0$ and $y_1 = 0$

$y = mx + 4 \Rightarrow 0 = mx - y + 4$

$$d = \frac{|m(0) + (-1)(0) + 4|}{\sqrt{m^2 + (-1)^2}} = \frac{4}{\sqrt{m^2 + 1}}$$

(b)

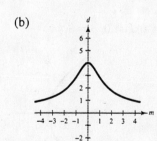

(c) The maximum distance of 4 occurs when the slope $m$ is 0 and the line through $(0, 4)$ is horizontal.

(d) The graph has a horizontal asymptote at $d = 0$. As the slope becomes larger, the distance between the origin and the line, $y = mx + 4$, becomes smaller and approaches 0.

# Section 6.2    Introduction to Conics:  Parabolas

■ A **parabola** is the set of all points *(x, y)* that are equidistant from a fixed line (**directrix**) and a fixed point (**focus**) not on the line.

■ The standard equation of a parabola with vertex *(h, k)* and:
(a) Vertical axis $x = h$ and directrix $y = k - p$ is:
$(x - h)^2 = 4p(y - k), p \neq 0$
(b) Horizontal axis $y = k$ and directrix $x = h - p$ is:
$(y - k)^2 = 4p(x - h), p \neq 0$

■ The tangent line to a parabola at a point *P* makes **equal angles** with:
(a) the line through *P* and the focus
(b) the axis of the parabola

**Solutions to Odd-Numbered Exercises**

**1.** $y^2 = -4x$
Vertex: $(0, 0)$
Opens to the left since $p$ is negative.
Matches graph (e).

**3.** $x^2 = -8y$
Vertex: $(0, 0)$
Opens downward since $p$ is negative.
Matches graph (d).

**5.** $(y - 1)^2 = 4(x - 3)$
Vertex: $(3, 1)$
Opens to the right since $p$ is positive.
Matches graph (a).

**7.** $y = \frac{1}{2}x^2$
$x^2 = 2y$
$x^2 = 4\left(\frac{1}{2}\right)y \Rightarrow h = 0, k = 0, p = \frac{1}{2}$
Vertex: $(0, 0)$
Focus: $\left(\frac{1}{2}, 0\right)$
Directrix: $y = -\frac{1}{2}$

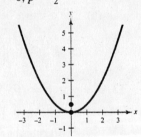

**9.** $y^2 = -6x$
$y^2 = 4\left(-\frac{3}{2}\right)x \Rightarrow h = 0, k = 0, p = -\frac{3}{2}$
Vertex: $(0, 0)$
Focus: $\left(-\frac{3}{2}, 0\right)$
Directrix: $x = \frac{3}{2}$

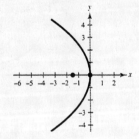

**11.** $x^2 + 8y = 0$
$x^2 = 4(-2)y \Rightarrow h = 0, k = 0, p = -2$
Vertex: $(0, 0)$
Focus: $(0, -2)$
Directrix: $y = 2$

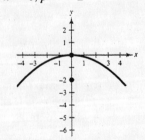

**13.** $(x - 1)^2 + 8(y + 2) = 0$

$(x - 1)^2 = 4(-2)(y + 2)$

$h = 1, k = -2, p = -2$

Vertex: $(1, -2)$

Focus: $(1, -4)$

Directrix: $y = 0$

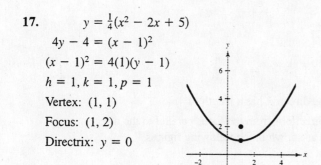

**15.** $\left(y + \frac{1}{2}\right)^2 = 2(x - 5)$

$= 4\left(\frac{1}{2}\right)(x - 5)$

$h = 5, k = -\frac{1}{2}, p = \frac{1}{2}$

Vertex: $\left(5, -\frac{1}{2}\right)$

Focus: $\left(\frac{11}{2}, -\frac{1}{2}\right)$

Directrix: $x = \frac{9}{2}$

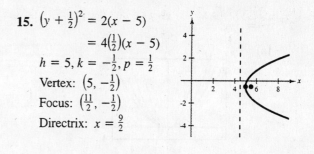

**17.** $y = \frac{1}{4}(x^2 - 2x + 5)$

$4y - 4 = (x - 1)^2$

$(x - 1)^2 = 4(1)(y - 1)$

$h = 1, k = 1, p = 1$

Vertex: $(1, 1)$

Focus: $(1, 2)$

Directrix: $y = 0$

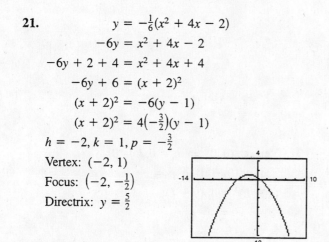

**19.** $y^2 + 6y + 8x + 25 = 0$

$y^2 + 6y + 9 = -8x - 25 + 9$

$(y + 3)^2 = 4(-2)(x + 2)$

$h = -2, k = -3, p = -2$

Vertex: $(-2, -3)$

Focus: $(-4, -3)$

Directrix: $x = 0$

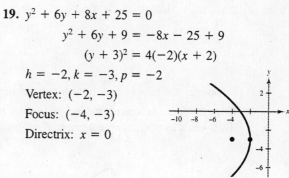

**21.** $y = -\frac{1}{6}(x^2 + 4x - 2)$

$-6y = x^2 + 4x - 2$

$-6y + 2 + 4 = x^2 + 4x + 4$

$-6y + 6 = (x + 2)^2$

$(x + 2)^2 = -6(y - 1)$

$(x + 2)^2 = 4\left(-\frac{3}{2}\right)(y - 1)$

$h = -2, k = 1, p = -\frac{3}{2}$

Vertex: $(-2, 1)$

Focus: $\left(-2, -\frac{1}{2}\right)$

Directrix: $y = \frac{5}{2}$

**23.** $y^2 + x + y = 0$

$y^2 + y + \frac{1}{4} = -x + \frac{1}{4}$

$\left(y + \frac{1}{2}\right)^2 = 4\left(-\frac{1}{4}\right)\left(x - \frac{1}{4}\right)$

$h = \frac{1}{4}, k = -\frac{1}{2}, p = -\frac{1}{4}$

Vertex: $\left(\frac{1}{4}, -\frac{1}{2}\right)$

Focus: $\left(0, -\frac{1}{2}\right)$

Directrix: $x = \frac{1}{2}$

To use a graphing calculator, enter:

$y_1 = -\frac{1}{2} + \sqrt{\frac{1}{4} - x}$

$y_2 = -\frac{1}{2} - \sqrt{\frac{1}{4} - x}$

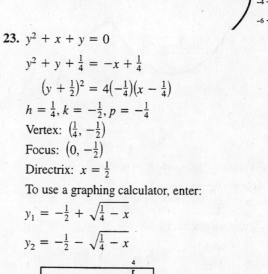

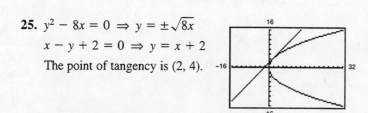

**25.** $y^2 - 8x = 0 \Rightarrow y = \pm\sqrt{8x}$

$x - y + 2 = 0 \Rightarrow y = x + 2$

The point of tangency is $(2, 4)$.

**27.** $x^2 = 4py \implies y = \dfrac{x^2}{4p}$

(a)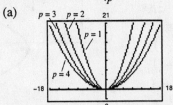

(b) $p = 1$: focus $(0, 1)$
$p = 2$: focus $(0, 2)$
$p = 3$: focus $(0, 3)$
$p = 4$: focus $(0, 4)$

As $p$ increases, the parabola opens wider.

(c)

| $p$-value | Length of chord |
|-----------|-----------------|
| 1 | 4 |
| 2 | 8 |
| 3 | 12 |
| 4 | 16 |

In general, the chord through the focus parallel to the directrix has a length of $4p$.

(d) Once the focus is located, move $2|p|$ units in both directions from the focus parallel to the directrix. This yields two points on the graph of the parabola, as shown in the following figures.

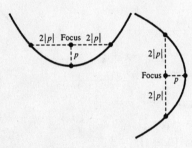

**29.** Vertex: $(0, 0) \implies h = 0, k = 0$

Graph opens upward.

$x^2 = 4py$

Point on graph: $(3, 6)$

$3^2 = 4p(6)$

$9 = 24p$

$\frac{3}{8} = p$

Thus, $x^2 = 4\left(\frac{3}{8}\right)y \implies y = \frac{2}{3}x^2$.

**31.** Vertex: $(0, 0) \implies h = 0, k = 0$

Focus: $\left(0, -\frac{3}{2}\right) \implies p = -\frac{3}{2}$

$(x - h)^2 = 4p(y - k)$

$x^2 = 4\left(-\frac{3}{2}\right)y$

$x^2 = -6y$

**33.** Vertex: $(0, 0) \implies h = 0, k = 0$

Focus: $(-2, 0) \implies p = -2$

$(y - k)^2 = 4p(x - h)$

$y^2 = 4(-2)x$

$y^2 = -8x$

**35.** Vertex: $(0, 0) \implies h = 0, k = 0$

Directrix: $y = -1 \implies p = 1$

$(x - h)^2 = 4p(y - k)$

$(x - 0)^2 = 4(1)(y - 0)$

$x^2 = 4y$ or $y = \frac{1}{4}x^2$

**37.** Vertex: $(0, 0) \Rightarrow h = 0, k = 0$

Directrix: $y = 2 \Rightarrow p = -2$

$(x - h)^2 = 4p(y - k)$

$(x - 0)^2 = 4(-2)(y - 0)$

$\quad x^2 = -8y$ or $y = -\frac{1}{8}x^2$

**39.** Vertex: $(0, 0) \Rightarrow h = 0, k = 0$

Horizontal axis and passes through the point $(4, 6)$

$(y - k)^2 = 4p(x - h)$

$(y - 0)^2 = 4p(x - 0)$

$\quad y^2 = 4px$

$\quad 6^2 = 4p(4)$

$\quad 36 = 16p \Rightarrow p = \frac{9}{4}$

$\quad y^2 = 4\left(\frac{9}{4}\right)x$

$\quad y^2 = 9x$

**41.** Vertex: $(3, 1)$ and opens downward. Passes through $(2, 0)$ and $(4, 0)$.

$y = -(x - 2)(x - 4)$

$\quad = -x^2 + 6x - 8$

$\quad = -(x - 3)^2 + 1$

$(x - 3)^2 = -(y - 1)$

**43.** Vertex: $(-2, 0)$ and opens to the right. Passes through $(0, 2)$.

$(y - 0)^2 = 4p(x + 2)$

$\quad 2^2 = 4p(0 + 2)$

$\quad \frac{1}{2} = p$

$\quad y^2 = 4\left(\frac{1}{2}\right)(x + 2)$

$\quad y^2 = 2(x + 2)$

**45.** Vertex: $(3, 2)$

Focus: $(1, 2)$

Horizontal axis

$p = 1 - 3 = -2$

$(y - 2)^2 = 4(-2)(x - 3)$

$(y - 2)^2 = -8(x - 3)$

**47.** Vertex: $(0, 4)$

Directrix: $y = 2$

Vertical axis

$p = 4 - 2 = 2$

$(x - 0)^2 = 4(2)(y - 4)$

$\quad x^2 = 8(y - 4)$

**49.** Focus: $(2, 2)$

Directrix: $x = -2$

Horizontal axis

Vertex: $(0, 2)$

$p = 2 - 0 = 2$

$(y - 2)^2 = 4(2)(x - 0)$

$(y - 2)^2 = 8x$

**51.** $(y - 3)^2 = 6(x + 1)$

For the upper half of the parabola:

$y - 3 = +\sqrt{6(x + 1)}$

$\quad y = \sqrt{6(x + 1)} + 3$

**53.** Vertex: $(0, 0) \Rightarrow h = 0, k = 0$

Focus: $(0, 3.5) \Rightarrow p = 3.5$

$(x - h)^2 = 4p(y - k)$

$(x - 0)^2 = 4(3.5)(y - 0)$

$\quad x^2 = 14y$ or $y = \frac{1}{14}x^2$

**55.** (a) Converting 16 meters to 1600 centimeters, and superimposing the coordinate plane over the parabola so that its vertex is $(0, 0)$, shows us that the points $(\pm 800, 3)$ are on the parabola.

$(x - 0)^2 = 4p(y - 0)$

$\quad x^2 = 4py$

At $(\pm 800, 3)$ we have:

$640,000 = 12p$

$\quad p = \dfrac{640,000}{12}$

$\quad x^2 = 4\left(\dfrac{640,000}{12}\right)y$

$\quad y = \dfrac{3x^2}{640,000}$

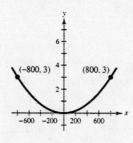

**—CONTINUED—**

**55.** —CONTINUED—

(b) Let $y = 1$, then:

$$1 = \frac{3x^2}{640,000}$$

$$x^2 = \frac{640,000}{3}$$

$$x = \frac{800\sqrt[3]{3}}{3} \approx 462 \text{ centimeters}$$

**57.** $R = 375x - \frac{3}{2}x^2$

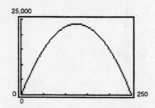

The revenue is maximum when $x = 125$ units.

**59.** $y = -0.08x^2 + x + 4$

(a)

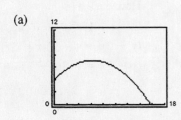

(b) The maximum occurs at the point $(6.25, 7.125)$ and the graph crosses the $x$-axis when $x \approx 15.69$ feet.

**61.** The slope of the line $y - y_1 = \frac{x_1}{2p}(x - x_1)$

is $m = \frac{x_1}{2p}$.

**63.** $x^2 = 2y \Rightarrow p = \frac{1}{2}$

Point: $(x_1, y_1) = (4, 8)$

Use: $y - y_1 = \frac{x_1}{2p}(x - x_1)$

$$y - 8 = \frac{4}{2(1/2)}(x - 4)$$

$$y - 8 = 4x - 16$$

$$y = 4x - 8 \Rightarrow 0 = 4x - y - 8$$

$x$-intercept: $(2, 0)$

**65.** $y = -2x^2 \Rightarrow x^2 = -\frac{1}{2}y \Rightarrow p = -\frac{1}{8}$

Point: $(x_1, y_1) = (-1, -2)$

Use: $y - y_1 = \frac{x_1}{2p}(x - x_1)$

$$y + 2 = \frac{-1}{2(-1/8)}(x + 1)$$

$$y + 2 = 4(x + 1)$$

$$y = 4x + 2 \Rightarrow 0 = 4x - y + 2$$

$x$-intercept: $\left(-\frac{1}{2}, 0\right)$

**67.** $f(x) = (x - 3)[x - (2 + i)][x - (2 - i)]$

$= (x - 3)(x - 2 - i)(x - 2 + i)$

$= (x - 3)(x^2 - 4x + 5)$

$= x^3 - 7x^2 + 17x - 15$

**69.** $g(x) = 6x^4 + 7x^3 - 29x^2 - 28x + 20$

Possible rational roots: $\pm 1, \pm 2, \pm 4, \pm 5, \pm 10, \pm 20,$
$\pm\frac{1}{2}, \pm\frac{5}{2}, \pm\frac{1}{3}, \pm\frac{2}{3}, \pm\frac{4}{3}, \pm\frac{5}{3}, \pm\frac{10}{3}, \pm\frac{20}{3}, \pm\frac{1}{6}, \pm\frac{5}{6}$

$x = \pm 2$ are both solutions.

| 2 | 6 | 7 | −29 | −28 | 20 |
|---|---|---|-----|-----|-----|
|   |   | 12 | 38 | 18 | −20 |
| −2 | 6 | 19 | 9 | −10 | 0 |
|   |   | −12 | −14 | 10 |   |
|   | 6 | 7 | −5 | 0 |   |

$g(x) = (x - 2)(x + 2)(6x^2 + 7x - 5)$

$= (x - 2)(x + 2)(2x - 1)(3x + 5)$

The zeros of $g(x)$ are $x = \pm 2$, $x = \frac{1}{2}$, $x = -\frac{5}{3}$.

# Section 6.3    Ellipses

- An **ellipse** is the set of all points *(x, y)* the sum of whose distances from two distinct fixed points **(foci)** is constant.
- The standard equation of an ellipse with center *(h, k)* and major and minor axes of lengths 2*a* and 2*b* is:

   (a) $\dfrac{(x-h)^2}{a^2} + \dfrac{(y-k)^2}{b^2} = 1$ if the major axis is horizontal.

   (b) $\dfrac{(x-h)^2}{b^2} + \dfrac{(y-k)^2}{a^2} = 1$ if the major axis is vertical.

- $c^2 = a^2 - b^2$ where *c* is the distance from the center to a focus.
- The eccentricity of an ellipse is $e = \dfrac{c}{a}$

## Solutions to Odd-Numbered Exercises

**1.** $\dfrac{x^2}{4} + \dfrac{y^2}{9} = 1$

Center: $(0, 0)$

$a = 3, b = 2$

Vertical major axis

Matches graph (b).

**3.** $\dfrac{x^2}{4} + \dfrac{y^2}{25} = 1$

Center: $(0, 0)$

$a = 5, b = 2$

Vertical major axis

Matches graph (d).

**5.** $\dfrac{(x-2)^2}{16} + (y+1)^2 = 1$

Center: $(2, -1)$

$a = 4, b = 1$

Horizontal major axis

Matches graph (a).

**7.** $\dfrac{x^2}{25} + \dfrac{y^2}{16} = 1$

Center: $(0, 0)$

$a = 5, b = 4, c = 3$

Foci: $(\pm 3, 0)$

Vertices: $(\pm 5, 0)$

$e = \dfrac{3}{5}$

**9.** $\dfrac{x^2}{16} + \dfrac{y^2}{25} = 1$

$a = 5, b = 4, c = 3$

Center: $(0, 0)$

Foci: $(0, \pm 3)$

Vertices: $(0, \pm 5)$

$e = \dfrac{3}{5}$

**11.** $\dfrac{x^2}{9} + \dfrac{y^2}{5} = 1$

Center: $(0, 0)$

$a = 3, b = \sqrt{5}, c = 2$

Foci: $(\pm 2, 0)$

Vertices: $(\pm 3, 0)$

$e = \dfrac{2}{3}$

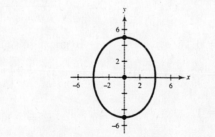

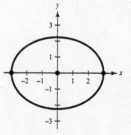

**13.** $\dfrac{(x-1)^2}{9} + \dfrac{(y-5)^2}{25} = 1$

$a = 5, b = 3, c = 4$

Center: $(1, 5)$

Foci: $(1, 9), (1, 1)$

Vertices: $(1, 10), (1, 0)$

$e = \dfrac{4}{5}$

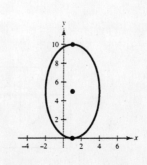

**15.**    $9x^2 + 4y^2 + 36x - 24y + 36 = 0$

$9(x^2 + 4x + 4) + 4(y^2 - 6y + 9) = -36 + 36 + 36$

$$\frac{(x+2)^2}{4} + \frac{(y-3)^2}{9} = 1$$

$a = 3, b = 2, c = \sqrt{5}$

Center: $(-2, 3)$

Foci: $\left(-2, 3 \pm \sqrt{5}\right)$

Vertices: $(-2, 6), (-2, 0)$

$e = \dfrac{\sqrt{5}}{3}$

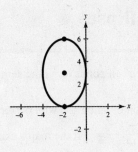

**17.**    $16x^2 + 25y^2 - 32x + 50y + 16 = 0$

$16(x^2 - 2x + 1) + 25(y^2 + 2y + 1) = -16 + 16 + 25$

$$\frac{(x-1)^2}{25/16} + (y+1)^2 = 1$$

$a = \dfrac{5}{4}, b = 1, c = \dfrac{3}{4}$

Center: $(1, -1)$

Foci: $\left(\dfrac{7}{4}, -1\right), \left(\dfrac{1}{4}, -1\right)$

Vertices: $\left(\dfrac{9}{4}, -1\right), \left(-\dfrac{1}{4}, -1\right)$

$e = \dfrac{3}{5}$

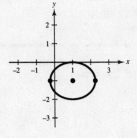

**19.** $5x^2 + 3y^2 = 15$

$$\frac{x^2}{3} + \frac{y^2}{5} = 1$$

Center: $(0, 0)$

$a = \sqrt{5}, b = \sqrt{3}, c = \sqrt{2}$

Foci: $\left(0, \pm\sqrt{2}\right)$

Vertices: $\left(0, \pm\sqrt{5}\right)$

To graph, solve for $y$.

$y^2 = \dfrac{15 - 5x^2}{3}$

$y_1 = \sqrt{\dfrac{15 - 5x^2}{3}}$

$y_2 = -\sqrt{\dfrac{15 - 5x^2}{3}}$

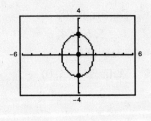

**21.**    $12x^2 + 20y^2 - 12x + 40y - 37 = 0$

$12\left(x^2 - x + \dfrac{1}{4}\right) + 20(y^2 + 2y + 1) = 37 + 3 + 20$

$$\frac{[x - (1/2)]^2}{5} + \frac{(y+1)^2}{3} = 1$$

$a = \sqrt{5}, b = \sqrt{3}, c = \sqrt{2}$

Center: $\left(\dfrac{1}{2}, -1\right)$

Foci: $\left(\dfrac{1}{2} \pm \sqrt{2}, -1\right)$

Vertices: $\left(\dfrac{1}{2} \pm \sqrt{5}, -1\right)$

$e = \dfrac{\sqrt{10}}{5}$

To graph, solve for $y$.

$(y + 1)^2 = 3\left[1 - \dfrac{(x - 0.5)^2}{5}\right]$

$y_1 = -1 + \sqrt{3\left[1 - \dfrac{(x - 0.5)^2}{5}\right]}$

$y_2 = -1 - \sqrt{3\left[1 - \dfrac{(x - 0.5)^2}{5}\right]}$

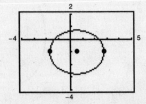

**23.** For the right half of the ellipse, solve for $x$ and use the positive square root.

$$\frac{(x-3)^2}{9} + \frac{y^2}{4} = 1$$

$$4(x-3)^2 + 9y^2 = 36$$

$$4(x-3)^2 = 36 - 9y^2$$

$$(x-3)^2 = \frac{9(4-y^2)}{4}$$

$$x - 3 = \frac{2}{3}\sqrt{4-y^2}$$

$$x = 3 + \frac{2}{3}\sqrt{4-y^2}$$

$$= \frac{3}{2}(2 + \sqrt{4-y^2})$$

**25.** Center: $(0, 0)$

$a = 2, b = 1$

Vertical major axis

$$\frac{(x-h)^2}{b^2} + \frac{(y-k)^2}{a^2} = 1$$

$$\frac{x^2}{1} + \frac{y^2}{4} = 1$$

**27.** Vertices: $(\pm 5, 0)$

$a = 5, c = 2 \Rightarrow b = \sqrt{21}$

Foci: $(\pm 2, 0)$

Horizontal major axis

Center: $(0, 0)$

$$\frac{(x-h)^2}{a^2} + \frac{(y-k)^2}{b^2} = 1$$

$$\frac{x^2}{25} + \frac{y^2}{21} = 1$$

**29.** Foci: $(\pm 5, 0) \Rightarrow c = 5$

Center: $(0, 0)$

Horizontal major axis

Major axis of length $12 \Rightarrow 2a = 12$

$$a = 6$$

$$6^2 - b^2 = 5^2 \Rightarrow b^2 = 11$$

$$\frac{(x-h)^2}{a^2} + \frac{(y-k)^2}{b^2} = 1$$

$$\frac{x^2}{36} + \frac{y^2}{11} = 1$$

**31.** Vertices: $(0, \pm 5) \Rightarrow a = 5$

Center: $(0, 0)$

Vertical major axis

$$\frac{(x-h)^2}{b^2} + \frac{(y-k)^2}{a^2} = 1$$

$$\frac{x^2}{b^2} + \frac{y^2}{25} = 1$$

Point: $(4, 2)$

$$\frac{4^2}{b^2} + \frac{2^2}{25} = 1$$

$$\frac{16}{b^2} = 1 - \frac{4}{25} = \frac{21}{25}$$

$$400 = 21b^2$$

$$\frac{400}{21} = b^2$$

$$\frac{x^2}{400/21} + \frac{y^2}{25} = 1$$

$$\frac{21x^2}{400} + \frac{y^2}{25} = 1$$

**33.** Center: $(2, 3)$

$a = 3, \quad b = 1$

Vertical major axis

$$\frac{(x-h)^2}{b^2} + \frac{(y-k)^2}{a^2} = 1$$

$$\frac{(x-2)^2}{1} + \frac{(y-3)^2}{9} = 1$$

**35.** Center: $(2, 2)$

$a = 3, \quad b = 2$

Horizontal major axis

$$\frac{(x-h)^2}{a^2} + \frac{(y-k)^2}{b^2} = 1$$

$$\frac{(x-2)^2}{9} + \frac{(y-2)^2}{4} = 1$$

**37.** Vertices: $(0, 2), (4, 2) \Rightarrow a = 2$

Minor axis of length $2 \Rightarrow b = 1$

Center: $(2, 2) = (h, k)$

$$\frac{(x - h)^2}{a^2} + \frac{(y - k)^2}{b^2} = 1$$

$$\frac{(x - 2)^2}{4} + \frac{(y - 2)^2}{1} = 1$$

**39.** Foci: $(0, 0), (0, 8) \Rightarrow c = 4$

Major axis of length $16 \Rightarrow a = 8$

$b^2 = a^2 - c^2 = 64 - 16 = 48$

Center: $(0, 4) = (h, k)$

$$\frac{(x - h)^2}{b^2} + \frac{(y - k)^2}{a^2} = 1$$

$$\frac{x^2}{48} + \frac{(y - 4)^2}{64} = 1$$

**41.** Vertices: $(3, 1), (3, 9) \Rightarrow a = 4$

Center: $(3, 5)$

Minor axis of length $6 \Rightarrow b = 3$

Vertical major axis

$$\frac{(x - h)^2}{b^2} + \frac{(y - k)^2}{a^2} = 1$$

$$\frac{(x - 3)^2}{9} + \frac{(y - 5)^2}{16} = 1$$

**43.** Center: $(0, 4)$

Vertices: $(-4, 4), (4, 4) \Rightarrow a = 4$

$a = 2c \Rightarrow 4 = 2c \Rightarrow c = 2$

$2^2 = 4^2 - b^2 \Rightarrow b^2 = 12$

Horizontal major axis

$$\frac{(x - h)^2}{a^2} + \frac{(y - k)^2}{b^2} = 1$$

$$\frac{x^2}{16} + \frac{(y - 4)^2}{12} = 1$$

**45.** (a) The length of the string is $2a$.

(b) The path is an ellipse because the sum of the distances from the two thumbtacks is always the length of the string, that is, it is constant.

**47.**

**49.** $\dfrac{x^2}{a^2} + \dfrac{y^2}{b^2} = 1$

(a) $a + b = 20 \Rightarrow b = 20 - a$

$A = \pi ab = \pi a(20 - a)$

(b) $264 = \pi a(20 - a)$

$0 = -\pi a^2 + 20\pi a - 264$

$0 = \pi a^2 - 20\pi a + 264$

$a = 14$ or $a = 6$. The equation of an ellipse with

an area of 264 is $\dfrac{x^2}{196} + \dfrac{y^2}{36} = 1.$

(c)

| $a$ | 8 | 9 | 10 | 11 | 12 | 13 |
|-----|------|------|------|------|------|------|
| $A$ | 301.6 | 311.0 | 314.2 | 311.0 | 301.6 | 285.9 |

The area is maximum when $a = 10$ and the ellipse is a circle.

(d)

The area is maximum (314.16) when $a = b = 10$ and the ellipse is a circle.

**51.** Vertices: $(\pm 5, 0) \Rightarrow a = 5$

Eccentricity: $\dfrac{3}{5} \Rightarrow c = \dfrac{3}{5}a = 3$

$b^2 = a^2 - c^2 = 25 - 9 = 16$

Center: $(0, 0) = (h, k)$

$\dfrac{(x - h)^2}{a^2} + \dfrac{(y - k)^2}{b^2} = 1$

$$\dfrac{x^2}{25} + \dfrac{y^2}{16} = 1$$

**53.**    $2a = 36.18 \Rightarrow a = 18.09$

$e = \dfrac{c}{a} = 0.97 \Rightarrow c = (18.09)(0.97) = 17.5473$

$b^2 = a^2 - c^2 = (18.09)^2 - (17.5473)^2 \approx 19.34$

The equation of the ellipse is:

$\dfrac{x^2}{(18.09)^2} + \dfrac{y^2}{19.34} = 1$

$\dfrac{x^2}{327.25} + \dfrac{y^2}{19.34} = 1$

**55.**    apogee $= 938 + 6378 = 7316$

perigee $= 212 + 6378 = 6590$

$$e = \dfrac{7316 - 6590}{7316 + 6590} \approx 0.052$$

**57.** False:  The graph of $\dfrac{x^2}{4} + y^4 = 1$ is not an ellipse.

The degree on $y$ is 4, not 2.

**59.** $\dfrac{x^2}{4} + \dfrac{y^2}{1} = 1$

$a = 2, b = 1, c = \sqrt{3}$

Points on the ellipse: $(\pm 2, 0), (0, \pm 1)$

Length of latus recta: $\dfrac{2b^2}{a} = \dfrac{2(1)^2}{2} = 1$

Additional points: $\left(-\sqrt{3}, \pm\dfrac{1}{2}\right), \left(\sqrt{3}, \pm\dfrac{1}{2}\right)$

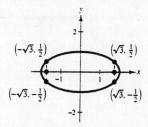

**61.** $9x^2 + 4y^2 = 36$

$\dfrac{x^2}{4} + \dfrac{y^2}{9} = 1$

$a = 3, b = 2, c = \sqrt{5}$

Points on the ellipse: $(\pm 2, 0), (0, \pm 3)$

Length of latus recta: $\dfrac{2b^2}{a} = \dfrac{2 \cdot 2^2}{3} = \dfrac{8}{3}$

Additional points: $\left(\pm\dfrac{4}{3}, -\sqrt{5}\right), \left(\pm\dfrac{4}{3}, \sqrt{5}\right)$

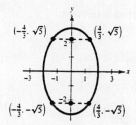

# Section 6.4    Hyperbolas

■ A **hyperbola** is the set of all points $(x, y)$ the difference of whose distances from two distinct fixed points (**foci**) is constant.

■ The standard equation of a hyperbola with center $(h, k)$ and transverse and conjugate axes of lengths $2a$ and $2b$ is:

(a) $\dfrac{(x - h)^2}{a^2} - \dfrac{(y - k)^2}{b^2} = 1$ if the traverse axis is horizontal.

(b) $\dfrac{(y - k)^2}{a^2} - \dfrac{(x - h)^2}{b^2} = 1$ if the traverse axis is vertical.

■ $c^2 = a^2 + b^2$ where $c$ is the distance from the center to a focus.

■ The asymptotes of a hyperbola are:

(a) $y = k \pm \dfrac{b}{a}(x - h)$ if the transverse axis is horizontal.

(b) $y = k \pm \dfrac{a}{b}(x - h)$ the transverse axis is vertical.

■ The eccentricity of a hyperbola is $e = \dfrac{c}{a}$.

■ To classify a nondegenerate conic from its general equation $Ax^2 + Cy^2 + Dx + Ey + F = 0$:
(a) If $A = C$ ($A \neq 0$, $C \neq 0$), then it is a circle.
(b) If $AC = 0$ ($A = 0$ or $C = 0$, but not both), then it is a parabola.
(c) If $AC > 0$, then it is an ellipse.
(d) If $AC < 0$, then it is a hyperbola.

## Solutions to Odd Numbered Exercises

1. $\dfrac{x^2}{16} - \dfrac{y^2}{4} = 1$

   Center: $(0, 0)$

   $a = 4, b = 2$

   Horizontal transverse axis

   Matches graph (b).

3. $\dfrac{y^2}{9} - \dfrac{x^2}{16} = 1$

   Center: $(0, 0)$

   $a = 3, b = 4$

   Vertical transverse axis

   Matches graph (e).

5. $\dfrac{(x - 1)^2}{16} - \dfrac{y^2}{4} = 1$

   Center: $(1, 0)$

   $a = 4, b = 2$

   Horizontal transverse axis

   Matches graph (a).

7. $x^2 - y^2 = 1$

   $a = 1, b = 1, c = \sqrt{2}$

   Center: $(0, 0)$

   Vertices: $(\pm 1, 0)$

   Foci: $(\pm \sqrt{2}, 0)$

   Asymptotes: $y = \pm x$

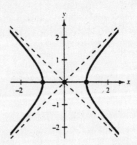

**9.** $\dfrac{y^2}{1} - \dfrac{x^2}{4} = 1$

$a = 1, b = 2, c = \sqrt{5}$

Center: $(0, 0)$

Vertices: $(0, \pm 1)$

Foci: $\left(0, \pm\sqrt{5}\right)$

Asymptotes: $y = \pm\dfrac{1}{2}x$

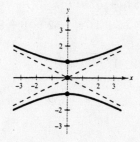

**11.** $\dfrac{y^2}{25} - \dfrac{x^2}{144} = 1$

$a = 5, b = 12, c = 13$

Center: $(0, 0)$

Vertices: $(0, \pm 5)$

Foci: $(0, \pm 13)$

Asymptotes: $y = \pm\dfrac{5}{12}x$

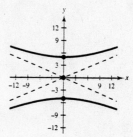

**13.** $\dfrac{(x - 1)^2}{4} - \dfrac{(y + 2)^2}{1} = 1$

$a = 2, b = 1, c = \sqrt{5}$

Center: $(1, -2)$

Vertices: $(-1, -2), (3, -2)$

Foci: $\left(1 \pm \sqrt{5}, -2\right)$

Asymptotes: $y = -2 \pm \dfrac{1}{2}(x - 1)$

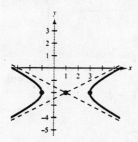

**15.** $(y + 6)^2 - (x - 2)^2 = 1$

$a = 1, b = 1, c = \sqrt{2}$

Center: $(2, -6)$

Vertices: $(2, -5), (2, -7)$

Foci: $\left(2, -6 \pm \sqrt{2}\right)$

Asymptotes: $y = -6 \pm (x - 2)$

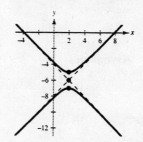

**17.**     $9x^2 - y^2 - 36x - 6y + 18 = 0$

$9\left(x^2 - 4x + 4\right) - \left(y^2 + 6y + 9\right) = -18 + 36 - 9$

$$\dfrac{(x - 2)^2}{1} - \dfrac{(y + 3)^2}{9} = 1$$

$a = 1, b = 3, c = \sqrt{10}$

Center: $(2, -3)$

Vertices: $(1, -3), (3, -3)$

Foci: $\left(2 \pm \sqrt{10}, -3\right)$

Asymptotes: $y = -3 \pm 3(x - 2)$

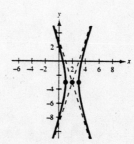

**19.**     $x^2 - 9y^2 + 2x - 54y - 80 = 0$

$(x^2 + 2x + 1) - 9(y^2 + 6y + 9) = 80 + 1 - 81$

$(x + 1)^2 - 9(y + 3)^2 = 0$

$$y + 3 = \pm\frac{1}{3}(x + 1)$$

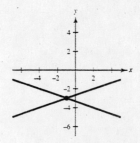

Degenerate hyperbola is two lines intersecting at $(-1, -3)$.

**21.** $2x^2 - 3y^2 = 6$

$\dfrac{x^2}{3} - \dfrac{y^2}{2} = 1$

$a = \sqrt{3}, b = \sqrt{2}, c = \sqrt{5}$

Center: $(0, 0)$

Vertices: $(\pm\sqrt{3}, 0)$

Foci: $(\pm\sqrt{5}, 0)$

Asymptotes: $y = \pm\sqrt{\dfrac{2}{3}}x$

To use a graphing calculator, solve first for $y$.

$y^2 = \dfrac{2x^2 - 6}{3}$

$\left.\begin{array}{l} y_1 = \sqrt{\dfrac{2x^2 - 6}{3}} \\[4mm] y_2 = -\sqrt{\dfrac{2x^2 - 6}{3}} \end{array}\right\}$ Hyperbola

$\left.\begin{array}{l} y_3 = \sqrt{\dfrac{2}{3}}x \\[4mm] y_4 = -\sqrt{\dfrac{2}{3}}x \end{array}\right\}$ Asymptotes

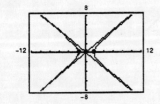

**23.**     $9y^2 - x^2 + 2x + 54y + 62 = 0$

$9(y^2 + 6y + 9) - (x^2 - 2x + 1) = -62 - 1 + 81$

$\dfrac{(y + 3)^2}{2} - \dfrac{(x - 1)^2}{18} = 1$

$a = \sqrt{2}, b = 3\sqrt{2}, c = 2\sqrt{5}$

Center: $(1, -3)$

Vertices: $(1, -3 \pm \sqrt{2})$

Foci: $(1, -3 \pm 2\sqrt{5})$

Asymptotes: $y = -3 \pm \dfrac{1}{3}(x - 1)$

To use a graphing calculator, solve for $y$ first.

$9(y + 3)^2 = 18 + (x - 1)^2$

$y = -3 \pm \sqrt{\dfrac{18 + (x - 1)^2}{9}}$

$\left.\begin{array}{l} y_1 = -3 + \dfrac{1}{3}\sqrt{18 + (x - 1)^2} \\[4mm] y_2 = -3 - \dfrac{1}{3}\sqrt{18 + (x - 1)^2} \end{array}\right\}$ Hyperbola

$\left.\begin{array}{l} y_3 = -3 + \dfrac{1}{3}(x - 1) \\[4mm] y_4 = -3 - \dfrac{1}{3}(x - 1) \end{array}\right\}$ Asymptotes

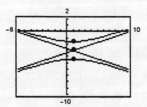

**25.** Vertices: $(0, \pm 3) \Rightarrow a = 3$

Solution point: $(-2, 5)$

Center: $(0, 0) = (h, k)$

$\dfrac{(y - k)^2}{a^2} - \dfrac{(x - h)^2}{b^2} = 1$

$\dfrac{y^2}{9} - \dfrac{x^2}{b^2} = 1 \Rightarrow b^2 = \dfrac{9x^2}{y^2 - 9} = \dfrac{9(-2)^2}{5^2 - 9} = \dfrac{36}{16} = \dfrac{9}{4}$

$\dfrac{y^2}{9} - \dfrac{x^2}{9/4} = 1$

**27.** Vertices: $(0, \pm 2) \Rightarrow a = 2$

Foci: $(0, \pm 4) \Rightarrow c = 4$

$b^2 = c^2 - a^2 = 16 - 4 = 12$

Center: $(0, 0) = (h, k)$

$\dfrac{(y - k)^2}{a^2} - \dfrac{(x - h)^2}{b^2} = 1$

$\dfrac{y^2}{4} - \dfrac{x^2}{12} = 1$

**29.** Vertices: $(\pm 1, 0) \Rightarrow a = 1$

Asymptotes: $y = \pm 3x \Rightarrow \dfrac{b}{a} = 3, b = 3$

Center: $(0, 0) = (h, k)$

$\dfrac{(x - h)^2}{a^2} - \dfrac{(y - k)^2}{b^2} = 1$

$\dfrac{x^2}{1} - \dfrac{y^2}{9} = 1$

**31.** Foci: $(0, \pm 8) \Rightarrow c = 8$

Asymptotes: $y = \pm 4x \Rightarrow \dfrac{a}{b} = 4 \Rightarrow a = 4b$

Center: $(0, 0) = (h, k)$

$c^2 = a^2 + b^2 \Rightarrow 64 = 16b^2 + b^2$

$\dfrac{64}{17} = b^2 \Rightarrow a^2 = \dfrac{1024}{17}$

$\dfrac{(y - k)^2}{a^2} - \dfrac{(x - h)^2}{b^2} = 1$

$\dfrac{y^2}{1024/17} - \dfrac{x^2}{64/17} = 1$

$\dfrac{17y^2}{1024} - \dfrac{17x^2}{65} = 1$

**33.** Vertices: $(0, 0), (0, 2) \Rightarrow a = 1$

Solution point: $\left( \sqrt{3}, 3 \right)$

Center: $(0, 1) = (h, k)$

$\dfrac{(y - k)^2}{a^2} - \dfrac{(x - h)^2}{b^2} = 1$

$(y - 1)^2 - \dfrac{x^2}{b^2} = 1 \Rightarrow 4 - \dfrac{3}{b^2} = 1 \Rightarrow b^2 = 1$

$(y - 1)^2 - x^2 = 1$

**35.** Center: $(3, 2) = (h, k)$

Vertices: $(1, 2), (5, 2) \Rightarrow a = 2$

Solution point: $(0, 0)$

$\dfrac{(x - h)^2}{a^2} - \dfrac{(y - k)^2}{b^2} = 1$

$\dfrac{(x - 3)^2}{4} - \dfrac{(y - 2)^2}{b^2} = 1 \Rightarrow \dfrac{9}{4} - \dfrac{4}{b^2} = 1 \Rightarrow b^2 = \dfrac{16}{5}$

$\dfrac{(x - 3)^2}{4} - \dfrac{(y - 2)^2}{16/5} = 1$

**37.** Vertices: $(2, 0), (6, 0) \Rightarrow a = 2$

Foci: $(0, 0), (8, 0) \Rightarrow c = 4$

$b^2 = c^2 - a^2 = 16 - 4 = 12$

Center: $(4, 0) = (h, k)$

$$\frac{(x - h)^2}{a^2} - \frac{(y - k)^2}{b^2} = 1$$

$$\frac{(x - 4)^2}{4} - \frac{y^2}{12} = 1$$

**39.** Vertices: $(4, 1), (4, 9) \Rightarrow a = 4$

Foci: $(4, 0), (4, 10) \Rightarrow c = 5$

$b^2 = c^2 - a^2 = 25 - 16 = 9$

Center: $(4, 5) = (h, k)$

$$\frac{(y - k)^2}{a^2} - \frac{(x - h)^2}{b^2} = 1$$

$$\frac{(y - 4)^2}{16} - \frac{(x - 4)^2}{9} = 1$$

**41.** Vertices: $(2, 3), (2, -3) \Rightarrow a = 3$

Solution point: $(0, 5)$

Center: $(2, 0) = (h, k)$

$$\frac{(y - k)^2}{a^2} - \frac{(x - h)^2}{b^2} = 1$$

$$\frac{y^2}{9} - \frac{(x - 2)^2}{b^2} = 1 \Rightarrow b^2 = \frac{9(x - 2)^2}{y^2 - 9} = \frac{9(-2)^2}{25 - 9} = \frac{36}{16} = \frac{9}{4}$$

$$\frac{y^2}{9} - \frac{(x - 2)^2}{9/4} = 1$$

**43.** Vertices: $(0, 2), (6, 2) \Rightarrow a = 3$

Asymptotes: $y = \frac{2}{3}x, y = 4 - \frac{2}{3}x$

$$\frac{b}{a} = \frac{2}{3} \Rightarrow b = 2$$

Center: $(3, 2) = (h, k)$

$$\frac{(x - h)^2}{a^2} - \frac{(y - k)^2}{b^2} = 1$$

$$\frac{(x - 3)^2}{9} - \frac{(y - 2)^2}{4} = 1$$

**45.** $x = 3 - \frac{2}{3}\sqrt{1 + (y - 1)^2}$ represents the left branch of the hyperbola.

**47.** Since $\overline{AB} = 1100$ feet and the sound takes one second longer to reach $B$ than $A$, the explosion must occur on the vertical line through $A$ and $B$ below $A$.

Foci: $(\pm 4400, 0) \Rightarrow c = 4400$

Center: $(0, 0) = (h, k)$

$$\frac{\overline{CE}}{1100} - \frac{\overline{AE}}{1100} = 5 \Rightarrow 2a = 5500, a = \frac{5500}{2} = 2750$$

$b^2 = c^2 - a^2 = (4400)^2 - (2750)^2 = 11,797,500$

$$\frac{x^2}{(2750)^2} - \frac{y^2}{11,797,500} = 1$$

$$y^2 = 11,797,500\left(\frac{x^2}{(2750)^2} - 1\right)$$

$$y^2 = 11,797,500\left(\frac{(4400)^2}{(2750)^2} - 1\right) = 18,404,100$$

$$y = -4290$$

The explosion occurs at $(4400, -4290)$.

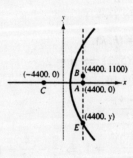

**49.** Center:  $(0, 0) = (h, k)$

Focus:  $(24, 0) \Rightarrow c = 24$

Solution point: $(24, 24)$

$24^2 = a^2 + b^2 \Rightarrow b^2 = 24^2 - a^2$

$$\frac{(x - h)^2}{a^2} - \frac{(y - k)^2}{b^2} = 1$$

$$\frac{x^2}{a^2} - \frac{y^2}{24^2 - a^2} = 1 \Rightarrow \frac{24^2}{a^2} - \frac{24^2}{24^2 - a^2} = 1$$

Solving yields $a^2 = \dfrac{(3 - \sqrt{5})24^2}{2} \approx 220.0124$  and $b^2 \approx 355.9876$.

Thus, we have $\dfrac{x^2}{220.0124} - \dfrac{y^2}{355.9876} = 1$.

The right vertex is at $(a, 0) \approx (14.83, 0)$.

**51.** $x^2 + y^2 - 6x + 4y + 9 = 0$

$A = 1, C = 1$

$A = C \Rightarrow$ Circle

**53.** $4x^2 - y^2 - 4x - 3 = 0$

$A = 4, C = -1$

$AC = 4(-1)$

$\quad = -4 < 0 \Rightarrow$ Hyperbola

**55.** $4x^2 + 3y^2 + 8x - 24y + 51 = 0$

$A = 4, C = 3$

$AC = 4(3) = 12 > 0 \Rightarrow$ Ellipse

**57.** $25x^2 - 10x - 200y - 119 = 0$

$A = 25, C = 0$

$AC = 25(0) = 0 \Rightarrow$ Parabola

**59.** $(x^3 - 3x^2) - (6 - 2x - 4x^2) = x^3 - 3x^2 - 6 + 2x + 4x^2$

$$= x^3 + x^2 + 2x - 6$$

**61.**

$$
\require{enclose}
\begin{array}{r}
x^2 - 2x + 1 + \dfrac{2}{x + 2} \\[2pt]
x + 2 \enclose{longdiv}{x^3 + 0x^2 - 3x + 4} \\
\underline{x^3 + 2x^2} \phantom{aaaaaaaaaa} \\
-2x^2 - 3x \phantom{aaaa} \\
\underline{-2x^2 - 4x} \phantom{aaaa} \\
x + 4 \\
\underline{x + 2} \\
2
\end{array}
$$

Thus, $\dfrac{x^3 - 3x + 4}{x + 2} = x^2 - 2x + 1 + \dfrac{2}{x + 2}$.

# Section 6.5    Rotation of Conics

■ The general second-degree equation $Ax^2 + Bxy + Cy^2 + Dx + Ey + F = 0$ can be rewritten as $A'(x')^2 + C'(y')^2 + D'x' + E'y' + F' = 0$ by rotating the coordinate axes through the angle $\theta$, where $\cot 2\theta = (A - C)/B$.

■ $x = x'\cos\theta - y'\sin\theta$
   $y = x'\sin\theta + y'\cos\theta$

■ The graph of the nondegenerate equation $Ax^2 + Bxy + Cy^2 + Dx + Ey + F = 0$ is:

   (a) An ellipse or circle if $B^2 - 4AC < 0$.

   (b) A parabola if $B^2 - 4AC = 0$.

   (c) A hyperbola if $B^2 - 4AC > 0$.

## Solutions to Odd-Numbered Exercises

**1.** $\theta = 90°$; Point: $(0, 3)$

$x' = x\cos\theta - y\sin\theta = 0(\cos 90°) - 3(\sin 90°) = -3$

$y' = x\sin\theta + y\cos\theta = 0(\sin 90°) + 3(\cos 90°) = 0$

Thus, $(x', y') = (-3, 0)$.

**3.** $\theta = 30°$; Point: $(1, 4)$

$x' = x\cos\theta - y\sin\theta = 1(\cos 30°) - 4(\sin 30°) = \dfrac{\sqrt{3}}{2} - \dfrac{4}{2} = \dfrac{1}{2}(\sqrt{3} - 4)$

$y' = x\sin\theta + y\cos\theta = 1(\sin 30°) + 4(\cos 30°) = \dfrac{1}{2} + \dfrac{4\sqrt{3}}{2} = \dfrac{1}{2}(1 + 4\sqrt{3})$

Thus, $(x', y') = \left(\dfrac{1}{2}(\sqrt{3} - 4), \dfrac{1}{2}(1 + 4\sqrt{3})\right)$.

**5.** $xy + 1 = 0$

$A = 0, B = 1, C = 0$

$\cot 2\theta = \dfrac{A - C}{B} = 0 \Rightarrow 2\theta = \dfrac{\pi}{2} \Rightarrow \theta = \dfrac{\pi}{4}$

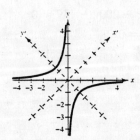

$x = x'\cos\dfrac{\pi}{4} - y'\sin\dfrac{\pi}{4}$        $y = x'\sin\dfrac{\pi}{4} + y'\cos\dfrac{\pi}{4}$

$= x'\left(\dfrac{\sqrt{2}}{2}\right) - y'\left(\dfrac{\sqrt{2}}{2}\right)$        $= x'\left(\dfrac{\sqrt{2}}{2}\right) + y'\left(\dfrac{\sqrt{2}}{2}\right)$

$= \dfrac{x' - y'}{\sqrt{2}}$             $= \dfrac{x' + y'}{\sqrt{2}}$

$$xy + 1 = 0$$

$$\left(\dfrac{x' - y'}{\sqrt{2}}\right)\left(\dfrac{x' + y'}{\sqrt{2}}\right) + 1 = 0$$

$$\dfrac{(y')^2}{2} - \dfrac{(x')^2}{2} = 1$$

**7.** $x^2 - 10xy + y^2 + 1 = 0$

$A = 1, B = -10, C = 1$

$\cot 2\theta = \dfrac{A-C}{B} = 0 \Rightarrow 2\theta = \dfrac{\pi}{2} \Rightarrow \theta = \dfrac{\pi}{4}$

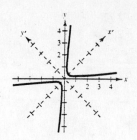

$x = x' \cos \dfrac{\pi}{4} - y' \sin \dfrac{\pi}{4}$ $\qquad\qquad$ $y = x' \sin \dfrac{\pi}{4} + y' \cos \dfrac{\pi}{4}$

$\quad = x'\left(\dfrac{\sqrt{2}}{2}\right) - y'\left(\dfrac{\sqrt{2}}{2}\right)$ $\qquad\qquad = x'\left(\dfrac{\sqrt{2}}{2}\right) + y'\left(\dfrac{\sqrt{2}}{2}\right)$

$\quad = \dfrac{x'-y'}{\sqrt{2}}$ $\qquad\qquad\qquad\quad = \dfrac{x'+y'}{\sqrt{2}}$

$x^2 - 10xy + y^2 + 1 = 0$

$\qquad \left(\dfrac{x'-y'}{\sqrt{2}}\right)^2 - 10\left(\dfrac{x'-y'}{\sqrt{2}}\right)\left(\dfrac{x'+y'}{\sqrt{2}}\right) + \left(\dfrac{x'+y'}{\sqrt{2}}\right)^2 + 1 = 0$

$\dfrac{(x')^2}{2} - x'y' + \dfrac{(y')^2}{2} - 5(x')^2 + 5(y')^2 + \dfrac{(x')^2}{2} + x'y' + \dfrac{(y')^2}{2} + 1 = 0$

$\qquad\qquad\qquad\qquad\qquad -4(x')^2 + 6(y')^2 = -1$

$\qquad\qquad\qquad\qquad\qquad\quad \dfrac{(x')^2}{1/4} - \dfrac{(y')^2}{1/6} = 1$

**9.** $xy - 2y - 4x = 0$

$A = 0, B = 1, C = 0$

$\cot 2\theta = \dfrac{A-C}{B} = 0 \Rightarrow 2\theta = \dfrac{\pi}{2} \Rightarrow \theta = \dfrac{\pi}{4}$

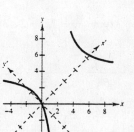

$x = x' \cos \dfrac{\pi}{4} - y' \sin \dfrac{\pi}{4}$ $\qquad\qquad$ $y = x' \sin \dfrac{\pi}{4} + y' \cos \dfrac{\pi}{4}$

$\quad = x'\left(\dfrac{\sqrt{2}}{2}\right) - y'\left(\dfrac{\sqrt{2}}{2}\right)$ $\qquad\qquad = x'\left(\dfrac{\sqrt{2}}{2}\right) + y'\left(\dfrac{\sqrt{2}}{2}\right)$

$\quad = \dfrac{x'-y'}{\sqrt{2}}$ $\qquad\qquad\qquad\quad = \dfrac{x'+y'}{\sqrt{2}}$

$xy - 2y - 4x = 0$

$\qquad \left(\dfrac{x'-y'}{\sqrt{2}}\right)\left(\dfrac{x'+y'}{\sqrt{2}}\right) - 2\left(\dfrac{x'+y'}{\sqrt{2}}\right) - 4\left(\dfrac{x'-y'}{\sqrt{2}}\right) = 0$

$\qquad \dfrac{(x')^2}{2} - \dfrac{(y')^2}{2} - \sqrt{2}x' - \sqrt{2}y' - 2\sqrt{2}x' + 2\sqrt{2}y' = 0$

$\left[(x')^2 - 6\sqrt{2}x' + (3\sqrt{2})^2\right] - \left[(y')^2 - 2\sqrt{2}y' + (\sqrt{2})^2\right] = 0 + (3\sqrt{2})^2 - (\sqrt{2})^2$

$\qquad\qquad\qquad (x' - 3\sqrt{2})^2 - (y' - \sqrt{2})^2 = 16$

$\qquad\qquad\qquad \dfrac{(x' - 3\sqrt{2})^2}{16} - \dfrac{(y' - \sqrt{2})^2}{16} = 1$

**11.** $5x^2 - 2xy + 5y^2 - 12 = 0$

$A = 5, B = -2, C = 5$

$$\cot 2\theta = \frac{A - C}{B} = 0 \Rightarrow 2\theta = \frac{\pi}{2} \Rightarrow \theta = \frac{\pi}{4}$$

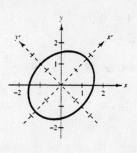

$x = x'\cos\dfrac{\pi}{4} - y'\sin\dfrac{\pi}{4}$ $\qquad$ $y = x'\sin\dfrac{\pi}{4} + y'\cos\dfrac{\pi}{4}$

$\quad = x'\left(\dfrac{\sqrt{2}}{2}\right) - y'\left(\dfrac{\sqrt{2}}{2}\right)$ $\qquad = x'\left(\dfrac{\sqrt{2}}{2}\right) + y'\left(\dfrac{\sqrt{2}}{2}\right)$

$\quad = \dfrac{x' - y'}{\sqrt{2}}$ $\qquad\qquad\quad = \dfrac{x' + y'}{\sqrt{2}}$

$$5x^2 - 2xy + 5y^2 - 12 = 0$$

$$5\left(\frac{x' - y'}{\sqrt{2}}\right)^2 - 2\left(\frac{x' - y'}{\sqrt{2}}\right)\left(\frac{x' + y'}{\sqrt{2}}\right) + 5\left(\frac{x' + y}{\sqrt{2}}\right)^2 - 12 = 0$$

$$\frac{5(x')^2}{2} - 5x'y' + \frac{5(y')^2}{2} - (x')^2 + (y')^2 + \frac{5(x')^2}{2} + 5x'y' + \frac{5(y')^2}{2} - 12 = 0$$

$$4(x')^2 + 6(y')^2 = 12$$

$$\frac{(x')^2}{3} + \frac{(y')^2}{2} = 1$$

**13.** $3x^2 - 2\sqrt{3}xy + y^2 + 2x + 2\sqrt{3}y = 0$

$A = 3, B = -2\sqrt{3}, C = 1$

$$\cot 2\theta = \frac{A - C}{B} = -\frac{1}{\sqrt{3}} \Rightarrow \theta = 60°$$

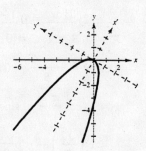

$x = x'\cos 60° - y'\sin 60°$ $\qquad\qquad$ $y = x'\sin 60° + y'\cos 60°$

$\quad = x'\left(\dfrac{1}{2}\right) - y'\left(\dfrac{\sqrt{3}}{2}\right) = \dfrac{x' - \sqrt{3}y'}{2}$ $\qquad = x'\left(\dfrac{\sqrt{3}}{2}\right) + y'\left(\dfrac{1}{2}\right) = \dfrac{\sqrt{3}x' - y'}{2}$

$$3x^2 - 2\sqrt{3}xy + y^2 + 2x + 2\sqrt{3}y = 0$$

$$3\left(\frac{x' - \sqrt{3}y'}{2}\right)^2 - 2\sqrt{3}\left(\frac{x' - \sqrt{3}y'}{2}\right)\left(\frac{\sqrt{3}x' + y'}{2}\right) + \left(\frac{\sqrt{3}x' + y'}{2}\right)^2 + 2\left(\frac{x' - \sqrt{3}y'}{2}\right) + 2\sqrt{3}\left(\frac{\sqrt{3}x' + y'}{2}\right) = 0$$

$$\frac{3(x')^2}{4} - \frac{6\sqrt{3}x'y'}{4} + \frac{9(y')^2}{4} - \frac{6(x')^2}{4} + \frac{4\sqrt{3}x'y'}{4} + \frac{6(y')^2}{4} + \frac{3(x')^2}{4} + \frac{2\sqrt{3}x'y'}{4} + \frac{(y')^2}{4}$$

$$+ x' - \sqrt{3}y' + 3x' + \sqrt{3}y' = 0$$

$$4(y')^2 + 4x' = 0$$

$$x' = -(y')^2$$

**15.** $9x^2 + 24xy + 16y^2 + 90x - 130y = 0$

$A = 9, B = 24, C = 16$

$\cot 2\theta = \dfrac{A - C}{B} = -\dfrac{7}{24} \Rightarrow \theta \approx 53.13°$

$\cos 2\theta = -\dfrac{7}{25}$

$\sin \theta = \sqrt{\dfrac{1 - \cos \theta}{2}} = \sqrt{\dfrac{1 - (-7/25)}{2}} = \dfrac{4}{5}$

$\cos \theta = \sqrt{\dfrac{1 + \cos 2\theta}{2}} = \sqrt{\dfrac{1 + (-7/25)}{2}} = \dfrac{3}{5}$

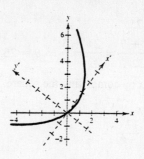

$x = x'\cos \theta - y'\sin \theta$ $\qquad\qquad$ $y = x'\sin \theta + y'\cos \theta$

$\quad = x'\left(\dfrac{3}{5}\right) - y'\left(\dfrac{4}{5}\right) = \dfrac{3x' - 4y'}{5}$ $\qquad$ $= x'\left(\dfrac{4}{5}\right) + y'\left(\dfrac{3}{5}\right)$

$\qquad\qquad\qquad\qquad\qquad\qquad\qquad\qquad\qquad = \dfrac{4x' + 3y'}{5}$

$$9x^2 + 24xy + 16y^2 + 90x - 130y = 0$$

$$9\left(\dfrac{3x' - 4y'}{5}\right)^2 + 24\left(\dfrac{3x' - 4y'}{5}\right)\left(\dfrac{4x' + 3y'}{5}\right) + 16\left(\dfrac{3x' - 4y'}{5}\right)^2 + 90\left(\dfrac{3x' - 4y'}{5}\right) - 130\left(\dfrac{4x' + 3y'}{5}\right) = 0$$

$$\dfrac{81(x')^2}{25} - \dfrac{216x'y'}{25} + \dfrac{144(y')^2}{25} + \dfrac{288(x')^2}{25} - \dfrac{168x'y'}{25} - \dfrac{288(y')^2}{25} + \dfrac{256(x')^2}{25} + \dfrac{384x'y'}{25}$$

$$+ \dfrac{144(y')^2}{25} + 54x' - 72y' - 104x' - 78y' = 0$$

$$25(x')^2 - 50x' - 150y' = 0$$

$$(x')^2 - 2x' + 1 = 6y' + 1$$

$$y' = \dfrac{(x')^2}{6} - \dfrac{x'}{3}$$

**17.** $x^2 + xy + y^2 = 10$

$\cot 2\theta = \dfrac{A - C}{B} = \dfrac{1 - 1}{0} = 0 \Rightarrow \theta = \dfrac{\pi}{4}$ or $45°$

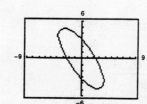

To graph the conic using a graphing calculator, we need to solve for $y$ in terms of $x$.

$y^2 + xy = 10 - x^2$

$y^2 + xy + \dfrac{x^2}{4} = 10 - x^2 + \dfrac{x^2}{4}$

$\left(y + \dfrac{x}{2}\right)^2 = \dfrac{40 - 3x^2}{4}$

$y = -\dfrac{x}{2} \pm \dfrac{\sqrt{40 - 3x^2}}{2}$

Enter $y_1 = \dfrac{-x + \sqrt{40 - 3x^2}}{2}$

and $y_2 = \dfrac{-x - \sqrt{40 - 3x^2}}{2}$.

**19.** $17x^2 + 32xy - 7y^2 = 75$

$$\cot 2\theta = \frac{A - C}{B} = \frac{17 + 7}{32} = \frac{24}{32} = \frac{3}{4} \Rightarrow \theta \approx 26.57°$$

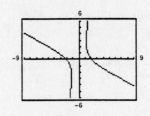

Solve for $y$ in terms of $x$ by completing the square.

$$-7y^2 + 32xy = -17x^2 + 75$$

$$y^2 - \frac{32}{7}xy = \frac{17}{7}x^2 - \frac{75}{7}$$

$$y^2 - \frac{32}{7}xy + \frac{256}{49}x^2 = \frac{119}{49}x^2 - \frac{525}{49} + \frac{256}{49}x^2$$

$$\left(y - \frac{16}{7}x\right)^2 = \frac{375x^2 - 525}{49}$$

$$y = \frac{16}{7}x \pm \sqrt{\frac{375x^2 - 525}{49}}$$

$$y = \frac{16x \pm 5\sqrt{15x^2 - 21}}{7}$$

Use $y_1 = \dfrac{16x + 5\sqrt{15x^2 - 21}}{7}$

and $y_2 = \dfrac{16x - 5\sqrt{15x^2 - 21}}{7}$.

**21.** $32x^2 + 50xy + 7y^2 = 52$

$$\cot 2\theta = \frac{A - C}{B} = \frac{32 - 7}{50} = \frac{1}{2} \Rightarrow \theta \approx 31.72°$$

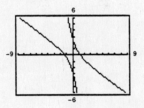

Solve for $y$ in terms of $x$ by completing the square.

$$7y^2 + 50xy = 52 - 32x^2$$

$$y^2 + \frac{50}{7}xy = \frac{52 - 32x^2}{7}$$

$$y^2 + \frac{50}{7}xy + \frac{625}{49}x^2 = \frac{52 - 32x^2}{7} + \frac{625x^2}{49}$$

$$\left(y + \frac{25}{7}x\right)^2 = \frac{364 + 401x^2}{49}$$

$$y = -\frac{25x}{7} \pm \frac{\sqrt{364 + 401x^2}}{7}$$

Enter $y_1 = \dfrac{-25x + \sqrt{364 + 401x^2}}{7}$

and $y_2 = \dfrac{-25x - \sqrt{364 + 401x^2}}{7}$.

**23.** $xy + 3 = 0$

$B^2 - 4AC = 1 \Rightarrow$ The graph is a hyperbola.

$$\cot 2\theta = \frac{A - C}{B} = 0 \Rightarrow \theta = 45°$$

Matches graph (e).

**25.** $-2x^2 + 3xy + 2y^2 + 3 = 0$

$B^2 - 4AC = (3)^2 - 4(-2)(2) = 25 \Rightarrow$ The graph is a hyperbola.

$\cot 2\theta = \dfrac{A - C}{B} = -\dfrac{4}{3} \Rightarrow \theta \approx -18.43°$

Matches graph (b).

**27.** $3x^2 + 2xy + y^2 - 10 = 0$

$B^2 - 4AC = (2)^2 - 4(3)(1) = -8 \Rightarrow$ The graph is an ellipse or circle.

$\cot 2\theta = \dfrac{A - C}{B} = 1 \Rightarrow \theta = 22.5°$

Matches graph (d).

**29.** $16x^2 - 24xy + 9y^2 - 30x - 40y = 0$

$B^2 - 4AC = (-24)^2 - 4(16)(9) = 0$

Parabola

**31.** $13x^2 - 8xy + 7y^2 - 45 = 0$

$B^2 - 4AC = (-8)^2 - 4(13)(7) = -300$

Ellipse or circle

**33.** $x^2 - 6xy - 5y^2 + 4x - 22 = 0$

$B^2 - 4AC = (-6)^2 - 4(1)(-5) = 56$

Hyperbola

**35.** $x^2 + 4xy + 4y^2 - 5x - y - 3 = 0$

$B^2 - 4AC = (4)^2 - 4(1)(4) = 0$

Parabola

**37.** $y^2 - 4x^2 = 0$

$\qquad y^2 = 4x^2$

$\qquad y = \pm 2x$

Two intersecting lines

**39.** $x^2 + 2xy + y^2 - 1 = 0$

$\qquad (x + y)^2 - 1 = 0$

$\qquad (x + y)^2 = 1$

$\qquad x + y = \pm 1$

$\qquad\qquad y = -x \pm 1$

Two parallel lines

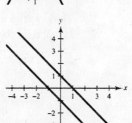

**41.**
$$-x^2 + y^2 + 4x - 6y + 4 = 0 \Rightarrow (y - 3)^2 - (x - 2)^2 = 1$$
$$\underline{x^2 + y^2 - 4x - 6y + 12 = 0 \Rightarrow (x - 2)^2 + (y - 3)^2 = 1}$$
$$2y^2 - 12y + 16 = 0$$
$$2(y - 2)(y - 4) = 0$$
$$y = 2 \text{ or } y = 4$$

For $y = 2$: $x^2 + 2^2 - 4x - 6(2) + 12 = 0$

$\qquad\qquad\qquad x^2 - 4x + 4 = 0$

$\qquad\qquad\qquad (x - 2)^2 = 0$

$\qquad\qquad\qquad\qquad x = 2$

For $y = 4$: $x^2 + 4^2 - 4x - 6(4) + 12 = 0$

$\qquad\qquad\qquad x^2 - 4x + 4 = 0$

$\qquad\qquad\qquad (x - 2)^2 = 0$

$\qquad\qquad\qquad\qquad x = 2$

The points of intersection are $(2, 2)$ and $(2, 4)$.

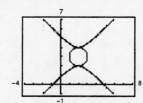

**43.**
$$-4x^2 - y^2 - 32x + 24y - 64 = 0$$
$$\underline{4x^2 + y^2 + 56x - 24y + 304 = 0}$$
$$24x \qquad\qquad + 240 = 0$$
$$x = -10$$

When $x = -10$: $4(-10)^2 + y^2 + 56(-10) - 24y + 304 = 0$
$$y^2 - 24y + 144 = 0$$
$$(y - 12)^2 = 0$$
$$y = 12$$

The point of intersection is $(-10, 12)$.
In standard form the equations are:
$$\frac{(x + 4)^2}{36} + \frac{(y - 12)^2}{144} = 1$$
$$\frac{(x + y)^2}{9} + \frac{(y - 12)^2}{36} = 1$$

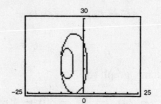

**45.**
$$x^2 - y^2 - 12x + 12y - 36 = 0$$
$$\underline{x^2 + y^2 - 12x - 12y + 36 = 0}$$
$$2x^2 - 24x = 0$$
$$2x(x - 12) = 0$$
$$x = 0 \text{ or } x = 12$$

When $x = 0$: $y^2 - 12y + 36 = 0$
$$(y - 6)^2 = 0$$
$$y = 6$$

When $x = 12$: $12^2 + y^2 - 12(12) - 12y + 36 = 0$
$$y^2 - 12y + 36 = 0$$
$$(y - 6)^2 = 0$$
$$y = 6$$

The points of intersection are $(0, 6)$ and $(12, 6)$.
In standard form the equations are:
$$\frac{(x - 6)^2}{36} + \frac{(y - 6)^2}{36} = 1$$
$$(x - 6)^2 + (y - 6)^2 = 36$$

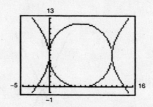

**47.**
$$-16x^2 - y^2 + 24y - 80 = 0$$
$$\underline{16x^2 + 25y^2 - 400 = 0}$$
$$24y^2 + 24y - 480 = 0$$
$$24(y + 5)(y - 4) = 0$$
$$y = -5 \text{ or } y = 4$$

When $y = -5$: $16x^2 + 25(-5)^2 - 400 = 0$
$$16x^2 = -225$$
No real solution

When $y = 4$: $16x^2 + 25(4)^2 - 400 = 0$
$$16x^2 = 0$$
$$x = 0$$

The point of intersection is $(0, 4)$.
In standard form the equations are:
$$\frac{x^2}{4} + \frac{(y - 12)^2}{64} = 1$$
$$\frac{x^2}{25} + \frac{y^2}{16} = 1$$

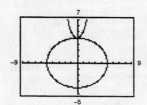

**49.** $x^2 + y^2 - 25 = 0 \Rightarrow y^2 = 25 - x^2$

$\qquad 9x - 4y^2 = 0 \Rightarrow 9x - 4(25 - x^2) = 0$

$\qquad\qquad\qquad\qquad 4x^2 + 9x - 100 = 0$

$\qquad\qquad\qquad\qquad (4x + 25)(x - 4) = 0$

$\qquad\qquad\qquad\qquad x = -\frac{25}{4} \text{ or } x = 4$

When $x = -\frac{25}{4}$: $\quad y^2 = 25 - \left(-\frac{25}{4}\right)^2$

$\qquad\qquad\qquad y^2 = -\frac{225}{16}$

$\qquad\qquad$ No real solution

When $x = 4$: $\quad y^2 = 25 - 4^2$

$\qquad\qquad\qquad y^2 = 9$

$\qquad\qquad\qquad y = \pm 3$

The points of intersection are $(4, 3)$, $(4, -3)$.

In standard form the equations are:

$x^2 + y^2 = 25$

$\qquad y^2 = \frac{9}{4}x$

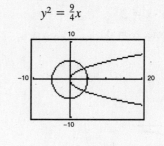

**51.** $\qquad\qquad\qquad x^2 + 2y^2 - 4x + 6y - 5 = 0$

$\qquad\qquad\qquad\qquad x + y + 5 = 0 \Rightarrow y = -x - 5$

$\qquad x^2 + 2(-x - 5)^2 - 4x + 6(-x - 5) - 5 = 0$

$\qquad x^2 + 2x^2 + 20x + 50 - 4x - 6x - 30 - 5 = 0$

$\qquad\qquad\qquad\qquad 3x^2 + 10x + 15 = 0$

No real solution

No points of intersection

In standard form we have:

$$\frac{(x - 2)^2}{27/2} + \frac{(y + 3/2)^2}{27/4} = 1$$

$$x + y = -5$$

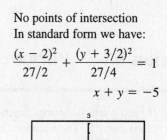

**53.** $\qquad\qquad\qquad xy + x - 2y + 3 = 0 \Rightarrow y = \dfrac{-x - 3}{x - 2}$

$\qquad\qquad\qquad x^2 + 4y^2 - 9 = 0$

$\qquad\qquad\qquad x^2 + 4\left(\dfrac{-x - 3}{x - 2}\right)^2 = 9$

$\qquad\qquad x^2(x - 2)^2 + 4(-x - 3)^2 = 9(x - 2)^2$

$\qquad x^2(x^2 - 4x + 4) + 4(x^2 + 6x + 9) = 9(x^2 - 4x + 4)$

$\qquad x^4 - 4x^3 + 4x^2 + 4x^2 + 24x + 36 = 9x^2 - 36x + 36$

$\qquad\qquad\qquad x^4 - 4x^3 - x^2 + 60x = 0$

$\qquad\qquad\qquad x(x + 3)(x^2 - 7x + 20) = 0$

$\qquad\qquad\qquad\qquad x = 0 \text{ or } x = -3$

Note: $x^2 - 7x + 20 = 0$ has no real solution.

When $x = 0$: $\quad y = \dfrac{-0 - 3}{0 - 2} = \dfrac{3}{2}$

When $x = -3$: $\quad y = \dfrac{-(-3) - 3}{-3 - 2} = 0$

The points of intersection are $\left(0, \dfrac{3}{2}\right)$, $(-3, 0)$.

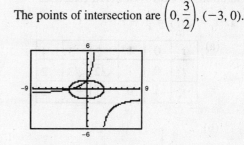

**55.** $(x')^2 + (y')^2 = (x \cos \theta + y \sin \theta)^2 + (y \cos \theta - x \sin \theta)^2$

$\qquad\qquad = x^2 \cos^2 \theta + 2xy \cos \theta \sin \theta + y^2 \sin^2 \theta + y^2 \cos^2 \theta - 2xy \cos \theta \sin \theta + x^2 \sin^2 \theta$

$\qquad\qquad = x^2(\cos^2 \theta + \sin^2 \theta) + y^2(\sin^2 \theta + \cos^2 \theta) = x^2 + y^2 = r^2$

**57.** $g(x) = \dfrac{2}{2-x}$

$y$-intercept:  (0, 1)

Vertical asymptote:  $x = 2$

Horizontal asymptote:  $y = 0$

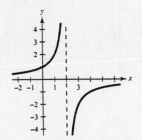

**59.** $h(t) = \dfrac{t^2}{2-t} = -t - 2 + \dfrac{4}{2-t}$

Intercept:  (0, 0)

Vertical asymptote:  $t = 2$

Slant asymptote:  $y = -t - 2$

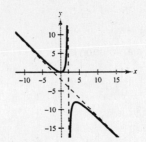

# Section 6.6    Parametric Equations

---

- If $f$ and $g$ are continuous functions of $t$ on an interval $I$, then the set of ordered pairs $(f(t),\ g(t))$ is a *plane curve C.* The equations $x = f(t)$ and $y = g(t)$ are *parametric equations* for $C$ and $t$ is the *parameter.*
- To eliminate the parameter:
  - (a) Solve for $t$ in one equation and substitute into the second equation.
  - (b) Use trigonometric identities.
- You should be able to find the parametric equations for a graph.

---

**Solutions to Odd-Numbered Exercises**

**1.** $x = \sqrt{t},\ y = 1 - t$

(a)

| $t$ | 0 | 1 | 2 | 3 | 4 |
|---|---|---|---|---|---|
| $x$ | 0 | 1 | $\sqrt{2}$ | $\sqrt{3}$ | 2 |
| $y$ | 1 | 0 | $-1$ | $-2$ | $-3$ |

(b)

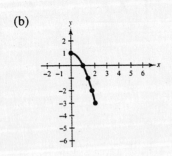

(c)  $x = \sqrt{t} \ \Rightarrow\ x^2 = t$

  $y = 1 - t \ \Rightarrow\ y = 1 - x^2$

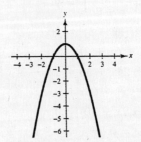

The graph of the parametric equations only shows the right half of the parabola, whereas the rectangular equation yields the entire parabola.

**3.** $x = t$
$\quad y = -2t \Rightarrow y = -2x$
$\qquad 2x + y = 0$

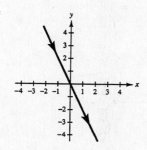

**5.** $x = 3t - 1 \Rightarrow t = \dfrac{x + 1}{3}$
$\quad y = 2t + 1 \Rightarrow y = 2\left(\dfrac{x + 1}{3}\right) + 1$
$\qquad 2x - 3y + 5 = 0$

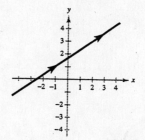

**7.** $x = \frac{1}{4}t \Rightarrow t = 4x$
$\quad y = t^2 \Rightarrow y = 16x^2$

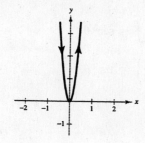

**9.** $x = t + 1 \Rightarrow t = x - 1$
$\quad y = t^2 \qquad \Rightarrow y = (x - 1)^2$

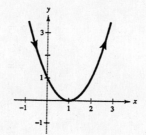

**11.** $x = t^3 \Rightarrow t = x^{1/3}$
$\quad y = \frac{1}{2}t^2 \Rightarrow y = \frac{1}{2}x^{2/3}$
$\qquad\qquad y = \frac{1}{2}\sqrt[3]{x^2}$

**13.** $\quad x = 2t \Rightarrow t = \dfrac{x}{2}$

$\quad y = |t - 2| \Rightarrow y = \left|\dfrac{x}{2} - 2\right|$

$\qquad\qquad y = \dfrac{1}{2}|x - 4|$

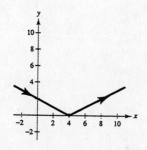

**15.** $x = 3\cos\theta \Rightarrow \left(\dfrac{x}{3}\right)^2 = \cos^2\theta$

$\quad y = 3\sin\theta \Rightarrow \left(\dfrac{y}{3}\right)^2 = \sin^2\theta$

$\quad \left(\dfrac{x}{3}\right)^2 + \left(\dfrac{y}{3}\right)^2 = 1$
$\qquad\quad x^2 + y^2 = 9$

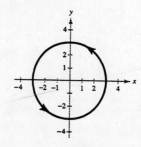

**17.** $x = 4 \sin 2\theta \Rightarrow \left(\dfrac{x}{4}\right)^2 = \sin^2 2\theta$

$y = 2 \cos 2\theta \Rightarrow \left(\dfrac{y}{2}\right)^2 = \cos^2 2\theta$

$\left(\dfrac{x}{4}\right)^2 + \left(\dfrac{y}{3}\right)^2 = 1$

$\dfrac{x^2}{16} + \dfrac{y^2}{4} = 1$

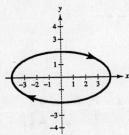

**19.** $x = 4 + 2 \cos \theta \Rightarrow \left(\dfrac{x-4}{2}\right)^2 = \cos^2 \theta$

$y = -1 + \sin \theta \Rightarrow (y + 1)^2 = \sin^2 \theta$

$\dfrac{(x-4)^2}{4} + \dfrac{(y+1)^2}{1} = 1$

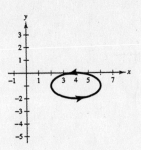

**21.** $x = 4 + 2 \cos \theta \Rightarrow \left(\dfrac{x-4}{2}\right)^2 = \cos^2 \theta$

$y = -1 + 4 \sin \theta \Rightarrow \left(\dfrac{y+1}{4}\right)^2 = \sin^2 \theta$

$\dfrac{(x-4)^2}{4} + \dfrac{(y+1)^2}{16} = 1$

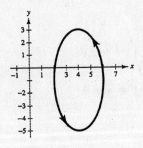

**23.** $x = 4 \sec \theta \Rightarrow \left(\dfrac{x}{4}\right)^2 = \sec^2 \theta$

$y = 3 \tan \theta \Rightarrow \left(\dfrac{y}{3}\right)^2 = \tan^2 \theta$

$1 + \left(\dfrac{y}{3}\right)^2 = \left(\dfrac{x}{4}\right)^2$

$\dfrac{x^2}{16} - \dfrac{y^2}{9} = 1$

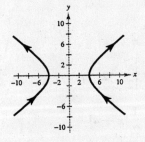

**25.** $x = e^{-1} \Rightarrow \dfrac{1}{x} = e^t$

$y = e^{3t} \Rightarrow y = (e^t)^3$

$y = \left(\dfrac{1}{x}\right)^3$

$y = \dfrac{1}{x^3}, \; x > 0, \, y > 0$

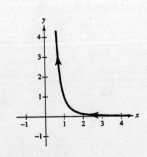

**27.** $x = t^3 \quad \Rightarrow x^{1/3} = t$

$y = 3 \ln t \Rightarrow y = \ln t^3$

$y = \ln(x^{1/3})^3$

$y = \ln x$

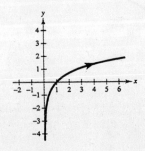

**29.** By eliminating the parameter, each curve becomes $y = 2x + 1$.

(a)   $x = t$

$y = 2t + 1$

There are no restrictions on $x$ and $y$.

Domain: $(-\infty, \infty)$

Orientation: Left to right

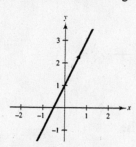

(b)   $x = \cos \theta \qquad \Rightarrow -1 \le x \le 1$

$y = 2 \cos \theta + 1 \Rightarrow -1 \le y \le 3$

The graph oscillates.

Domain: $[-1, 1]$

Orientation: Depends on $\theta$

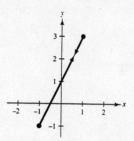

(c)   $x = e^{-t} \qquad \Rightarrow x > 0$

$y = 2e^{-t} + 1 \Rightarrow y > 1$

Domain: $(0, \infty)$

Orientation: Downward or right to left

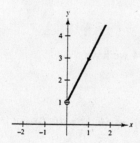

(d)   $x = e^{t} \qquad \Rightarrow x > 0$

$y = 2e^{t} + 1 \Rightarrow y > 1$

Domain: $(0, \infty)$

Orientation: Upward or left to right

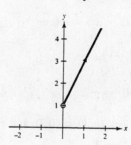

**31.** (a)   $x = \cos \theta$

$y = \sin \theta$

$x^2 + y^2 = 1$

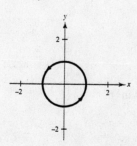

Oriented counterclockwise

(b)   $x = \sin \theta$

$y = \cos \theta$

$x^2 + y^2 = 1$

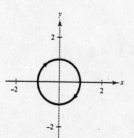

Oriented clockwise

**—CONTINUED—**

**31. —CONTINUED—**

(c)  $x = \sin^2 \theta$

$y = \cos^2 \theta$

$x + y = 1, 0 \le x \le 1, 0 \le y \le 1$

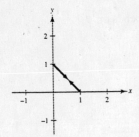

Oscillates

(d)  $x = -\cos \theta$

$y = \sin \theta$

$x^2 + y^2 = 1$

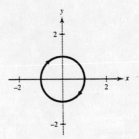

Oriented clockwise

**33.**  $x = x_1 + t(x_2 - x_1), y = y_1 + t(y_2 - y_1)$

$\dfrac{x - x_1}{x_2 - x_1} = t$

$y = y_1 + \left(\dfrac{x - x_1}{x_2 - x_1}\right)(y_2 - y_1)$

$y - y_1 = \dfrac{y_2 - y_1}{x_2 - x_1}(x - x_1) = m(x - x_1)$

**35.**  $x = h + a \cos \theta, y = k + b \sin \theta$

$\dfrac{x - h}{a} = \cos \theta, \dfrac{y - k}{b} = \sin \theta$

$\dfrac{(x - h)^2}{a^2} + \dfrac{(y - k)^2}{b^2} = 1$

**37.** From Exercise 33 we have:

$x = 0 + t(5 - 0) = 5t$

$y = 0 + t(-2 - 0) = -2t$

**39.** From Exercise 34 we have:

$x = 2 + 4 \cos \theta$

$y = 1 + 4 \sin \theta$

**41.** Vertices:  $(\pm 5, 0) \Rightarrow (h, k) = (0, 0)$ and $a = 5$

Foci:  $(\pm 4, 0) \Rightarrow c = 4$

$c^2 = a^2 - b^2 \Rightarrow 16 = 25 - b^2 \Rightarrow b = 3$

From Exercise 35 we have:

$x = 5 \cos \theta$

$y = 3 \sin \theta$

**43.** Vertices:  $(\pm 4, 0) \Rightarrow (h, k) = (0, 0)$ and $a = 4$

Foci:  $(\pm 5, 0) \Rightarrow c = 5$

$c^2 = a^2 + b^2 \Rightarrow 25 = 16 + b^2 \Rightarrow b = 3$

From Exercise 36 we have:

$x = 4 \sec \theta$

$y = 3 \tan \theta$

**45.**  $y = 3x - 2$

Examples

$x = t, \qquad y = 3t - 2$

$x = \dfrac{t}{3}, \qquad y = t - 2$

$x = \dfrac{t + 2}{3}, \quad y = t$

$x = 2t, \qquad y = 6t - 2$

**47.**  $y = x^3$

Examples

$x = t, \qquad y = t^3$

$x = \sqrt[3]{t}, \quad y = t$

$x = \tan t, \quad y = \tan^3 t$

**49.** $x = 2(\theta - \sin \theta)$

$y = 2(1 - \cos \theta)$

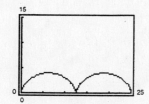

**51.** $x = \theta - \dfrac{3}{2}\sin \theta$

$y = 1 - \dfrac{3}{2}\cos \theta$

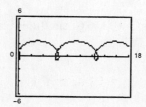

**53.** $x = 3\cos^3 \theta$

$y = 3\sin^3 \theta$

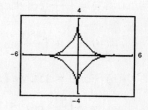

**55.** $x = 2\cot \theta$

$y = 2\sin^2 \theta$

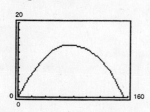

**57.** $x = 2\cos\theta \Rightarrow -2 \le x \le 2$

$y = \sin 2\theta \Rightarrow -1 \le y \le 1$

Matches graph (b).

Domain: $[-2, 2]$

Range: $[-1, 1]$

**59.** $x = \dfrac{1}{2}(\cos \theta + \theta \sin \theta)$

$y = \dfrac{1}{2}(\sin \theta - \theta \cos \theta)$

Matches graph (d).

Domain: $(-\infty, \infty)$

Range: $(-\infty, \infty)$

**61.** $x = (v_0 \cos \theta)t$

$y = h + (v_0 \sin \theta)t - 16t^2$

(a) $\theta = 20°$, $v_0 = 88$ ft/sec
Maximum height: 14.2 ft
Range: 155.6 ft

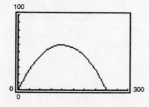

(b) $\theta = 20°$, $v_0 = 132$ ft/sec
Maximum height: 31.8 ft
Range: 350.0 ft

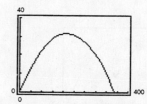

(c) $\theta = 245°$, $v_0 = 88$ ft/sec
Maximum height: 60.5 ft
Range: 242.0 ft

(d) $\theta = 45°$, $v_0 = 132$ ft/sec
Maximum height: 136.1 ft
Range: 544.5 ft

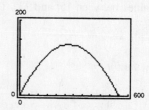

**63. (a)** 100 miles per hour $= 100\left(\dfrac{5280}{3600}\right)$ ft/sec $= \dfrac{440}{3}$ ft/sec

$$x = \left(\frac{440}{3} \cos \theta\right)t \approx (146.67 \cos \theta)t$$

$$y = 3 + \left(\frac{440}{3} \sin \theta\right)t - 16t^2 \approx 3 + (146.67 \sin \theta)t - 16t^2$$

**(b)** For $\theta = 15°$, we have:

$$x = \left(\frac{440}{3} \cos 15°\right)t \approx 141.7t$$

$$y = 3 + \left(\frac{440}{3} \sin 15°\right)t - 16t^2 \approx 3 + 38.0t - 16t^2$$

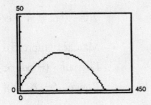

The ball hits the ground inside the ballpark, so it is not a home run.

**(c)** For $\theta = 23°$, we have:

$$x = \left(\frac{440}{3} \cos 23°\right)t \approx 135.0t$$

$$y = 3 + \left(\frac{440}{3} \sin 23°\right)t - 16t^2 \approx 3 + 57.3t - 16t^2$$

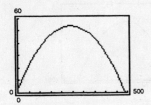

The ball easily clears the 10-foot fence at 400 feet so it is a home run.

**(d)** Find $\theta$ so that $y = 10$ when $x = 400$ by graphing the parametric equations for $\theta$ values between $15°$ and $23°$. This occurs when $\theta \approx 19.4°$.

**65.** $x = (v_0 \cos \theta)t \Rightarrow t = \dfrac{x}{v_0 \cos \theta}$

$y = h + (v_0 \sin \theta)t - 16t^2$

$= h + (v_0 \sin \theta)\left(\dfrac{x}{v_0 \cos \theta}\right) - 16\left(\dfrac{x}{v_0 \cos \theta}\right)^2$

$= h + (\tan \theta)x - \dfrac{16x^2}{v_0{}^2 \cos^2 \theta}$

$= -\dfrac{16 \sec^2 \theta}{v_0{}^2}x^2 + (\tan \theta)x + h$

**67.** When the circle has rolled $\theta$ radians, the center is at $(a\theta, a)$.

$$\sin \theta = \sin(180° - \theta)$$

$$= \frac{|AC|}{b} = \frac{|BD|}{b} \Rightarrow |BD| = b \sin \theta$$

$$\cos \theta = -\cos(180° - \theta)$$

$$= \frac{|AP|}{-b} \Rightarrow |AP| = -b \cos \theta$$

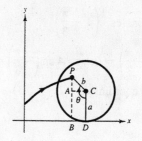

Therefore, $x = a\theta - b \sin \theta$ and $y = a - b \cos \theta$.

**69.** $x = t$

$y = t^2 + 1 \Rightarrow y = x^2 + 1$

$x = 3t$

$y = 9t^2 + 1 \Rightarrow y = x^2 + 1$

True

**71.**

$$\begin{array}{rcl} 5x - 7y = & 11 & \Rightarrow \\ -3x + y = & -13 & \Rightarrow \end{array} \quad \begin{array}{rcl} 5x - 7y = & 11 \\ \underline{-21x + 7y = -91} \\ -16x \quad\quad = -80 \\ x = 5 \end{array}$$

$$5(5) - 7y = 11 \Rightarrow y = 2$$

Solution: $(5, 2)$

**73.**

$$\begin{array}{rcl} 3a - 2b + c = & 8 & \Rightarrow \\ 2a + b - 3c = & -3 & \Rightarrow \end{array} \quad \begin{array}{rcl} 9a - 6b + 3c = 24 \\ \underline{2a + b - 3c = -3} \\ 11a - 5b \quad\quad = 21 \end{array}$$

$$\begin{array}{rcl} 2a + b - 3c = & -3 & \Rightarrow 6a + 3b - 9c = -9 \\ a - 3b + 9c = & 16 & \Rightarrow \underline{a - 3b + 9c = 16} \\ & & \quad 7a \quad\quad\quad = 7 \\ & & \quad a \quad\quad\quad = 1 \end{array}$$

$$a = 11(1) - 5b = 21 \Rightarrow b = -2$$

$$3(1) - 2(-2) + c = 8 \Rightarrow c = 1$$

Solution: $(1, -2, 1)$

# Section 6.7    Polar Coordinates

---

- In polar coordinates you do not have unique representation of points. The point $(r, \theta)$ can be repsresented by $(r, \theta \pm 2n\pi)$ or by $(-r, \theta \pm (2n + 1)\pi)$ where $n$ is any integer. The pole is represented by $(0, \theta)$ where $\theta$ is any angle.

- To convert from polar coordinates to rectangular coordinates, use the following relationships.

  $x = r \cos \theta$

  $y = r \sin \theta$

- To convert from rectangular coordinates to polar coordinates, use the following relationships.

  $r = \pm \sqrt{x^2 + y^2}$

  $\tan \theta = y/x$

  If $\theta$ is in the same quadrant as the point $(x, y)$, then $r$ is positive. If $\theta$ is in the opposite quadrant as the point $(x, y)$, then $r$ is negative.

- You should be able to convert rectangular equations to polar form and vice versa.

---

**Solutions to Odd-Numbered Exercises**

**1.** Polar coordinates: $\left(4, \frac{3\pi}{6}\right)$

$x = 4 \cos\left(\frac{3\pi}{6}\right) = 0, y = 4 \sin\left(\frac{3\pi}{6}\right) = 4$

Rectangular coordinates: $(0, 4)$

**3.** Polar coordinates: $\left(-1, \frac{5\pi}{4}\right)$

$x = -1 \cos\left(\frac{5\pi}{4}\right) = \frac{\sqrt{2}}{2}, y = -1 \sin\left(\frac{5\pi}{4}\right) = \frac{\sqrt{2}}{2}$

Rectangular coordinates: $\left(\frac{\sqrt{2}}{2}, \frac{\sqrt{2}}{2}\right)$

**5.** Polar coordinates: $\left(4, -\dfrac{\pi}{3}\right)$

$x = 4\cos\left(-\dfrac{\pi}{3}\right) = 2, y = 4\sin\left(-\dfrac{\pi}{3}\right) = -2\sqrt{3}$

Rectangular coordinates: $(2, -2\sqrt{3})$

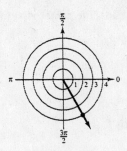

**7.** Polar coordinates: $\left(0, -\dfrac{7\pi}{6}\right)$

$x = 0\cos\left(-\dfrac{7\pi}{6}\right) = 0, y = 0\sin\left(-\dfrac{7\pi}{6}\right) = 0$

Rectangular coordinates: $(0, 0)$

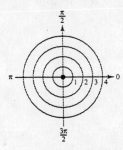

**9.** Polar coordinates: $\left(\sqrt{2}, 2.36\right)$

$x = \sqrt{2}\cos(2.36) \approx -1.004$

$y = \sqrt{2}\sin(2.36) \approx 0.996$

Rectangular coordinates: $(-1.004, 0.996)$

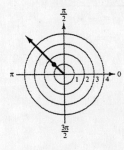

**11.** Polar coordinates: $\left(2, \dfrac{3\pi}{4}\right)$

$\left(2 < \dfrac{3\pi}{4}\right) \blacktriangleright$ Rec

$\approx (-1.4142, 1.4142)$

$= \left(-\sqrt{2}, \sqrt{2}\right)$

**13.** Polar coordinates: $(-4.5, 1.3)$

$(-4.5 < 1.3) \blacktriangleright$ Rec

$\approx (-1.204, -4.336)$

**15.** Rectangular coordinates: $(1, 1)$

$r = \pm\sqrt{2}, \tan\theta = 1, \theta = \dfrac{\pi}{4}$ or $\dfrac{5\pi}{4}$

Polar coordinates: $\left(\sqrt{2}, \dfrac{\pi}{4}\right), \left(-\sqrt{2}, \dfrac{5\pi}{4}\right)$

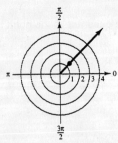

**17.** Rectangular coordinates: $(-6, 0)$

$r = \pm 6, \tan\theta = 0, \theta = 0$ or $\pi$

Polar coordinates: $(6, \pi), (-6, 0)$

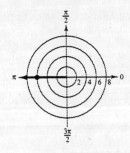

**19.** Rectangular coordinates: $(-3, 4)$

$r = \pm\sqrt{9 + 16} = \pm 5$, $\tan\theta = -\dfrac{4}{3}$, $\theta \approx 2.214,\ 5.356$

Polar coordinates: $(5, 2.214)$, $(-5, 5.356)$

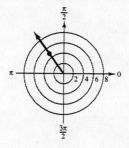

**21.** Rectangular coordinates: $\left(-\sqrt{3},\ -\sqrt{3}\right)$

$r = \pm\sqrt{3 + 3} = \pm\sqrt{6}$, $\tan\theta = 1$, $\theta = \dfrac{\pi}{4}$ or $\dfrac{5\pi}{4}$

Polar coordinates: $\left(\sqrt{6}, \dfrac{5\pi}{4}\right),\ \left(-\sqrt{6}, \dfrac{\pi}{4}\right)$

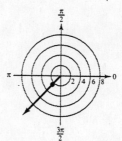

**23.** Rectangular coordinates: $(4, 6)$

$r = \pm\sqrt{4^2 + 6^2} = \pm\sqrt{52} = \pm 2\sqrt{13}$

$\tan\theta = \dfrac{6}{4} \Rightarrow \theta \approx 0.983,\ 4.124$

Polar coordinates: $\left(2\sqrt{13},\ 0.983\right),\ \left(-2\sqrt{13},\ 4.124\right)$

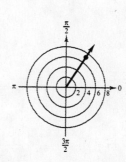

**25.** Rectangular: $(3, -2)$

$(3, -2) \blacktriangleright$ Pol

$\approx (3.606, -0.588)$

**27.** Rectangular: $\left(\sqrt{3}, 2\right)$

$\left(\sqrt{3}, 2\right) \blacktriangleright$ Pol

$\approx (2.646, 0.857)$

**29.** Rectangular: $\left(\frac{5}{2}, \frac{4}{3}\right)$

$\left(\frac{5}{2}, \frac{4}{3}\right) \blacktriangleright$ Pol

$\approx (2.833, 0.490)$

**31.** True, $|r_1| = |r_2|$

**33.** $x^2 + y^2 = 9$

$r = 3$

**35.** $x^2 + y^2 - 2ax = 0$

$r^2 - 2ar\cos\theta = 0$

$r(r - 2a\cos\theta) = 0$

$r = 2a\cos\theta$

**37.** $y = 4$

$r\sin\theta = 4$

$r = 4\csc\theta$

**39.** $x = 10$

$r\cos\theta = 10$

$r = 10\sec\theta$

**41.** $3x - y + 2 = 0$

$3r\cos\theta - r\sin\theta + 2 = 0$

$r(3\cos\theta - \sin\theta) = -2$

$r = \dfrac{-2}{3\cos\theta - \sin\theta}$

**43.** $xy = 4$

$(r\cos\theta)(r\sin\theta) = 4$

$r^2 = 4\sec\theta\csc\theta = 8\csc 2\theta$

**45.** $\left(x^2 + y^2\right)^2 - 9\left(x^2 - y^2\right) = 0$

$\left(r^2\right) - 9\left(r^2\cos^2\theta - r^2\sin^2\theta\right) = 0$

$r^2\left[r^2 - 9(\cos 2\theta)\right] = 0$

$r^2 = 9\cos 2\theta$

**47.** $r = 4\sin\theta$

$r^2 = 4r\sin\theta$

$x^2 + y^2 = 4y$

$x^2 + y^2 - 4y = 0$

**49.**
$$\theta = \frac{\pi}{6}$$

$$\tan \theta = \frac{\sqrt{3}}{3}$$

$$\frac{y}{x} = \frac{\sqrt{3}}{3}$$

$$y = \frac{\sqrt{3}}{3}x$$

$$\sqrt{3}\,x - 3y = 0$$

**51.**
$$r = 2 \csc \theta$$

$$r \sin \theta = 2$$

$$y = 2$$

**53.**
$$r = 2 \sin 3\,\theta$$

$$r = 2(3 \sin \theta - 4 \sin^3 \theta)$$

$$r^4 = 6r^3 \sin \theta - 8r^3 \sin^3 \theta$$

$$(x^2 + y^2)^2 = 6(x^2 + y^2)y - 8y^3$$

$$(x^2 + y^2)^2 = 6x^2 y - 2y^3$$

**55.**
$$r = \frac{6}{2 - \sin \theta}$$

$$r(2 - \sin \theta) = 6$$

$$2r = 6 + r \sin \theta$$

$$2(\pm \sqrt{x^2 + y^2}) = 6 + 3y$$

$$4(x^2 + y^2) = (6 + 3y)^2$$

$$4x^2 + 4y^2 = 36 + 36y + 9y^2$$

$$4x^2 - 5y^2 - 36y - 36 = 0$$

**57.**
$$r = 3$$

$$r^2 = 9$$

$$x^2 + y^2 = 9$$

**59.**
$$\theta = \frac{\pi}{4}$$

$$\tan \theta = \tan \frac{\pi}{4}$$

$$\frac{y}{x} = 1$$

$$y = x$$

$$x - y = 0$$

**61.**
$$r = 3 \sec \theta$$

$$r \cos \theta = 3$$

$$x = 3$$

$$x - 3 = 0$$

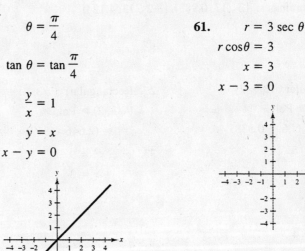

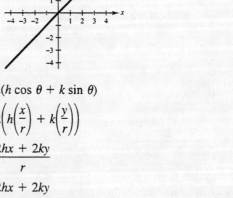

**63.**
$$r = 2(h \cos \theta + k \sin \theta)$$

$$r = 2\left(h\left(\frac{x}{r}\right) + k\left(\frac{y}{r}\right)\right)$$

$$r = \frac{2hx + 2ky}{r}$$

$$r^2 = 2hx + 2ky$$

$$x^2 + y^2 = 2hx + 2ky$$

$$x^2 - 2hx + y^2 - 2ky = 0$$

$$(x^2 - 2hx + h^2) + (y^2 - 2ky + k^2) = h^2 + k^2$$

$$(x - h)^2 + (y - k)^2 = h^2 + k^2$$

Center: *(h, k)*

Radius: $\sqrt{h^2 + k^2}$

**65.** (a) $(r_1, \theta_1) = (x_1, y_1)$ where $x_1 = r_1 \cos \theta_1$ and $y_1 = r_1 \sin \theta_1$.

$(r_2, \theta_2) = (x_2, y_2)$ where $x_2 = r_2 \cos \theta_2$ and $y_2 = r_2 \sin \theta_2$.

$$d = \sqrt{(x_1 - x_2)^2 + (y_1 - y_2)^2}$$
$$= \sqrt{x_1^2 - 2x_1 x_2 + x_2^2 + y_1^2 - 2y_1 y_2 + y_2^2}$$
$$= \sqrt{(x_1^2 + y_1^2) + (x_2^2 + y_2 + y_2^2) - 2(x_1 x_2 + y_1 y_2)}$$
$$= \sqrt{r_1^2 + r_2^2 - 2\left(r_1 r_2 \cos \theta_1 \cos \theta_2 + r_1 r_2 \sin \theta_1 \sin \theta_2\right)}$$
$$= \sqrt{r_1^2 + r_2^2 - 2 r_1 r_2 \cos(\theta_1 - \theta_2)}$$

(b) If $\theta_1 = \theta_2$, then
$$d = \sqrt{r_1^2 + r_2^2 - 2 r_1 r_2}$$
$$= \sqrt{(r_1 - r_2)^2}$$
$$= |r_1 - r_2|.$$
This represents the distance between two points on the line $\theta = \theta_1 = \theta_2$.

(c) If $\theta_1 - \theta_2 = 90°$, then
$$d = \sqrt{r_1^2 + r_2^2}.$$
This is the result of the Pythagorean Theorem.

(d) The results should be the same. For example, use the points

$$\left(3, \frac{\pi}{6}\right) \text{ and } \left(4, \frac{\pi}{3}\right).$$

The distance is $d \approx 2.053$. Now use the representations

$$\left(-3, \frac{7\pi}{6}\right) \text{ and } \left(-4, \frac{4\pi}{3}\right).$$

The distance is still $d \approx 2.053$.

**67.**  $5x - 7y = -11$

$-3x + y = -3$

By Cramer's Rule we have:

$$x = \frac{\begin{vmatrix} -11 & -7 \\ -3 & 1 \end{vmatrix}}{\begin{vmatrix} 5 & -7 \\ -3 & 1 \end{vmatrix}} = \frac{-32}{-16} = 2$$

$$y = \frac{\begin{vmatrix} 5 & -11 \\ -3 & -3 \end{vmatrix}}{\begin{vmatrix} 5 & -7 \\ -3 & 1 \end{vmatrix}} = \frac{-48}{-16} = 3$$

Solution: $(2, 3)$

**69.**  $3a - 2b + c = 0$

$2a + b - 3c = 0$

$a - 3b + 9c = 8$

$$\begin{vmatrix} 3 & -2 & 1 \\ 2 & 1 & -3 \\ 1 & -3 & 9 \end{vmatrix} = 35$$

By Cramer's Rule we have:

$$x = \frac{\begin{vmatrix} 0 & -2 & 1 \\ 0 & 1 & -3 \\ 8 & -3 & 9 \end{vmatrix}}{35} = \frac{40}{35} = \frac{8}{7}$$

$$y = \frac{\begin{vmatrix} 3 & 0 & 1 \\ 2 & 0 & -3 \\ 1 & 8 & 9 \end{vmatrix}}{35} = \frac{88}{35}$$

$$z = \frac{\begin{vmatrix} 3 & -2 & 0 \\ 2 & 1 & 0 \\ 1 & -3 & 8 \end{vmatrix}}{35} = \frac{56}{35} = \frac{8}{5}$$

Solution: $\left(\frac{8}{7}, \frac{88}{35}, \frac{8}{5}\right)$

**71.**   $x + y + z - 3w = -8$
$3x - y - 2z + w = 7$
$-x + y - z + 2w = -2$
$2y + w = -6$

$$\begin{vmatrix} 1 & 1 & 1 & -3 \\ 3 & -1 & -2 & 1 \\ -1 & 1 & -1 & 2 \\ 0 & 2 & 0 & 1 \end{vmatrix} = 20$$

By Cramer's Rule we have:

$$x = \dfrac{\begin{vmatrix} -8 & 1 & 1 & -3 \\ 7 & -1 & -2 & 1 \\ -2 & 1 & -1 & 2 \\ -6 & 2 & 0 & 1 \end{vmatrix}}{20} = \dfrac{20}{20} = 1$$

$$z = \dfrac{\begin{vmatrix} 1 & 1 & -8 & -3 \\ 3 & -1 & 7 & 1 \\ -1 & 1 & -2 & 2 \\ 0 & 2 & -6 & 1 \end{vmatrix}}{20} = \dfrac{20}{20} = 1$$

$$y = \dfrac{\begin{vmatrix} 1 & -8 & 1 & -3 \\ 3 & 7 & -2 & 1 \\ -1 & -2 & -1 & 2 \\ 0 & -6 & 0 & 1 \end{vmatrix}}{20} = \dfrac{-80}{20} = -4$$

$$w = \dfrac{\begin{vmatrix} 1 & 1 & 1 & -8 \\ 3 & -1 & -2 & 7 \\ -1 & 1 & -1 & -2 \\ 0 & 2 & 0 & -6 \end{vmatrix}}{20} = \dfrac{40}{20} = 2$$

Solution: $(1, -4, 1, 2)$

# Section 6.8    Graphs of Polar Equations

■   When graphing polar equations:

1. Test for symmetry
   (a) $\theta = \pi/2$: Replace $(r, \theta)$ by $(r, \pi - \theta)$ or $(-r, -\theta)$.
   (b) Polar axis:  Replace $(r, \theta)$ by $(r, -\theta)$ or $(-r, \pi - \theta)$.
   (c) Pole:  Replace $(r, \theta)$ by $(r, \pi + \theta)$ or $(-r, \theta)$.
   (d) $r = f(\sin \theta)$ is symmetric with repsect to the line $\theta = \pi/2$.
   (e) $r = f(\cos \theta)$ is symmetric with respect to the polar axis.

2. Find the $\theta$ values for which $|r|$ is maximum.

3. Find the $\theta$ values for which $r = 0$.

4. Know the different types of polar graphs.
   (a) Limacons
       $r = a \pm b \cos \theta$
       $r = a \pm b \sin \theta$

   (b) Rose Curves, $n \geq 2$
       $r = a \cos n\theta$
       $r = a \sin n\theta$

   (c) Circles
       $r = a \cos \theta$
       $r = a \sin \theta$
       $r = a$

   (d) Lemniscates
       $r^2 = a^2 \cos 2\theta$
       $r^2 = a^2 \sin 2\theta$

5. Plot additional points.

**Solutions to Odd-Numbered Exercises**

**1.** $r = 3 \cos 2\theta$

Rose curve with 4 petals

**3.** $r = 2 - \cos \theta$

Convex limaçon

**5.** $r = 6 \sin 2\theta$

Rose curve with 4 petals

**7.** $r = 10 + 6 \cos \theta$

$\theta = \dfrac{\pi}{2}$: $\quad -r = 10 + 6 \cos(-\theta)$

$\qquad\qquad -r = 10 + 6 \cos \theta$

$\qquad$ Not an equivalent equation

Polar $\quad r = 10 + 6 \cos(-\theta)$

axis: $\quad\; r = 10 + 6 \cos \theta$

$\qquad$ Equivalent equation

Pole: $\quad -r = 10 + 6 \cos \theta$

$\qquad$ Not an equivalent equation

*Answer:* Symmetric with respect to polar axis

**9.** $r = \dfrac{2}{1 + \sin \theta}$

$\theta = \dfrac{\pi}{2}$: $\quad r = \dfrac{2}{1 + \sin(\pi - \theta)}$

$\qquad\qquad r = \dfrac{2}{1 + \sin \pi \cos \theta - \cos \pi \sin \theta}$

$\qquad\qquad r = \dfrac{2}{1 + \sin \theta}$

$\qquad$ Equivalent equation

Polar

axis: $\quad r = \dfrac{2}{1 + \sin(-\theta)}$

$\qquad\qquad r = \dfrac{2}{1 - \sin \theta}$

$\qquad$ Not an equivalent equation

Pole: $\quad -r = \dfrac{2}{1 + \sin \theta}$

*Answer:* Symmetric with respect to $\theta = \pi/2$

**11.** $r = 4 \sec \theta \csc \theta$

$\theta = \dfrac{\pi}{2}$: $\quad -r = 4 \sec(-\theta) \csc(-\theta)$

$\qquad\qquad -r = -4 \sec \theta \csc \theta$

$\qquad\qquad\; r = 4 \sec \theta \csc \theta$

$\qquad$ Equivalent equation

Polar $\quad -r = 4 \sec(\pi - \theta) \csc(\pi - \theta)$

axis: $\quad -r = 4(-\sec \theta) \csc \theta$

$\qquad\qquad r = 4 \sec \theta \csc \theta$

$\qquad$ Equivalent equation

Pole: $\quad r = 4 \sec(\pi + \theta) \csc(\pi + \theta)$

$\qquad\qquad r = 4(-\sec \theta)(-\csc \theta)$

$\qquad\qquad r = 4 \sec \theta \csc \theta$

$\qquad$ Equivalent equation

*Answer:* Symmetric with respect to $\theta = \pi/2$, polar axis, and pole

**13.** $|r| = |10(1 - \sin \theta)| = 10|1 - \sin \theta| \leq 10(2) = 20$

$|1 - \sin \theta| = 2$

$\quad 1 - \sin \theta = 2 \qquad$ or $\quad 1 - \sin \theta = -2$

$\qquad \sin \theta = -1 \qquad\qquad\qquad \sin \theta = 3$

$\qquad\quad \theta = \dfrac{3\pi}{2} \qquad\qquad$ Not possible

Maximum: $|r| = 20$ when $\theta = \dfrac{3\pi}{2}$.

$\quad 0 = 10(1 - \sin \theta)$

$\sin \theta = 1$

$\quad \theta = \dfrac{\pi}{2}$

Zero: $r = 0$ when $\theta = \dfrac{\pi}{2}$.

**15.** $|r| = |4 \cos 3\theta| = 4|\cos 3\theta| \leq 4$

$|\cos 3\theta| = 1$

$\cos 3\theta = \pm 1$

$\theta = 0, \dfrac{\pi}{3}, \dfrac{2\pi}{3}$

Maximum: $|r| = 4$ when $\theta = 0, \dfrac{\pi}{3}, \dfrac{2\pi}{3}$.

$0 = 4 \cos 3\theta$

$\cos 3\theta = 0$

$\theta = \dfrac{\pi}{6}, \dfrac{\pi}{2}, \dfrac{5\pi}{6}$

Zero: $r = 0$ when $\theta = \dfrac{\pi}{6}, \dfrac{\pi}{2}, \dfrac{5\pi}{6}$.

**17.** Circle: $r = 5$

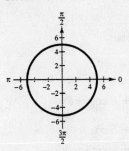

**19.** Circle: $r = \dfrac{\pi}{6}$

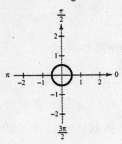

**21.** $r = 3 \sin \theta$

Symmetric with respect to $\theta = \dfrac{\pi}{2}$

Circle with a radius of $\dfrac{3}{2}$

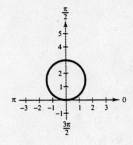

**23.** $r = 3 - 3 \cos \theta$

Symmetric with respect to polar axis

$\dfrac{a}{b} = \dfrac{3}{3} = 1 \Rightarrow$ Cardioid

$|r| = 6$ when $\theta = \pi$.

$r = 0$ when $\pi = 0$.

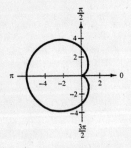

**25.** $r = 4 + 4 \sin \theta$

Symmetric with respect to

$\theta = \dfrac{\pi}{2}$

$\dfrac{a}{b} = \dfrac{4}{4} = 1 \Rightarrow$ Cardioid

$|r| = 8$ when $\theta = \dfrac{\pi}{2}$.

$r = 0$ when $\theta = \dfrac{3\pi}{2}$.

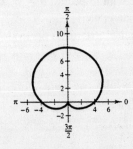

**27.** $r = 3 - 2 \cos \theta$

Symmetric with respect to polar axis

$\dfrac{a}{b} = \dfrac{3}{2} > 1 \implies$ Dimpled limaçon

$|r| = 5$ when $\theta = \pi$.

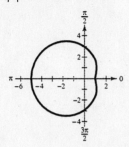

**29.** $r = 2 + \sin \theta$

Symmetric with respect to $\theta = \dfrac{\pi}{2}$

$\dfrac{a}{b} = \dfrac{2}{1} \geq 2 \implies$ Convex limaçon

$|r| = 3$ when $\theta = \dfrac{\pi}{2}$.

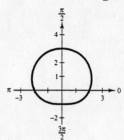

**31.** $r = 2 + 4 \sin \theta$

Symmetric with respect to $\theta = \dfrac{\pi}{2}$

$\dfrac{a}{b} = \dfrac{2}{4} < 1 \implies$ Limaçon with inner loop

$|r| = 6$ when $\theta = \dfrac{\pi}{2}$.

$r = 0$ when $\theta = \dfrac{7\pi}{6}, \dfrac{11\pi}{6}$.

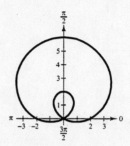

**33.** $r = 3 - 4 \cos \theta$

Symmetric with respect to polar axis

$\dfrac{a}{b} = \dfrac{2}{4} < 1 \implies$ Limaçon with inner loop

$|r| = 7$ when $\theta = \pi$.

$r = 0$ when $\cos \theta = \dfrac{3}{4}$ or

$\theta \approx 0.723, 5.560$

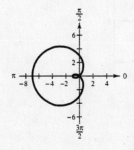

**35.** $r = 2 \cos 3\theta$

Symmetric with respect to polar axis

Rose curve ($n = 3$) with 3 petals

$|r| = 2$ when $\theta = 0, \dfrac{2\pi}{3}, \dfrac{4\pi}{3}$.

$r = 0$ when $\theta = \dfrac{\pi}{6}, \dfrac{\pi}{2}, \dfrac{5\pi}{6}$.

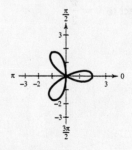

**37.** $r = 3 \sin 2\theta$

Symmetric with respect to $\theta = \dfrac{\pi}{2}$

Rose curve ($n = 2$) with 4 petals

$|r| = 3$ when $\theta = \dfrac{\pi}{4}, \dfrac{3\pi}{4}, \dfrac{5\pi}{4}, \dfrac{7\pi}{4}$.

$r = 0$ when $\theta = 0, \dfrac{\pi}{2}, \pi$.

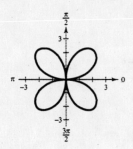

**39.** $r = 2 \sec \theta$

$r = \dfrac{2}{\cos \theta}$

$r \cos \theta = 2$

$x = 2 \Rightarrow$ Line

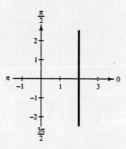

**41.** $r = \dfrac{3}{\sin \theta - 2 \cos \theta}$

$r(\sin \theta - 2 \cos \theta) = 3$

$y - 2x = 3$

$y = 2x + 3 \Rightarrow$ Line

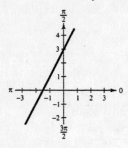

**43.** $r^2 = 4 \cos 2\theta$

Symmetric with respect to polar axis

Lemniscate

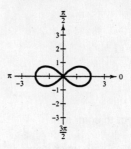

**45.** $r = \dfrac{\theta}{2}$

Symmetric with respect to $\theta = \dfrac{\pi}{2}$

Spiral

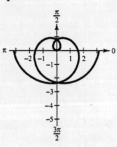

**47.** $r = 6 \cos \theta$

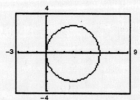

**49.** $r = 3(2 - \sin \theta)$

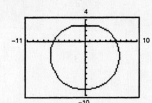

**51.** $r = 4 \sin \theta \cos^2 \theta$

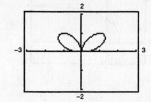

**53.** $r = 2 \csc \theta + 5 = \dfrac{2}{\sin \theta} + 5$

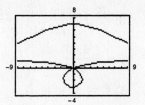

**55.** $r = 3 - 4 \cos \theta$
$0 \leq \theta < 2\pi$

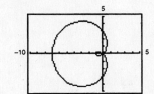

**57.** $r = 2 + \sin \theta$
$0 \leq \theta < 2\pi$

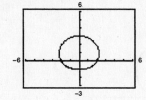

**59.** $r = 2 \cos\left(\dfrac{3\theta}{2}\right)$

$0 \leq \theta < 4\pi$

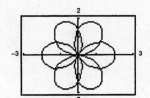

**61.** $r^2 = 4 \sin 2\theta$
$0 \leq \theta < \pi$

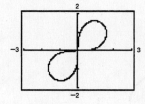

**63.**
$$r = 2 - \sec \theta = 2 - \frac{1}{\cos \theta}$$
$$r \cos \theta = 2 \cos \theta - 1$$
$$r(r \cos \theta) = 2r \cos \theta - r$$
$$\left(\pm \sqrt{x^2 + y^2}\right)x = 2x - \left(\pm \sqrt{x^2 + y^2}\right)$$
$$\left(\pm \sqrt{x^2 + y^2}\right)(x + 1) = 2x$$
$$\left(\pm \sqrt{x^2 + y^2}\right) = \frac{2x}{x + 1}$$
$$x^2 + y^2 = \frac{4x^2}{(x + 1)^2}$$
$$y^2 = \frac{4x^2}{(x + 1)^2} - x^2$$
$$= \frac{4x^2 - x^2(x + 1)^2}{(x + 1)^2} = \frac{4x^2 - x^2(x^2 + 2x + 1)}{(x + 1)^2}$$
$$= \frac{-x^4 - 2x^3 + 3x^2}{(x + 1)^2} = \frac{-x^2(x^2 + 2x - 3)}{(x + 1)^2}$$
$$y = \pm \sqrt{\frac{x^2(3 - 2x - x^2)}{(x + 1)^2}} = \pm \left|\frac{x}{x + 1}\right| \sqrt{3 - 2x - x^2}$$

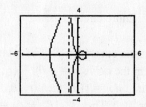

The graph has an asymptote at $x = -1$.

**65.** $r = \dfrac{2}{\theta}$

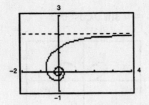

$\theta = \dfrac{2}{r} = \dfrac{2 \sin \theta}{r \sin \theta} = \dfrac{2 \sin \theta}{y}$

$y = \dfrac{2 \sin \theta}{\theta}$

As $\theta \Rightarrow 0$, $y \Rightarrow 2$.

**67.** $r = 4 \sin \theta$

(a) $0 \le \theta \le \dfrac{\pi}{2}$ 

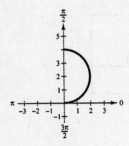

Right half of the circle

(b) $\dfrac{\pi}{2} \le \theta \le \pi$

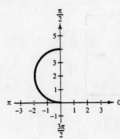

Left half of the circle

(c) $-\dfrac{\pi}{2} \le \theta \le \dfrac{\pi}{2}$

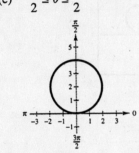

Entire circle

(d) $\dfrac{\pi}{4} \le \theta \le \dfrac{3\pi}{4}$

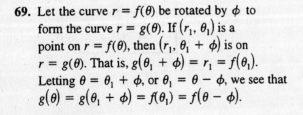

Top half of the circle

**69.** Let the curve $r = f(\theta)$ be rotated by $\phi$ to form the curve $r = g(\theta)$. If $(r_1, \theta_1)$ is a point on $r = f(\theta)$, then $(r_1, \theta_1 + \phi)$ is on $r = g(\theta)$. That is, $g(\theta_1 + \phi) = r_1 = f(\theta_1)$. Letting $\theta = \theta_1 + \phi$, or $\theta_1 = \theta - \phi$, we see that $g(\theta) = g(\theta_1 + \phi) = f(\theta_1) = f(\theta - \phi)$.

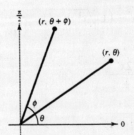

**71.** (a) $r = 2 - \sin\left(\theta - \dfrac{\pi}{4}\right)$

$= 2 - \dfrac{\sqrt{2}}{2}(\sin \theta - \cos \theta)$

(b) $r = 2 - \sin\left(\theta - \dfrac{\pi}{2}\right)$

$= 2 + \cos \theta$

(c) $r = 2 - \sin(\theta - \pi)$

$= 2 + \sin \theta$

(d) $r = 2 - \sin\left(\theta - \dfrac{3\pi}{2}\right)$

$= 2 - \cos \theta$

**73.** (a)  $r = 1 - \sin\theta$

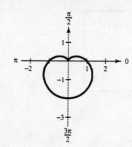

(b)  $r = 1 - \sin\left(\theta - \dfrac{\pi}{4}\right)$

Rotate the graph in part (a) through the angle $\dfrac{\pi}{4}$.

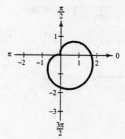

**75.**  $r = 2 + k\cos\theta$

    $k = 0$:  $r = 2$

        Circle

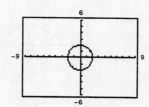

$k = 2$:  $r = 2 + 2\cos\theta$

    Cardioid

    $k = 1$:  $r = 2 + \cos\theta$

        Convex limaçon

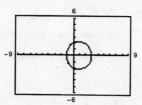

$k = 3$:  $r = 2 + 3\cos\theta$

    Limaçon with
    inner loop

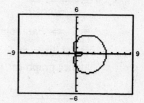

# Section 6.9    Polar Equations of Conics

- The graph of a polar equation of the form

$$r = \frac{ep}{1 \pm e\cos\theta} \quad \text{or} \quad r = \frac{ep}{1 \pm e\sin\theta}$$

  is a conic, *where $e > 0$ is the eccentricity and $|p|$ is the distance between the focus (pole) and the directrix.*

  (a) If $e < 1$, the graph is an ellipse.
  (b) If $e = 1$, the graph is a parabola.
  (c) If $e > 1$, the graph is a hyperbola.

- Guidelines for finding polar equations of conics:

  (a) Horizontal directrix above the pole: $r = \dfrac{ep}{1 + e\sin\theta}$

  (b) Horizontal directrix below the pole: $r = \dfrac{ep}{1 - e\sin\theta}$

  (c) Vertical directrix to the right of the pole: $r = \dfrac{ep}{1 + e\cos\theta}$

  (d) Vertical directrix to the left of the pole: $r = \dfrac{ep}{1 - e\cos\theta}$

**Solutions to Odd-Numbered Exercises**

1. $r = \dfrac{2e}{1 + e\cos\theta}$

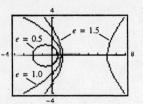

  (a) $e = 1$, $r = \dfrac{2}{1 + \cos\theta}$, parabola

  (b) $e = 0.5$, $r = \dfrac{1}{1 + 0.5\cos\theta} = \dfrac{2}{2 + \cos\theta}$, ellipse

  (c) $e = 1.5$, $r = \dfrac{3}{1 + 1.5\cos\theta} = \dfrac{6}{2 + 3\cos\theta}$, hyperbola

3. $r = \dfrac{2e}{1 - e\sin\theta}$

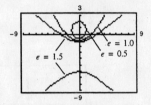

  (a) $e = 1$, $r = \dfrac{2}{1 - \sin\theta}$, parabola

  (b) $e = 0.5$, $r = \dfrac{1}{1 - 0.5\sin\theta} = \dfrac{2}{2 - \sin\theta}$, ellipse

  (c) $e = 1.5$, $r = \dfrac{3}{1 - 1.5\sin\theta} = \dfrac{6}{2 - 3\sin\theta}$, hyperbola

5. $r = \dfrac{4}{1 - \cos\theta}$

$e = 1 \Rightarrow$ Parabola

Vertical directrix to the left of the pole

Matches graph (b).

7. $r = \dfrac{3}{1 + 2\sin\theta}$

$e = 2 \Rightarrow$ Hyperbola

Matches graph (d).

9. $r = \dfrac{2}{1 - \cos\theta}$

$e = 1$, the graph is a parabola.

Vertex: $(1, \pi)$

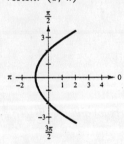

11. $r = \dfrac{5}{1 + \sin\theta}$

$e = 1$, the graph is a parabola.

Vertex: $\left(\dfrac{5}{2}, \dfrac{\pi}{2}\right)$

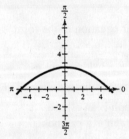

13. $r = \dfrac{2}{2 - \cos\theta} = \dfrac{1}{1 - \frac{1}{2}\cos\theta}$

$e = \dfrac{1}{2} < 1$, the graph is an ellipse.

Vertices: $(2, 0)$, $\left(\dfrac{2}{3}, \pi\right)$

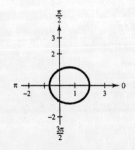

15. $r = \dfrac{4}{2 + \sin\theta} = \dfrac{2}{1 + \frac{1}{2}\sin\theta}$

$e = \dfrac{1}{2} < 1$, the graph is an ellipse.

Vertices: $\left(\dfrac{4}{3}, \dfrac{\pi}{2}\right)$, $\left(4, \dfrac{3\pi}{2}\right)$

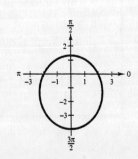

**17.** $r = \dfrac{3}{2 + 4 \sin \theta} = \dfrac{3/2}{1 + 2 \sin \theta}$

$e = 2 > 1$, the graph is a hyperbola.

Vertices: $\left(\dfrac{1}{2}, \dfrac{\pi}{2}\right), \left(-\dfrac{3}{2}, \dfrac{3\pi}{2}\right)$

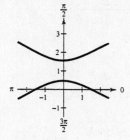

**19.** $r = \dfrac{3}{2 - 6 \cos \theta} = \dfrac{3/2}{1 - 3 \cos \theta}$

$e = 3 > 1$, the graph is a hyperbola.

Vertices: $\left(-\dfrac{3}{4}, 0\right), \left(\dfrac{3}{8}, \pi\right)$

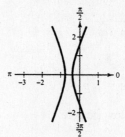

**21.** $r = \dfrac{6}{2 - \cos \theta} = \dfrac{3}{1 - \frac{1}{2} \cos \theta}$

$e = \dfrac{1}{2} < 1$, the graph is an ellipse.

Vertices: $(6, 0), (2, \pi)$

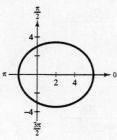

**23.** $r = \dfrac{-1}{1 - \sin \theta}$

$e = 1 \Rightarrow$ Parabola

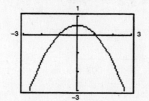

**25.** $r = \dfrac{3}{-4 + 2 \cos \theta}$

$e = \dfrac{1}{2} \Rightarrow$ Ellipse

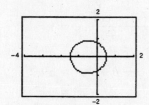

**27.** $r = \dfrac{2}{1 - \cos(\theta - \pi/4)}$

Rotate the graph in Exercise 9 through the angle $\dfrac{\pi}{4}$.

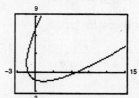

**29.** $r = \dfrac{4}{2 + \sin(\theta + \pi/6)}$

Rotate the graph in Exercise 15 through the angle $-\pi/6$.

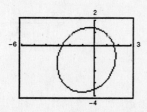

**31.** Parabola:  $e = 1$

Directrix:  $x = -1$

Vertical directrix to the left of the pole

$$r = \dfrac{1(1)}{1 - 1 \cos \theta} = \dfrac{1}{1 - \cos \theta}$$

**33.** Ellipse:   $e = \dfrac{1}{2}$

Directrix:  $y = 1$

$p = 1$

Horizontal directrix above the pole

$r = \dfrac{\frac{1}{2}(1)}{1 + \frac{1}{2}\sin\theta} = \dfrac{1}{2 + \sin\theta}$

**37.** Parabola

Vertex: $\left(1, -\dfrac{\pi}{2}\right) \Rightarrow e = 1, p = 2$

Horizontal directrix below the pole

$r = \dfrac{1(2)}{1 - 1\sin\theta} = \dfrac{2}{1 - \sin\theta}$

**41.** Ellipse:  Vertices $(2, 0), (8, \pi)$

Center: $(3, \pi); c = 3, a = 5, e = \dfrac{3}{5}$

Vertical directrix to the right of the pole

$r = \dfrac{\frac{3}{5}p}{1 + \frac{3}{5}\cos\theta} = \dfrac{3p}{5 + 3\cos\theta}$

$2 = \dfrac{3p}{5 + 3\cos 0}$

$p = \dfrac{16}{3}$

$r = \dfrac{3\left(\frac{16}{3}\right)}{5 + 3\cos\theta} = \dfrac{16}{5 + 3\cos\theta}$

**45.** Hyperbola:  Vertices $\left(1, \dfrac{3\pi}{2}\right), \left(9, \dfrac{3\pi}{2}\right)$

Center: $\left(5, \dfrac{3\pi}{2}\right); c = 5, a = 4, e = \dfrac{5}{4}$

Horizontal directrix below the pole

$r = \dfrac{5/4p}{1 - 5/4\sin\theta} = \dfrac{5p}{4 - 5\sin\theta}$

$1 = \dfrac{5p}{4 - 5\sin(3\pi/2)}$

$p = \dfrac{9}{5}$

$r = \dfrac{5(9/5)}{4 - 5\sin\theta} = \dfrac{9}{4 - 5\sin\theta}$

**35.** Hyperbola:  $e = 2$

Directrix:   $x = 1$

$p = 1$

Vertical directrix to the right of the pole

$r = \dfrac{2(1)}{1 + 2\cos\theta} = \dfrac{2}{1 + 2\cos\theta}$

**39.** Parabola

Vertex: $(5, \pi) \Rightarrow e = 1, p = 10$

Vertical directrix to the left of the pole

$r = \dfrac{1(10)}{1 - 1\cos\theta} = \dfrac{10}{1 - \cos\theta}$

**43.** Ellipse:  Vertices $(20, 0), (4, \pi)$

Center: $(8, 0); c = 8, a = 12, e = \dfrac{2}{3}$

Vertical directrix to the left of the pole

$r = \dfrac{\frac{2}{3}p}{1 - \frac{2}{3}\cos\theta} = \dfrac{2p}{3 - 2\cos\theta}$

$20 = \dfrac{2p}{3 - 2\cos 0}$

$p = 10$

$r = \dfrac{2(10)}{3 - 2\cos\theta} = \dfrac{20}{3 - 2\cos\theta}$

**47.**

$\dfrac{x^2}{a^2} + \dfrac{y^2}{b^2} = 1$

$\dfrac{r^2\cos^2\theta}{a^2} + \dfrac{r^2\sin^2\theta}{b^2} = 1$

$\dfrac{r^2\cos^2\theta}{a^2} + \dfrac{r^2(1 - \cos^2\theta)}{b^2} = 1$

$r^2b^2\cos^2\theta + r^2a^2 - r^2a^2\cos^2\theta = a^2b^2$

$r^2(b^2 - a^2)\cos^2\theta + r^2a^2 = a^2b^2$

$b^2 - a^2 = -c^2$

$-r^2c^2\cos^2\theta + r^2a^2 = a^2b^2$

$-r^2\left(\dfrac{c}{a}\right)^2\cos^2\theta + r^2 = b^2, e = \dfrac{c}{a}$

$-r^2e^2\cos^2\theta + r^2 = b^2$

$r^2(1 - e^2\cos^2\theta) = b^2$

$r^2 = \dfrac{b^2}{1 - e^2\cos^2\theta}$

**49.** $\dfrac{x^2}{169} + \dfrac{y^2}{144} = 1$

$a = 13, b = 12, c = 5, e = \dfrac{5}{13}$

$r^2 = \dfrac{144}{1 - \left(\frac{25}{169}\right)\cos^2\theta} = \dfrac{24{,}336}{169 - 25\cos^2\theta}$

**51.** $\dfrac{x^2}{9} - \dfrac{y^2}{16} = 1$

$a = 3, b = 4, c = 5, e = \dfrac{5}{3}$

$r^2 = \dfrac{-16}{1 - \left(\frac{25}{9}\right)\cos^2\theta} = \dfrac{144}{25\cos^2\theta - 9}$

**53.** One focus: $\left(5, \dfrac{\pi}{2}\right)$

Vertices: $\left(4, \dfrac{\pi}{2}\right), \left(4, -\dfrac{\pi}{2}\right)$

$a = 4, c = 5 \Rightarrow b = 3$ and $e = \dfrac{5}{4}$

$$\dfrac{y^2}{16} - \dfrac{x^2}{9} = 1$$

$$\dfrac{r^2\sin^2\theta}{16} - \dfrac{r^2\cos^2\theta}{9} = 1$$

$$9r^2\sin^2\theta - 16r^2(1 - \sin^2\theta) = 144$$

$$25r^2\sin^2\theta - 16r^2 = 144$$

$$r^2 = \dfrac{144}{25\sin^2\theta - 16} = \dfrac{-144}{16 - 25\sin^2\theta}$$

**55.** When $\theta = 0$, $r = c + a = ea + a = a(1 + e)$. Therefore,

$$a(1 + e) = \dfrac{ep}{1 - e\cos\theta}$$

$$a(1 + e)(1 - e) = ep$$

$$a(1 - e^2) = ep.$$

Thus, $r = \dfrac{ep}{1 - e\cos\theta} = \dfrac{(1 - e^2)a}{1 - e\cos\theta}.$

**57.** $r = \dfrac{[1 - (0.0167)^2](92.957 \times 10^6)}{1 - 0.0167\cos\theta}$

$\approx \dfrac{9.2931 \times 10^7}{1 - 0.0167\cos\theta}$

Perihelion distance:
$r = 92.957 \times 10^6(1 - 0.0167) \approx 9.1405 \times 10^7$

Aphelion distance:
$r = 92.957 \times 10^6(1 + 0.0167) \approx 9.4509 \times 10^7$

**59.** $r = \dfrac{[1 - (0.2481)^2](5.9 \times 10^9)}{1 - 0.2481\cos\theta}$

$\approx \dfrac{5.5368 \times 10^9}{1 - 0.2481\cos\theta}$

Perihelion distance:
$r = 5.9 \times 10^9(1 - 0.2481) \approx 4.4362 \times 10^9$

Aphelion distance:
$r = 5.9 \times 10^9(1 + 0.2481) \approx 7.3638 \times 10^9$

**61.** Vertex: $\left(4100, \dfrac{\pi}{2}\right)$

Focus: $(0, 0)$

$e = 1, p = 8200$

$r = \dfrac{ep}{1 + e\sin\theta} = \dfrac{8200}{1 + \sin\theta}$

When $\theta = 30°$, $r = 8200/1.5 \approx 5466.67$.
Distance between the surface of the earth and the satellite is $r - 4000 \approx 1467$ miles.

**63.** $f(x) = 4^{x-2}$

Exponential function

| $x$ | 0 | 1 | 2 | 3 |
|-----|-----|-----|-----|-----|
| $y$ | $\frac{1}{16}$ | $\frac{1}{4}$ | 1 | 4 |

**65.** $h(t) = \log_4(t - 2)$

Logarithmic function

$4^y = t - 2$

$4^y + 2 = t$

| $t$ | $2\frac{1}{4}$ | 3 | 6 |
|-----|-----|-----|-----|
| $y$ | $-1$ | 0 | 1 |

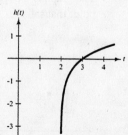

Vertical asymptote: $t = 2$

# Review Exercises for Chapter 6

**Solutions to Odd-Numbered Exercises**

**1.** $m = \tan 120°$
$= -\sqrt{3}$

**3.** $y = -x + 10$
$m = -1 = \tan \theta$
$\theta = \arctan(-1) = 135°$

**5.** $(1, 2) \Rightarrow x_1 = 1, y_1 = 2$
$x - y - 3 = 0 \Rightarrow A = 1, B = -1, C = -3$
$d = \dfrac{|1(1) + (-1)(2) + (-3)|}{\sqrt{1^2 + (-1)^2}} = \dfrac{4}{\sqrt{2}} = 2\sqrt{2}$

**7.** $4x^2 + y^2 = 4$
$\dfrac{x^2}{1} + \dfrac{y^2}{4} = 1$
Ellipse with center $(0, 0)$ and a vertical major axis
Matches graph (g).

**9.** $4x^2 - y^2 = 4$
$\dfrac{x^2}{1} - \dfrac{y^2}{4} = 1$
Hyperbola with center $(0, 0)$ and a horizontal transverse axis
Matches graph (f).

**11.** $x^2 + 4y^2 = 4$
$\dfrac{x^2}{4} + \dfrac{y^2}{1} = 1$
Ellipse with center $(0, 0)$ and a horizontal major axis.
Matches graph (d)

**13.** $x^2 = -6y$
$y = -\dfrac{1}{6}x^2$
Parabola with vertex $(0, 0)$ and opening downward
Matches graph (b).

**15.** $x^2 - 5y^2 = -5$
$\dfrac{y^2}{1} - \dfrac{x^2}{5} = 1$
Hyperbola with center $(0, 0)$ and a vertical transverse axis
Matches graph (h).

**17.** $4x - y^2 = 0$
$y^2 = 4x$
The graph is a parabola.
Vertex: $(0, 0)$

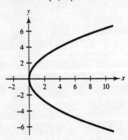

**19.** $x^2 - 6x + 2y + 9 = 0$
$(x - 3)^2 = -2y$
The graph is a parabola.
Vertex: $(3, 0)$

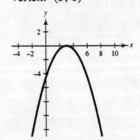

**21.** $x^2 + y^2 - 2x - 4y + 5 = 0$
$(x - 1)^2 + (y - 2)^2 = 0$
The graph is a degenerate circle. $(1, 2)$ is the only point that satisfies this equation.

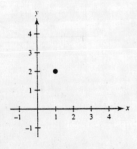

**23.** $4x^2 + y^2 = 16$
$\dfrac{x^2}{4} + \dfrac{y^2}{16} = 1$
The graph is an ellipse.
Center: $(0, 0)$
Vertices: $(0, \pm 4)$

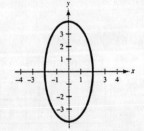

**25.** $x^2 + 9y^2 + 10x - 18y + 25 = 0$

$$(x + 5)^2 + 9(y - 1)^2 = 9$$

$$\frac{(x + 5)^2}{9} + \frac{(y - 1)^2}{1} = 1$$

The graph is an ellipse.

Center: $(-5, 1)$

Vertices: $(-8, 1)\ (-2, 1)$

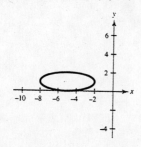

**27.** $5y^2 - 4x^2 = 20$

$$\frac{y^2}{4} - \frac{x^2}{5} = 1$$

The graph is a hyperbola.

Center: $(0, 0)$

Vertices: $(\pm 2, 0)$

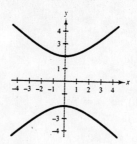

**29.** $B^2 - 4AC = 2^2 - 4(1)(1) = 0$

The graph is a parabola.

$$\cot 2\theta = \frac{A - C}{B} = 0 \Rightarrow 2\theta = \frac{\pi}{2} \Rightarrow \theta = \frac{\pi}{4}$$

$$x = x' \cos \frac{\pi}{4} - y' \sin \frac{\pi}{4} = \frac{x' - y'}{\sqrt{2}}$$

$$y = x' \sin \frac{\pi}{4} + y' \cos \frac{\pi}{4} = \frac{x' + y'}{\sqrt{2}}$$

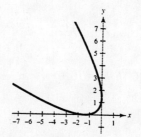

$$\left(\frac{x' - y'}{\sqrt{2}}\right)^2 + \left(\frac{x' + y'}{\sqrt{2}}\right)^2 + 2\left(\frac{x' - y'}{\sqrt{2}}\right)\left(\frac{x' + y'}{\sqrt{2}}\right) + 2\sqrt{2}\left(\frac{x' - y'}{\sqrt{2}}\right) - 2\sqrt{2}\left(\frac{x' + y'}{\sqrt{2}}\right) + 2 = 0$$

$$2(x')^2 - 4y' + 2 = 0$$

$$(x')^2 = 2y' - 1$$

Vertex: $(x', y') = \left(0, \dfrac{1}{2}\right)$, $\theta = 45°$

**31.** $AC = 3(2) = 6 > 0$

The graph is an ellipse.

$$3x^2 + 2y^2 - 12x + 12y + 29 = 0$$

$$\frac{(x - 2)^2}{1/3} + \frac{(y + 3)^2}{1/2} = 1$$

Center: $(2, -3)$

Vertices: $\left(2, -3 \pm \dfrac{\sqrt{2}}{2}\right)$

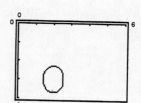

To use a graphing calculator, we need to solve for $y$ in terms of $x$.

$$2(y + 3)^2 = 1 - 3(x - 2)^2|$$

$$y = -3 \pm \sqrt{\frac{1 - (x - 2)^2}{2}}$$

**33.** $x^2 - 10xy + y^2 + 1 = 0$

Since $B^2 - 4AC = (-10)^2 - 4(1)(1) > 0$, the graph is a hyperbola.

To use a graphing calculator, we need to solve for $y$ in terms of $x$.

$(y^2 - 10xy + 25x^2) = -x^2 - 1 + 25x^2$

$(y - 5x)^2 = 24x^2 - 1$

$y = 5x \pm \sqrt{24x^2 - 1}$

**35.** Vertex: $(4, 2) = (h, k)$

Focus: $(4, 0) \Rightarrow p = -2$

$(x - h)^2 = 4p(y - k)$

$(x - 4)^2 = -8(y - 2)$

**37.** Vertex: $(0, 2) = (h, k)$

Directrix: $x = -3 \Rightarrow p = 3$

$(y - k^2) = 4p(x - h)$

$(y - 2)^2 = 12x$

**39.** Vertices: $(-3, 0), (7, 0) \Rightarrow a = 5$

$\qquad (h, k) = (2, 0)$

Foci: $(0, 0), (4, 0) \Rightarrow c = 2$

$b^2 = a^2 - c^2 = 25 - 4 = 21$

$\dfrac{(x - h)^2}{a^2} + \dfrac{(y - k)^2}{b^2} = 1$

$\dfrac{(x - 2)^2}{25} + \dfrac{y^2}{21} = 1$

**41.** Vertices: $(0, \pm 6) \Rightarrow a = 6, (h, k) = (0, 0)$

Passes through $(2, 2)$

$\dfrac{(x - h)^2}{b^2} + \dfrac{(y - k)^2}{a^2} = 1$

$\dfrac{x^2}{b^2} + \dfrac{y^2}{36} = 1 \Rightarrow b^2 = \dfrac{36(4)}{36 - 4} = \dfrac{36x^2}{36 - y^2} = \dfrac{9}{2}$

$\dfrac{x^2}{9/2} + \dfrac{y^2}{36} = 1$

$\dfrac{2x^2}{9} + \dfrac{y^2}{36} = 1$

**43.** Vertices: $(0, \pm 1) \Rightarrow a = 1, (h, k) = (0, 0)$

Foci: $(0, \pm 3) \Rightarrow c = 3$

$b^2 = c^2 - a^2 = 9 - 1 = 8$

$\dfrac{(y - k)^2}{a^2} - \dfrac{(x - h)^2}{b^2} = 1$

$y^2 - \dfrac{x^2}{8} = 1$

**45.** Foci: $(0, 0), (8, 0) \Rightarrow c = 4, (h, k) = (4, 0)$

Asymptotes: $y = \pm 2(x - 4) \Rightarrow \dfrac{b}{a} = 2, b = 2a$

$b^2 = c^2 - a^2 \Rightarrow 4a^2 = 16 - a^2 \Rightarrow a^2 = \dfrac{16}{5},$

$b^2 = \dfrac{64}{5}$

$\dfrac{(x - h)^2}{a^2} - \dfrac{(y - k)^2}{b^2} = 1$

$\dfrac{(x - 4)^2}{16/5} - \dfrac{y^2}{64/5} = 1$

$\dfrac{5(x - 4)^2}{16} - \dfrac{5y^2}{64} = 1$

**47.** Parabola

Opens downward

Vertex: $(0, 12)$

$(x - h)^2 = 4p(y - k)$

$\qquad x^2 = 4p(y - 12)$

Solution points: $(\pm 4, 10)$

$\quad 16 = 4p(10 - 12)$

$\quad 16 = -8p$

$\quad -2 = p$

$\quad x^2 = -8(y - 12)$

To find the $x$-intercepts, let $y = 0$.

$x^2 = 96$

$\quad x = \pm\sqrt{96} = \pm 4\sqrt{6}$

At the base,z the archway is $2\left(4\sqrt{6}\right) = 8\sqrt{6}$ meters wide.

**49.** $BD = AD + 6\left(\dfrac{1100}{5280}\right)$

$\quad CD = AD + 8\left(\dfrac{1100}{5280}\right)$

$\quad 2a = CD - BD = 2\left(\dfrac{1100}{5280}\right)$

$\quad a = \dfrac{5}{24}, c = 2 \Rightarrow b^2 = \dfrac{2279}{576}$

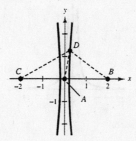

Thus, we have $\dfrac{576x^2}{25} - \dfrac{576y^2}{2279} = 1$ OR:

$\quad CD = AD + 8\left(\dfrac{1100}{5280}\right)$

$\quad BD = AD + 6\left(\dfrac{1100}{5280}\right)$

$\quad 2a = BD - AD = 6\left(\dfrac{1100}{5280}\right)$

$\quad a = 3\left(\dfrac{5}{24}\right) = \dfrac{5}{8}, c = 1 \Rightarrow b^2 = \dfrac{39}{64}$

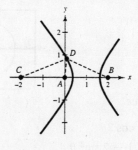

Center: $(1, 0)$

$\dfrac{64(x - 1)^2}{25} - \dfrac{64y^2}{39} = 1$

$\quad CD - 8\left(\dfrac{1100}{5280}\right) = AD$

$\qquad BD = CD - 8\left(\dfrac{1100}{5280}\right) + 6\left(\dfrac{1100}{5280}\right) = CD - 2\left(\dfrac{1100}{5280}\right)$

Thus, the friend to the west hears the explosion two seconds after the friend to the east.

**51.** $x = 2t \Rightarrow \dfrac{x}{2} = t$

$y = 4t \Rightarrow y = 4\left(\dfrac{x}{2}\right) = 2x$

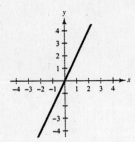

**53.** $x = 1 + 4t,\ y = 2 - 3t$

$t = \dfrac{x - 1}{4}$

$y = 2 - 3\left(\dfrac{x - 1}{4}\right)$

$3x + 4y = 11$

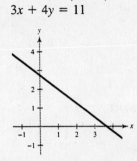

**55.** $x = \dfrac{1}{t},\ y = t^2$

$t = \dfrac{1}{x}$

$y = \dfrac{1}{x^2}$

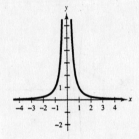

**57.** $x = 6\cos\theta,\ y = 6\sin\theta$

$\cos\theta = \dfrac{x}{6},\ \sin\theta = \dfrac{y}{6}$

$\dfrac{x^2}{36} + \dfrac{y^2}{36} = 1$

$x^2 + y^2 = 36$

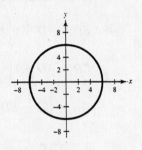

**59.** $(h, k) = (-3, 4)$

$2a = 8 \Rightarrow a = 4$

$2b = 6 \Rightarrow b = 3$

$\dfrac{(x + 3)^2}{16} + \dfrac{(y - 4)^2}{9} = 1$

$x = -3 + 4\cos\theta$

$y = 4 + 3\sin\theta$

This solution is not unique.

**61.** $x = \cos 3\theta + 5\cos\theta$

$y = \sin 3\theta + 5\sin\theta$

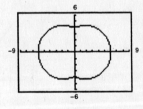

**63.**    $r = 3\cos\theta$

$r^2 = 3r\cos\theta$

$x^2 + y^2 = 3x$

**65.**    $r^2 = \cos 2\theta$

$r^2 = 1 - 2\sin^2\theta$

$r^4 = r^2 - 2r^2\sin^2\theta$

$(x^2 + y^2)^2 = x^2 + y^2 - 2y^2$

$(x^2 + y^2)^2 - x^2 + y^2 = 0$

**67.** $(x^2 + y^2)^2 = ax^2y$

$(r^2)^2 = ar^2\cos^2\theta\, r\sin\theta$

$r = a\cos^2\theta\sin\theta$

**69.** $r = 4$

Circle of radius 4 centered at the pole

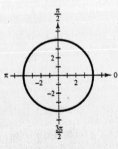

**71.** $r = 4\sin 2\theta$

Symmetric with respect to $\theta = \pi/2$

Rose curve ($n = 2$) with 4 petals

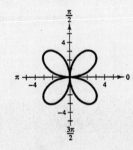

**73.** $r = -2 - 2 \cos \theta$

Symmetric with respect to polar axis

$\dfrac{a}{b} = \dfrac{2}{2} = 1 \implies$ Cardioid

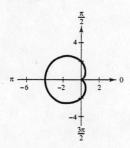

**75.** $r = 4 - 3 \cos \theta$

Symmetric with respect to polar axis

$\dfrac{a}{b} = \dfrac{4}{3} > 0 \implies$ Dimpled limaçon

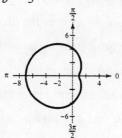

**77.** $r = -3 \cos 2\theta$

Symmetric with respect to the polar axis, $\theta = \pi/2$, and the pole

Rose curve with 4 petals

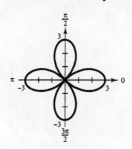

**79.** $r = \dfrac{2}{1 - \sin \theta}$, $e = 1$

Parabola symmetric with $\theta = \pi/2$ and the vertex at $(1, 3\pi/2)$

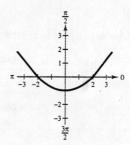

**81.** $r^2 = 4 \sin^2 2\theta \implies r = \pm 2 \sin 2\theta$

Symmetric with respect to $\theta = \pi/2$, polar axis, and pole

Rose curve ($n = 2$) with 4 petals

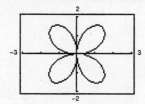

**83.** $r = \dfrac{3}{\cos(\theta - \pi/4)}$

The graph is a line.

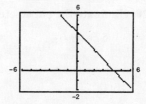

**85.** Center: $(5, \pi/2)$

Solution point: $(0, 0) \implies$ Radius $= 5 \implies a = 10$

$r = a \sin \theta$

$r = 10 \sin \theta$

**87.** Parabola: $r = \dfrac{ep}{1 - e \cos \theta}$, $e = 1$

Vertex: $(2, \pi)$

Focus: $(0, 0) \implies p = 4$

$r = \dfrac{4}{1 - \cos \theta}$

**89.** Ellipse: $r = \dfrac{ep}{1 - e \cos \theta}$;   Vertices: $(5, 0), (1, \pi) \implies a = 3$;   One focus: $(0, 0) \implies c = 2$

$e = \dfrac{c}{a} = \dfrac{2}{3}, p = \dfrac{5}{2}$

$r = \dfrac{(2/3)(5/2)}{1 - (2/3) \cos \theta} = \dfrac{5/3}{1 - (2/3) \cos \theta} = \dfrac{5}{3 - 2 \cos \theta}$

# ❑ Practice Test for Chapter 6

1. Find the angle, $\theta$, between the lines $3x + 4y = 12$ and $4x - 3y = 12$.

2. Find the distance between the point $(5, -9)$ and the line $3x - 7y = 21$.

3. Find the vertex, focus and directrix of the parabola $x^2 - 6x - 4y + 1 = 0$.

4. Find an equation of the parabola with its vertex at $(2, -5)$ and focus at $(2, -6)$.

5. Find the center, foci, vertices, and eccentricity of the ellipse $x^2 + 4y^2 - 2x + 32y + 61 = 0$.

6. Find an equation of the ellipse with vertices $(0, \pm 6)$ and eccentricity $e = \frac{1}{2}$.

7. Find the center, vertices, foci, and asymptotes of the hyperbola $16y^2 - x^2 - 6x - 128y + 231 = 0$.

8. Find an equation of the hyperbola with vertices at $(\pm 3, 2)$ and foci at $(\pm 5, 2)$.

9. Rotate the axes to eliminate the $xy$-term. Sketch the graph of the resulting equation, showing both sets of axes.

   $5x^2 + 2xy + 5y^2 - 10 = 0$

10. Use the discriminant to determine whether the graph of the equation is a parabola, ellipse, or hyperbola.

    (a) $6x^2 - 2xy + y^2 = 0$            (b) $x^2 + 4xy + 4y^2 - x - y + 17 = 0$

11. Convert the polar point $\left(\sqrt{2}, (3\pi)/4\right)$ to rectangular coordinates.

12. Convert the rectangular point $\left(\sqrt{3}, -1\right)$ to polar coordinates.

13. Convert the rectangular equation $4x - 3y = 12$ to polar form.

14. Convert the polar equation $r = 5 \cos \theta$ to rectangular form.

15. Sketch the graph of $r = 1 - \cos \theta$.

16. Sketch the graph of $r = 5 \sin 2\theta$.

17. Sketch the graph of $r = \dfrac{3}{6 - \cos \theta}$.

18. Find a polar equation of the parabola with its vertex at $\left(6, \pi/2\right)$ and focus at $(0, 0)$.

**For Exercises 19 and 20, eliminate the parameter and write the corresponding rectangular equation.**

19. $x = 3 - 2 \sin \theta, y = 1 + 5 \cos \theta$            20. $x = e^{2t}, y = e^{4t}$

# Practice Test Solutions

## ❑ Chapter P Practice Test Solutions

**1.** $\dfrac{|-42| - 20}{15 - |-4|} = \dfrac{42 - 20}{15 - 4} = \dfrac{22}{11} = 2$

**2.** $|x - 7| \leq 4$

**3.** (a)

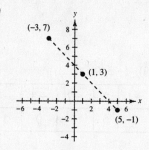

(b) $d = \sqrt{[5 - (-3)]^2 + (-1 - 7)^2}$

$= \sqrt{(8)^2 + (-8)^2}$

$= \sqrt{64 + 64}$

$= \sqrt{128}$

$= 8\sqrt{2}$

(c) $\left( \dfrac{-3 + 5}{2}, \dfrac{7 + (-1)}{2} \right) = (1, 3)$

**4.** $5x + 4 = 7x - 8$

$4 + 8 = 7x - 5x$

$12 = 2x$

$x = 6$

**5.** $\dfrac{x}{3} - 5 = \dfrac{x}{5} + 1$

$15\left( \dfrac{x}{3} - 5 \right) = 15\left( \dfrac{x}{5} + 1 \right)$

$5x - 75 = 3x + 15$

$2x = 90$

$x = 45$

**6.** $\dfrac{3x + 1}{6x - 7} = \dfrac{2}{5}$

$5(3x + 1) = 2(6x - 7)$

$15x + 5 = 12x - 14$

$3x = -19$

$x = -\dfrac{19}{3}$

**7.** $(x - 3)^2 + 4 = (x + 1)^2$

$x^2 - 6x + 9 + 4 = x^2 + 2x + 1$

$-8x = -12$

$x = \dfrac{-12}{-8}$

$x = \dfrac{3}{2}$

**8.** $3x - 5y = 15$

Line

$x$-intercept: $(5, 0)$

$y$-intercept: $(0, -3)$

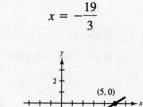

**9.** $y = \sqrt{9 - x}$

Domain: $(-\infty, 9]$

$x$-intercept: $(9, 0)$

$y$-intercept: $(0, 3)$

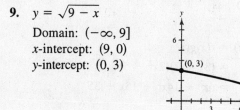

**10.** $m = \dfrac{-1 - 4}{3 - 2} = -5$

$y - 4 = -5(x - 2)$

$y - 4 = -5x + 10$

$y = -5x + 14$

**11.** $y = \dfrac{4}{3}x - 3$

**12.** $f(x - 3) = (x - 3)^2 - 2(x - 3) + 1$

$= x^2 - 6x + 9 - 2x + 6 + 1$

$= x^2 - 8x + 16$

**13.** $f(3) = 12 - 11 = 1$

$\dfrac{f(x) - f(3)}{x - 3} = \dfrac{(4x - 11) - 1}{x - 3}$

$= \dfrac{4x - 12}{x - 3}$

$= \dfrac{4(x - 3)}{x - 3} = 4, \ x \neq 3$

**14.** $f(x) = \sqrt{36 - x^2} = \sqrt{(6 + x)(6 - x)}$

Domain: $[\,6, 6]$ because $(6 + x)(6 - x) \geq 0$ on this interval

Range: $[0, 6]$

**15.** (a) $6x - 5y + 4 = 0$

$$y = \frac{6x + 4}{5} \text{ is a function of } x.$$

(b) $x^2 + y^2 = 9$

$\quad y = \pm\sqrt{9 - x^2}$ is not a function of $x$.

(c) $y^3 = x^2 + 6$

$\quad y = \sqrt[3]{x^2 + 6}$ is a function of $x$.

**16.** Parabola

Vertex: $(0, -5)$

Intercepts: $(0, -5)$, $\left(\pm\sqrt{5}, 0\right)$

$y$-axis symmetry

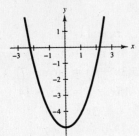

**17.** Intercepts: $(0, 3)$, $(-3, 0)$

| $x$ | 0 | 1 | −1 | 2 | −2 | −3 | −4 |
|---|---|---|---|---|---|---|---|
| $y$ | 3 | 4 | 2 | 5 | 1 | 0 | 1 |

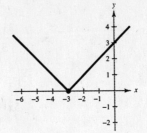

**18.** (a) $f(x + 2)$

Horizontal shift two units to the left

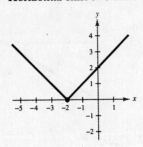

(b) $-f(x) + 2$

Reflection in the $x$-axis and a vertical shift two units upward

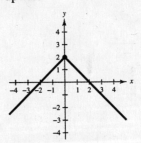

**19.** (a) $(g - f)(x) = g(x) - f(x)$

$\qquad = (2x^2 - 5) - (3x + 7)$

$\qquad = 2x^2 - 3x - 12$

(b) $(fg)(x) = f(x)g(x)$

$\qquad = (3x + 7)(2x^2 - 5)$

$\qquad = 6x^3 + 14x^2 - 15x - 35$

**20.** $f(g(x)) = f(2x + 3)$

$\qquad = (2x + 3)^2 - 2(2x + 3) + 16$

$\qquad = 4x^2 + 12x + 9 - 4x - 6 + 16$

$\qquad = 4x^2 + 8x + 19$

**21.** $\qquad f(x) = x^3 + 7$

$\qquad y = x^3 + 7$

$\qquad x = y^3 + 7$

$\qquad x - 7 = y^3$

$\qquad \sqrt[3]{x - 7} = y$

$\qquad f^{-1}(x) = \sqrt[3]{x - 7}$

**22. (a)** $f(x) = |x - 6|$ does not have an inverse. Its graph does not pass the horizontal line test.

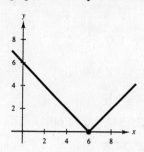

**(b)**    $f(x) = ax + b, a \neq 0$ does have an inverse.

$$y = ax + b$$

$$x = ay + b$$

$$\frac{x - b}{a} = y$$

$$f^{-1}(x) = \frac{x - b}{a}$$

**(c)**    $f(x) = x^3 - 19$  does have an inverse.

$$y = x^3 - 19$$

$$x = y^3 - 19$$

$$x + 19 = y^3$$

$$\sqrt[3]{x + 19} = y$$

$$f^{-1}(x) = \sqrt[3]{x + 19}$$

**23.** $[x - (-3)]^2 + (y - 5)^2 = 6^2$

$(x + 3)^2 + (y - 5)^2 = 36$

**24.** (5, 32) and (9, 44)

$$m = \frac{44 - 32}{9 - 5} = \frac{12}{4} = 3$$

$$y - 32 = 3(x - 5)$$

$$y - 32 = 3x - 15$$

$$y = 3x + 17$$

When $x = 20$, $y = 3(20) + 17$

$$y = \$77.$$

**25.**

$$60,000 = xy$$

$$y = \frac{60,000}{x}$$

$$2x + 2y = 1100$$

$$2x + 2\left(\frac{60,000}{x}\right) = 1100$$

$$x + \frac{60,000}{x} = 550$$

$$x^2 + 60,000 = 550x$$

$$x^2 - 550x + 60,000 = 0$$

$$(x - 150)(x - 400) = 0$$

$$x = 150 \quad \text{or} \quad x = 400$$

$$y = 400 \qquad y = 150$$

Length:  400 feet

Width:  150 feet

# ❑ Chapter 1  Practice Test Solutions

**1.** $350° = 350\left(\dfrac{\pi}{180}\right) = \dfrac{35\pi}{18}$

**2.** $\dfrac{5\pi}{9} = \dfrac{5\pi}{9} \cdot \dfrac{180}{\pi} = 100°$

**3.** $135° \, 14' \, 12'' = \left(135 + \dfrac{14}{60} + \dfrac{12}{3600}\right)°$

$\approx 135.2367°$

**4.** $-22.569° = -(22° + 0.569(60)')$

$= -22° \, 34.14'$

$= -(22° \, 34' + 0.14(60)'')$

$\approx -22° \, 34' \, 8''$

**5.** $\cos\theta = \dfrac{2}{3}$

$x = 2, \ r = 3, \ y = \pm\sqrt{9-4} = \pm\sqrt{5}$

$\tan\theta = \dfrac{y}{x} = \pm\dfrac{\sqrt{5}}{2}$

**6.** $\sin\theta = 0.9063$

$\theta = \arcsin(0.9063)$

$\theta = 65° = \dfrac{13\pi}{36}$

or

$\theta = 180° - 65° = 115° = \dfrac{23\pi}{36}$

**7.** $\tan 20° = \dfrac{35}{x}$

$x = \dfrac{35}{\tan 20°} \approx 96.1617$

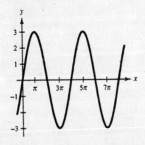

**8.** $\theta = \dfrac{6\pi}{5}$, $\theta$ is in Quadrant III.

Reference angle: $\dfrac{6\pi}{5} - \pi = \dfrac{\pi}{5}$ or $36°$

**9.** $\csc 3.92 = \dfrac{1}{\sin 3.92} \approx -1.4242$

**10.** $\tan\theta = 6 = \dfrac{6}{1}$, $\theta$ lies in Quadrant III.

$y = -6, \ x = -1, \ r = \sqrt{36+1} = \sqrt{37}$,

so $\sec\theta = \dfrac{\sqrt{37}}{-1} \approx -6.0828$.

**11.** Period: $4\pi$

Amplitude: 3

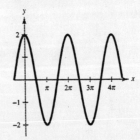

**12.** Period: $2\pi$

Amplitude: 2

**13.** Period: $\dfrac{\pi}{2}$

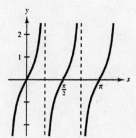

**14.** Period: $2\pi$

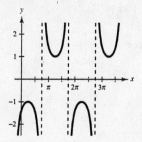

**15.**

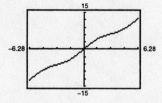

**16.**

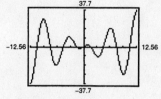

**17.**  $\theta = \arcsin 1$

$\sin \theta = 1$

$\theta = \dfrac{\pi}{2}$

**18.**  $\theta = \arctan(-3)$

$\tan \theta = -3$

$\theta \approx -1.249 \approx -71.565°$

**19.** $\sin\left(\arccos \dfrac{4}{\sqrt{35}}\right)$

$\sin \theta = \dfrac{\sqrt{19}}{\sqrt{35}} \approx 0.7368$

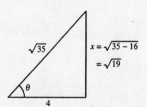

**20.** $\cos\left(\arcsin \dfrac{x}{4}\right)$

$\cos \theta = \dfrac{\sqrt{16 - x^2}}{4}$

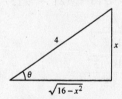

**21.** Given $A = 40°$, $c = 12$

$B = 90° - 40° = 50°$

$\sin 40° = \dfrac{a}{12}$

$a = 12 \sin 40° \approx 7.713$

$\cos 40° = \dfrac{b}{12}$

$b = 12 \cos 40° \approx 9.193$

**22.** Given $B = 6.84°$, $a = 21.3$

$A = 90° - 6.84° = 83.16°$

$\sin 83.16° = \dfrac{21.3}{c}$

$c = \dfrac{21.3}{\sin 83.16°} \approx 21.453$

$\tan 83.16° = \dfrac{21.3}{b}$

$b = \dfrac{21.3}{\tan 83.16°} \approx 2.555$

**23.** Given $a = 5, b = 9$

$c = \sqrt{25 + 81} = \sqrt{106} \approx 10.296$

$\tan A = \frac{5}{9}$

$A = \arctan \frac{5}{9} \approx 29.055°$

$B = 90° - 29.055° = 60.945°$

**24.** $\sin 67° = \dfrac{x}{20}$

$x = 20 \sin 67° \approx 18.41$ feet

**25.** $\tan 5° = \dfrac{250}{x}$

$x = \dfrac{250}{\tan 5°}$

$\approx 2857.513$ feet

$\approx 0.541$ mi

# ❑ Chapter 2  Practice Test Solutions

**1.** $\tan x = \dfrac{4}{11}$, $\sec x < 0 \implies x$ is in Quadrant III.

$y = -4, x = -11, r = \sqrt{16 + 121} = \sqrt{137}$

$\sin x = -\dfrac{4}{\sqrt{137}} = -\dfrac{4\sqrt{137}}{137}$          $\csc x = -\dfrac{\sqrt{137}}{4}$

$\cos x = -\dfrac{11}{\sqrt{137}} = -\dfrac{11\sqrt{137}}{137}$          $\sec x = -\dfrac{\sqrt{137}}{11}$

$\tan x = \dfrac{4}{11}$          $\cot x = \dfrac{11}{4}$

**2.** $\dfrac{\sec^2 x + \csc^2 x}{\csc^2 x(1 + \tan^2 x)} = \dfrac{\sec^2 x + \csc^2 x}{\csc^2 x + (\csc^2 x)\tan^2 x} = \dfrac{\sec^2 x + \csc^2 x}{\csc^2 x + \dfrac{1}{\sin^2 x} \cdot \dfrac{\sin^2 x}{\cos^2 x}}$

$= \dfrac{\sec^2 x + \csc^2 x}{\csc^2 x + \dfrac{1}{\cos^2 x}} = \dfrac{\sec^2 x + \csc^2 x}{\csc^2 x + \sec^2 x} = 1$

**3.** $\ln|\tan\theta| - \ln|\cot\theta| = \ln\left|\dfrac{\tan\theta}{\cot\theta}\right| = \ln\left|\dfrac{\sin\theta/\cos\theta}{\cos\theta/\sin\theta}\right| = \ln\left|\dfrac{\sin^2\theta}{\cos^2\theta}\right| = \ln|\tan^2\theta| = 2\ln|\tan\theta|$

**4.** $\cos\left(\dfrac{\pi}{2} - x\right) = \dfrac{1}{\csc x}$ is true since $\cos\left(\dfrac{\pi}{2} - x\right) = \sin x = \dfrac{1}{\csc x}$.

**5.** $\sin^4 x + (\sin^2 x)\cos^2 x = \sin^2 x(\sin^2 x + \cos^2 x) = \sin^2 x(1) = \sin^2 x$

**6.** $(\csc x + 1)(\csc x - 1) = \csc^2 x - 1 = \cot^2 x$

**7.** $\dfrac{\cos^2 x}{1 - \sin x} \cdot \dfrac{1 + \sin x}{1 + \sin x} = \dfrac{\cos^2 x(1 + \sin x)}{1 - \sin^2 x} = \dfrac{\cos^2 x(1 + \sin x)}{\cos^2 x} = 1 + \sin x$

**8.** $\dfrac{1 + \cos\theta}{\sin\theta} + \dfrac{\sin\theta}{1 + \cos\theta} = \dfrac{(1 + \cos\theta)^2 + \sin^2\theta}{\sin\theta(1 + \cos\theta)}$

$$= \dfrac{1 + 2\cos\theta + \cos^2\theta + \sin^2\theta}{\sin\theta(1 + \cos\theta)} = \dfrac{2 + 2\cos\theta}{\sin\theta(1 + \cos\theta)} = \dfrac{2}{\sin\theta} = 2\csc\theta$$

**9.** $\tan^4 x + 2\tan^2 x + 1 = (\tan^2 x + 1)^2 = (\sec^2 x)^2 = \sec^4 x$

**10.** (a) $\sin 105° = \sin(60° + 45°) = \sin 60° \cos 45° + \cos 60° \sin 45°$

$$= \dfrac{\sqrt{3}}{2} \cdot \dfrac{\sqrt{2}}{2} + \dfrac{1}{2} \cdot \dfrac{\sqrt{2}}{2} = \dfrac{\sqrt{2}}{4}\left(\sqrt{3} + 1\right)$$

(b) $\tan 15° = \tan(60° - 45°) = \dfrac{\tan 60° - \tan 45°}{1 + \tan 60° \tan 45°}$

$$= \dfrac{\sqrt{3} - 1}{1 + \sqrt{3}} \cdot \dfrac{1 - \sqrt{3}}{1 - \sqrt{3}} = \dfrac{2\sqrt{3} - 1 - 3}{1 - 3} = \dfrac{2\sqrt{3} - 4}{-2} = 2 - \sqrt{3}$$

**11.** $(\sin 42°)\cos 38° - (\cos 42°)\sin 38° = \sin(42° - 38°) = \sin 4°$

**12.** $\tan\left(\theta + \dfrac{\pi}{4}\right) = \dfrac{\tan\theta + \tan(\pi/4)}{1 - (\tan\theta)\tan(\pi/4)} = \dfrac{\tan\theta + 1}{1 - \tan\theta(1)} = \dfrac{1 + \tan\theta}{1 - \tan\theta}$

**13.** $\sin(\arcsin x - \arccos x) = \sin(\arcsin x)\cos(\arccos x) - \cos(\arcsin x)\sin(\arccos x)$

$$= (x)(x) - \left(\sqrt{1 - x^2}\right)\left(\sqrt{1 - x^2}\right) = x^2 - (1 - x^2) = 2x^2 - 1$$

**14.** (a) $\cos(120°) = \cos[2(60°)] = 2\cos^2 60° - 1 = 2\left(\dfrac{1}{2}\right)^2 - 1 = -\dfrac{1}{2}$

(b) $\tan(300°) = \tan[2(150°)] = \dfrac{2\tan 150°}{1 - \tan^2 150°} = \dfrac{-2\sqrt{3}/3}{1 - (1/3)} = -\sqrt{3}$

**15.** (a) $\sin 22.5° = \sin\dfrac{45°}{2} = \sqrt{\dfrac{1 - \cos 45°}{2}} = \sqrt{\dfrac{1 - \sqrt{2}/2}{2}} = \dfrac{\sqrt{2 - \sqrt{2}}}{2}$

(b) $\tan\dfrac{\pi}{12} = \tan\dfrac{\pi/6}{2} = \dfrac{\sin(\pi/6)}{1 + \cos(\pi/6)} = \dfrac{1/2}{1 + \sqrt{3}/2} = \dfrac{1}{2 + \sqrt{3}} = 2 - \sqrt{3}$

**16.** $\sin\theta = \dfrac{4}{5}$, $\theta$ lies in Quadrant II $\implies$ $\cos\theta = -\dfrac{3}{5}$.

$$\cos\dfrac{\theta}{2} = \sqrt{\dfrac{1 + \cos\theta}{2}} = \sqrt{\dfrac{1 - 3/5}{2}} = \sqrt{\dfrac{2}{10}} = \dfrac{1}{\sqrt{5}} = \dfrac{\sqrt{5}}{5}$$

1. $(\sin^2 x)\cos^2 x = \dfrac{1 - \cos 2x}{2} \cdot \dfrac{1 + \cos 2x}{2} = \dfrac{1}{4}[1 - \cos^2 2x] = \dfrac{1}{4}\left[1 - \dfrac{1 + \cos 4x}{2}\right]$

$\qquad\qquad\quad = \dfrac{1}{8}[2 - (1 + \cos 4x)] = \dfrac{1}{8}[1 - \cos 4x]$

18. $6(\sin 5\theta)\cos 2\theta = 6\left\{\dfrac{1}{2}[\sin(5\theta + 2\theta) + \sin(5\theta - 2\theta)]\right\} = 3[\sin 7\theta + \sin 3\theta]$

19. $\sin(x + \pi) + \sin(x - \pi) = 2\left(\sin\dfrac{[(x + \pi) + (x - \pi)]}{2}\right)\cos\dfrac{[(x + \pi) - (x - \pi)]}{2} = 2\sin x \cos\pi = -2\sin x$

20. $\dfrac{\sin 9x + \sin 5x}{\cos 9x - \cos 5x} = \dfrac{2 \sin 7x \cos 2x}{-2 \sin 7x \sin 2x} = -\dfrac{\cos 2x}{\sin 2x} = -\cot 2x$

21. $\dfrac{1}{2}[\sin(u + v) - \sin(u - v)] = \dfrac{1}{2}\{(\sin u)\cos v + (\cos u)\sin v - [(\sin u)\cos v - (\cos u)\sin v]\}$

$\qquad\qquad\qquad\qquad\qquad = \dfrac{1}{2}[2(\cos u)\sin v] = (\cos u)\sin v$

22. $4 \sin^2 x = 1$

$\qquad \sin^2 x = \dfrac{1}{4}$

$\qquad \sin x = \pm\dfrac{1}{2}$

$\qquad \sin x = \dfrac{1}{2} \qquad$ or $\quad \sin x = -\dfrac{1}{2}$

$\qquad x = \dfrac{\pi}{6}$ or $\dfrac{5\pi}{6} \qquad x = \dfrac{7\pi}{6}$ or $\dfrac{11\pi}{6}$

23. $\tan^2\theta + (\sqrt{3} - 1)\tan\theta - \sqrt{3} = 0$

$\qquad (\tan\theta - 1)(\tan\theta + \sqrt{3}) = 0$

$\qquad \tan\theta = 1 \qquad$ or $\quad \tan\theta = -\sqrt{3}$

$\qquad \theta = \dfrac{\pi}{4}$ or $\dfrac{5\pi}{4} \qquad \theta = \dfrac{2\pi}{3}$ or $\dfrac{5\pi}{3}$

24. $\qquad\qquad\qquad \sin 2x = \cos x$

$\qquad 2(\sin x)\cos x - \cos x = 0$

$\qquad\quad \cos x(2 \sin x - 1) = 0$

$\qquad \cos x = 0 \qquad$ or $\quad \sin x = \dfrac{1}{2}$

$\qquad x = \dfrac{\pi}{2}$ or $\dfrac{3\pi}{2} \qquad x = \dfrac{\pi}{6}$ or $\dfrac{5\pi}{6}$

25. $\tan^2 x - 6 \tan x + 4 = 0$

$\qquad\qquad \tan x = \dfrac{-(-6) \pm \sqrt{(-6)^2 - 4(1)(4)}}{2(1)}$

$\qquad\qquad \tan x = \dfrac{6 \pm \sqrt{20}}{2} = 3 \pm \sqrt{5}$

$\qquad \tan x = 3 + \sqrt{5} \qquad$ or $\quad \tan x = 3 - \sqrt{5}$

$\qquad x \approx 1.3821$ or $4.5237 \qquad x = 0.6524$ or $3.7940$

# ❑ Chapter 3  Practice Test Solutions

**1.** $C = 180° − (40° + 12°) = 128°$

$$a = \sin 40° \left( \frac{100}{\sin 12°} \right) \approx 309.164$$

$$c = \sin 128° \left( \frac{100}{\sin 12°} \right) \approx 379.012$$

**2.** $\sin A = 5 \left( \dfrac{\sin 150°}{20} \right) = 0.125$

$A \approx 7.181°$

$B \approx 180° − (150° + 7.181°) = 22.819°$

$b = \sin 22.819° \left( \dfrac{20}{\sin 150°} \right) \approx 15.513$

**3.** Area $= \frac{1}{2} ab \sin C$

$\quad = \frac{1}{2}(3)(5)\sin 130°$

$\quad \approx 5.745$ square units

**4.** $h = b \sin A$

$\quad = 35 \sin 22.5°$

$\quad \approx 13.394$

$a = 10$

Since $a < h$ and $A$ is acute, the triangle has no solution.

**5.** $\cos A = \dfrac{(53)^2 + (38)^2 − (49)^2}{2(53)(38)} \approx 0.4598$

$\quad A \approx 62.627°$

$\cos B = \dfrac{(49)^2 + (38)^2 − (53)^2}{2(49)(38)} \approx 0.2782$

$\quad B \approx 73.847°$

$\quad C \approx 180° − (62.627° + 73.847°)$

$\quad\quad = 43.526°$

**6.** $\quad c^2 = (100)^2 + (300)^2 − 2(100)(300)\cos 29°$

$\quad\quad \approx 47522.8176$

$\quad c \approx 218$

$\cos A = \dfrac{(300)^2 + (218)^2 − (100)^2}{2(300)(218)} \approx 0.97495$

$\quad A \approx 12.85°$

$\quad B \approx 180° − (12.85° + 29°) = 138.15°$

**7.** $\quad s = \dfrac{a + b + c}{2} = \dfrac{4.1 + 6.8 + 5.5}{2} = 8.2$

Area $= \sqrt{s(s − a)(s − b)(s − c)}$

$\quad = \sqrt{8.2(8.2 − 4.1)(8.2 − 6.8)(8.2 − 5.5)}$

$\quad \approx 11.273$ square units

**8.** $x^2 = (40)^2 + (70)^2 − 2(40)(70)\cos 168°$

$\quad \approx 11977.6266$

$\quad x \approx 190.442$ miles

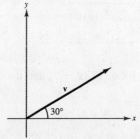

**9.** $\mathbf{w} = 4(3\mathbf{i} + \mathbf{j}) − 7(−\mathbf{i} + 2\mathbf{j})$

$\quad = 19\mathbf{i} − 10\mathbf{j}$

**10.** $\dfrac{\mathbf{v}}{\|\mathbf{v}\|} = \dfrac{5\mathbf{i} + 3\mathbf{j}}{\sqrt{25 + 9}} = \dfrac{5}{\sqrt{34}}\mathbf{i} − \dfrac{3}{\sqrt{34}}\mathbf{j}$

$\quad = \dfrac{5\sqrt{34}}{34}\mathbf{i} − \dfrac{3\sqrt{34}}{34}\mathbf{j}$

**11.** $\quad \mathbf{u} = 6\mathbf{i} + 5\mathbf{j} \quad\quad \mathbf{v} = 2\mathbf{i} − 3\mathbf{j}$

$\mathbf{u} \cdot \mathbf{v} = 6(2) + 5(−3) = −3$

$\|\mathbf{u}\| = \sqrt{61} \quad\quad \|\mathbf{v}\| = \sqrt{13}$

$\cos \theta = \dfrac{−3}{\sqrt{61}\sqrt{13}}$

$\quad \theta \approx 96.116°$

**12.** $\quad 4(\mathbf{i} \cos 30° + \mathbf{j} \sin 30°)$

$\quad = 4 \left( \dfrac{\sqrt{3}}{2}\mathbf{i} + \dfrac{1}{2}\mathbf{j} \right)$

$\quad = \langle 2\sqrt{3}, 2 \rangle$

**13.** $\text{proj}_{\mathbf{v}}\mathbf{u} = \left(\dfrac{\mathbf{u} \cdot \mathbf{v}}{\|\mathbf{v}\|^2}\right)\mathbf{v} = \dfrac{-10}{20}\langle -2, 4\rangle = \langle 1, -2\rangle$

**14.** $\mathbf{u} = 7\cos 35°\mathbf{i} + 7\sin 35°\mathbf{j}$

$\mathbf{v} = 4\cos 123°\mathbf{i} + 4\sin 123°\mathbf{j}$

$\mathbf{u} + \mathbf{v} = (7\cos 35° + 4\cos 123°)\mathbf{i} + (7\sin 35° + 4\sin 123°)\mathbf{j}$

$\approx \langle 3.56, 7.37\rangle$

**15.** Answer is not unique. Two possibilities are: $\langle 10, 3\rangle$ and $\langle -10, -3\rangle$

# ❑ Chapter 4  Practice Test Solutions

**1.** $4 + \sqrt{-81} - 3i^2 = 4 + 9i + 3 = 7 + 9i$

**2.** $\dfrac{3 + i}{5 - 4i} \cdot \dfrac{5 + 4i}{5 + 4i} = \dfrac{15 + 12i + 5i + 4i^2}{25 + 16} = \dfrac{11 + 17i}{41} = \dfrac{11}{41} + \dfrac{17}{41}i$

**3.** $x = \dfrac{-(-4) \pm \sqrt{(-4)^2 - 4(1)(7)}}{2(1)} = \dfrac{4 \pm \sqrt{-12}}{2} = \dfrac{4 \pm 2\sqrt{3}i}{2} = 2 \pm \sqrt{3}i$

**4.** False: $\sqrt{-6}\sqrt{-6} = \left(\sqrt{6}i\right)\left(\sqrt{6}i\right) = \sqrt{36}i^2 = 6(-1) = -6$

**5.** $b^2 - 4ac = (-8)^2 - 4(3)(7) = -20 < 0$

Two complex solutions.

**6.** $x^4 + 13x^2 + 36 = 0$

$(x^2 + 4)(x^2 + 9) = 0$

$x^2 + 4 = 0 \implies x^2 = -4 \implies x = \pm 2i$

$x^2 + 9 = 0 \implies x^2 = -9 \implies x = \pm 3i$

**7.** $f(x) = (x - 3)[x - (-1 + 4i)][x - (-1 - 4i)]$

$= (x - 3)[(x + 1) - 4i][(x + 1) + 4i]$

$= (x - 3)[(x + 1)^2 - 16i^2]$

$= (x - 3)(x^2 + 2x + 17)$

$= x^3 - x^2 + 11x - 51$

**8.** Since $x = 4 + i$ is a zero, so is its conjugate $4 - i$.

$[x - (4 + i)][x - (4 - i)] = [(x - 4) - i][(x - 4) + i]$

$= (x - 4)^2 - i^2$

$= x^2 - 8x + 17$

$$
\begin{array}{r}
x - 2 \\
x^2 - 8x + 17 \overline{\smash{\big)}\ x^3 - 10x^2 + 33x - 34} \\
\underline{x^3 - 8x^2 + 17x} \\
-2x^2 + 16x - 34 \\
\underline{-2x^2 + 16x - 34} \\
0
\end{array}
$$

Thus, $f(x) = (x^2 - 8x + 17)(x - 2)$ and the zeros of $f(x)$ are: $4 \pm i, 2$

**9.** $r = \sqrt{25 + 25} = \sqrt{50} = 5\sqrt{2}$

$\tan \theta = \dfrac{-5}{5} = -1$

Since $z$ is in Quadrant IV,

$\theta = 315°$

$z = 5\sqrt{2}(\cos 315° + i \sin 315°).$

**10.** $\cos 225° = -\dfrac{\sqrt{2}}{2},\ \sin 225° = -\dfrac{\sqrt{2}}{2}$

$z = 6\left(-\dfrac{\sqrt{2}}{2} - i\dfrac{\sqrt{2}}{2}\right)$

$\quad = -3\sqrt{2} - 3\sqrt{2}i$

**11.** $[7(\cos 23° + i \sin 23°)][4(\cos 7° + i \sin 7°] = 7(4)[\cos(23° + 7°) + i \sin(23° + 7°)]$

$$= 28(\cos 30° + i \sin 30°)$$

$$= 14\sqrt{3} + 14i$$

**12.** $\dfrac{9\left(\cos\dfrac{5\pi}{4} + i \sin\dfrac{5\pi}{4}\right)}{3(\cos \pi + i \sin \pi)} = \dfrac{9}{3}\left[\cos\left(\dfrac{5\pi}{4} - \pi\right) + i \sin\left(\dfrac{5\pi}{4} - \pi\right)\right] = 3\left(\cos\dfrac{\pi}{4} + i \sin\dfrac{\pi}{4}\right) = \dfrac{3\sqrt{2}}{2} + \dfrac{3\sqrt{2}}{2}i$

**13.** $(2 + 2i)^8 = \left[2\sqrt{2}(\cos 45° + i \sin 45°)\right]^8 = \left(2\sqrt{2}\right)^8[\cos(8)(45°) + i \sin(8)(45°)]$

$$= 4096[\cos 360° + i \sin 360°] = 4096$$

**14.** $z = 8\left(\cos\dfrac{\pi}{3} + i \sin\dfrac{\pi}{3}\right), n = 3$

The cube roots of $z$ are: $\sqrt[3]{8}\left[\cos\dfrac{(\pi/3) + 2\pi k}{3} + i \sin\dfrac{(\pi/3) + 2\pi k}{3}\right], k = 0, 1, 2.$

For $k = 0,\quad \sqrt[3]{8}\left[\cos\dfrac{\pi/3}{3} + i \sin\dfrac{\pi/3}{3}\right] = 2\left(\cos\dfrac{\pi}{9} + i \sin\dfrac{\pi}{9}\right).$

For $k = 1,\ \sqrt[3]{8}\left[\cos\dfrac{\pi/3 + 2\pi}{3} + i \sin\dfrac{\pi/3 + 2\pi}{3}\right] = 2\left(\cos\dfrac{7\pi}{9} + i \sin\dfrac{7\pi}{9}\right).$

For $k = 2,\ \sqrt[3]{8}\left[\cos\dfrac{\pi/3 + 4\pi}{3} + i \sin\dfrac{\pi/3 + 4\pi}{3}\right] = 2\left(\cos\dfrac{13\pi}{9} + i \sin\dfrac{13\pi}{9}\right).$

**15.** $x^4 = -i = 1\left(\cos\dfrac{3\pi}{2} + i \sin\dfrac{3\pi}{2}\right)$

The fourth roots are: $\sqrt[4]{1}\left[\cos\dfrac{(3\pi/2) + 2\pi k}{4} + i \sin\dfrac{(3\pi/2) + 2\pi k}{4}\right], k = 0, 1, 2, 3$

For $k = 0, \cos\left(\dfrac{3\pi/2}{4}\right) + i \sin\left(\dfrac{3\pi/2}{4}\right) = \cos\dfrac{3\pi}{8} + i \sin\dfrac{3\pi}{8}.$

For $k = 1, \cos\left(\dfrac{3\pi/2 + 2\pi}{4}\right) + i \sin\left(\dfrac{3\pi/2 + 2\pi}{4}\right) = \cos\dfrac{7\pi}{8} + i \sin\dfrac{7\pi}{8}.$

For $k = 2, \cos\left(\dfrac{3\pi/2 + 4\pi}{4}\right) + i \sin\left(\dfrac{3\pi/2 + 4\pi}{4}\right) = \cos\dfrac{11\pi}{8} + i \sin\dfrac{11\pi}{8}.$

For $k = 3, \cos\left(\dfrac{3\pi/2 + 6\pi}{4}\right) + i \sin\left(\dfrac{3\pi/2 + 6\pi}{4}\right) = \cos\dfrac{15\pi}{8} + i \sin\dfrac{15\pi}{8}.$

# ❑ Chapter 5 Practice Test Solutions

**1.** $x^{3/5} = 8$

$x = 8^{5/3} = \left(\sqrt[3]{8}\right)^5 = 2^5 = 32$

**2.** $3^{x-1} = \frac{1}{81}$

$3^{x-1} = 3^{-4}$

$x - 1 = -4$

$x = -3$

**3.** $f(x) = 2^{-x} = \left(\frac{1}{2}\right)^x$

| $x$ | $-2$ | $-1$ | $0$ | $1$ | $2$ |
|---|---|---|---|---|---|
| $f(x)$ | $4$ | $2$ | $1$ | $\frac{1}{2}$ | $\frac{1}{4}$ |

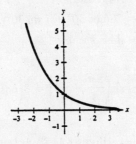

**4.** $g(x) = e^x + 1$

| $x$ | $-2$ | $-1$ | $0$ | $1$ | $2$ |
|---|---|---|---|---|---|
| $g(x)$ | $1.14$ | $1.37$ | $2$ | $3.72$ | $8.39$ |

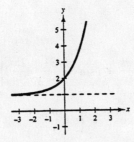

**5.** $A = P\left(1 + \frac{r}{n}\right)^{nt}$ OR $A = Pe^{rt}$

(a) $A = 5000\left(1 + \frac{0.09}{12}\right)^{12(3)} \approx \$6543.23$

(b) $A = 5000\left(1 + \frac{0.09}{4}\right)^{4(3)} \approx \$6530.25$

(c) $A = 5000e^{(0.09)(3)} \approx \$6549.82$

**6.** $7^{-2} = \frac{1}{49}$

$\log_7 \frac{1}{49} = -2$

**7.** $x - 4 = \log_2 \frac{1}{64}$

$2^{x-4} = \frac{1}{64}$

$2^{x-4} = 2^{-6}$

$x - 4 = -6$

$x = -2$

**8.** $\log_b \sqrt[4]{\frac{8}{25}} = \frac{1}{4} \log_b \frac{8}{25}$

$= \frac{1}{4}[\log_b 8 - \log_b 25]$

$= \frac{1}{4}[\log_b 2^3 - \log_b 5^2]$

$= \frac{1}{4}[3 \log_b 2 - 2 \log_b 5]$

$= \frac{1}{4}[3(0.3562) - 2(0.8271)]$

$= -0.1464$

**9.** $5 \ln x - \frac{1}{2} \ln y + 6 \ln z = \ln x^5 - \ln \sqrt{y} + \ln z^6 = \ln\left(\frac{x^5 z^6}{\sqrt{y}}\right)$

**10.** $\log_9 28 = \dfrac{\log 28}{\log 9} \approx 1.5166$

**11.** $\log N = 0.6646$

$N = 10^{0.6646} \approx 4.62$

**12.**

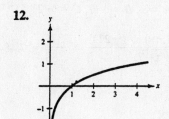

**13.** Domain:

$$x^2 - 9 > 0$$
$$(x + 3)(x - 3) > 0$$
$$x < -3 \text{ or } x > 3$$

**14.**

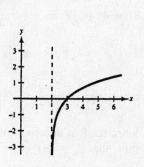

**15.** $\dfrac{\ln x}{\ln y} \neq \ln(x - y)$ since $\dfrac{\ln x}{\ln y} = \log_y x.$

**16.** $5^3 = 41$

$x = \log_5 41 = \dfrac{\ln 41}{\ln 5} \approx 2.3074$

**17.** $x - x^2 = \log_5 \frac{1}{25}$

$5^{x - x^2} = \frac{1}{25}$

$5^{x - x^2} = 5^{-2}$

$x - x^2 = -2$

$0 = x^2 - x - 2$

$0 = (x + 1)(x - 2)$

$x = -1 \text{ or } x = 2$

**18.** $\log_2 x + \log_2(x - 3) = 2$

$\log_2[x(x - 3)] = 2$

$x(x - 3) = 2^2$

$x^2 - 3x = 4$

$x^2 - 3x - 4 = 0$

$(x + 1)(x - 4) = 0$

$x = 4$

$x = -1 \text{ (extraneous)}$

$x = 4$ is the only solution.

**19.** $\dfrac{e^x + e^{-x}}{3} = 4$

$e^x(e^x + e^{-x}) = 12e^x$

$e^{2x} + 1 = 12e^x$

$e^{2x} - 12e^x + 1 = 0$

$e^x = \dfrac{12 \pm \sqrt{144 - 4}}{2}$

$e^x \approx 11.9161 \qquad \text{or} \qquad e^x \approx 0.0839$

$x = \ln 11.9161 \qquad\qquad x = \ln 0.0839$

$x \approx 2.478 \qquad\qquad\qquad x \approx -2.478$

**20.** $A = Pe^{et}$

$12{,}000 = 6000e^{0.13t}$

$2 = e^{0.13t}$

$0.13t = \ln 2$

$t = \dfrac{\ln 2}{0.13}$

$t \approx 5.3319$ years or 5 years 4 months

## ❑ Chapter 6  Practice Test Solutions

**1.** $3x + 4y = 12 \implies y = -\frac{3}{4}x + 3 \implies m_1 = -\frac{3}{4}$

$4x - 3y = 12 \implies y = \frac{4}{3}x - 4 \implies m_2 = \frac{4}{3}$

$\tan\theta = \left| \dfrac{\frac{4}{3} - \left(-\frac{3}{4}\right)}{1 + \left(\frac{4}{3}\right)\left(-\frac{3}{4}\right)} \right| = \left| \dfrac{\frac{25}{12}}{0} \right|$

Since $\tan\theta$ is undefined, the lines are perpendicular (note that $m_2 = -1/m_1$) and $\theta = 90°$.

**2.** $x_1 = 5, x_2 = -9, A = 3, B = -7, C = -21$

$d = \dfrac{|3(5) + (-7)(-9) + (-21)|}{\sqrt{3^2 + (-7)^2}} = \dfrac{57}{\sqrt{58}} \approx 7.484$

**3.** $x^2 - 6x - 4y + 1 = 0$

$x^2 - 6x + 9 = 4y - 1 + 9$

$(x - 3)^2 = 4y + 8$

$(x - 3)^2 = 4(1)(y + 2) \implies p = 1$

Vertex: $(3, -2)$

Focus: $(3, -1)$

Directrix: $y = -3$

**4.** Vertex: $(2, -5)$

Focus: $(2, -6)$

Vertical axis; opens downward with $p = -1$

$(x - h)^2 = 4p(y - k)$

$(x - 2)^2 = 4(-1)(y + 5)$

$x^2 - 4x + 4 = -4y - 20$

$x^2 - 4x + 4y + 24 = 0$

**5.** $\qquad x^2 + 4y^2 - 2x + 32y + 61 = 0$

$(x^2 - 2x + 1) + 4(y^2 + 8y + 16) = -61 + 1 + 64$

$(x - 1)^2 + 4(y + 4)^2 = 4$

$\dfrac{(x - 1)^2}{4} + \dfrac{(y + 4)^2}{1} = 1$

$a = 2, b = 1, c = \sqrt{3}$

Horizontal major axis

Center: $(1, -4)$

Foci: $\left(1 \pm \sqrt{3}, -4\right)$

Vertices: $(3, -4), (-1, -4)$

Eccentricity: $e = \dfrac{\sqrt{3}}{2}$

**6.** Vertices: $(0, \pm 6)$

Eccentricity: $e = \dfrac{1}{2}$

Center: $(0, 0)$

Vertical major axis

$a = 6, e = \dfrac{c}{a} = \dfrac{c}{6} = \dfrac{1}{2} \implies c = 3$

$b^2 = (6)^2 - (3)^2 = 27$

$\dfrac{x^2}{27} + \dfrac{y^2}{36} = 1$

**7.** $\qquad 16y^2 - x^2 - 6x - 128y + 231 = 0$

$16(y^2 - 8y + 16) - (x^2 + 6x + 9) = -231 + 256 - 9$

$16(y - 4)^2 - (x + 3)^2 = 16$

$\dfrac{(y - 4)^2}{1} - \dfrac{(x + 3)^2}{16} = 1$

$a = 1, b = 4, c = \sqrt{17}$

Center: $(-3, 4)$

Vertical transverse axis

Vertices: $(-3, 5), (-3, 3)$

Foci: $\left(-3, 4 \pm \sqrt{17}\right)$

Asymptotes: $y = 4 \pm \dfrac{1}{4}(x + 3)$

**8.** Vertices: $(\pm 3, 2)$

Foci: $(\pm 5, 2)$

Center: $(0, 2)$

Horizontal transverse axis

$a = 3, c = 5, b = 4$

$\dfrac{(x - 0)^2}{9} - \dfrac{(y - 2)^2}{16} = 1$

$\dfrac{x^2}{9} - \dfrac{(y - 2)^2}{16} = 1$

**9.** $5x^2 + 2xy + 5y^2 - 10 = 0$

$A = 5, B = 2, C = 5$

$\cot 2\theta = \dfrac{5 - 5}{2} = 0$

$2\theta = \dfrac{\pi}{2} \Rightarrow \theta = \dfrac{\pi}{4}$

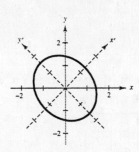

$x = x' \cos \dfrac{\pi}{4} - y' \sin \dfrac{\pi}{4} \qquad\qquad x = x' \cos \dfrac{\pi}{4} + y' \sin \dfrac{\pi}{4}$

$= \dfrac{x' - y'}{\sqrt{2}} \qquad\qquad\qquad\qquad = \dfrac{x' + y'}{\sqrt{2}}$

$$5\left(\dfrac{x' - y'}{\sqrt{2}}\right)^2 + 2\left(\dfrac{x' - y'}{\sqrt{2}}\right)\left(\dfrac{x' + y'}{\sqrt{2}}\right) + 5\left(\dfrac{x' + y'}{\sqrt{2}}\right)^2 - 10 = 0$$

$$\dfrac{5(x')^2}{2} - \dfrac{10x'y'}{2} + \dfrac{5(y')^2}{2} + (x')^2 - (y')^2 + \dfrac{5(x')^2}{2} + \dfrac{10x'y'}{2} + \dfrac{5(y')^2}{2} - 10 = 0$$

$$6(x')^2 + 4(y')^2 - 10 = 0$$

$$\dfrac{3(x')^2}{5} + \dfrac{2(y')^2}{5} = 1$$

$$\dfrac{(x')^2}{5/3} + \dfrac{(y')^2}{5/2} = 1$$

Ellipse centered at the origin

**10.** (a) $6x^2 - 2xy + y^2 = 0$

$A = 6, B = -2, C = 1$

$B^2 - 4AC = (-2)^2 - 4(6)(1) = -20 < 0$

Ellipse

(b) $x^2 + 4xy + 4y^2 - x - y + 17 = 0$

$A = 1, B = 4, C = 4$

$B^2 - 4AC = (4)^2 - 4(1)(4) = 0$

Parabola

**11.** Polar: $\left(\sqrt{2}, \dfrac{3\pi}{4}\right)$

$x = \sqrt{2} \cos \dfrac{3\pi}{4} = \sqrt{2}\left(-\dfrac{1}{\sqrt{2}}\right) = -1$

$y = \sqrt{2} \sin \dfrac{3\pi}{4} = \sqrt{2}\left(\dfrac{1}{\sqrt{2}}\right) = 1$

Rectangular: $(-1, 1)$

**12.** Rectangular: $\left(\sqrt{3}, -1\right)$

$r = \pm\sqrt{(\sqrt{3})^2 + (-1)^2} = \pm 2$

$\tan \theta = \dfrac{\sqrt{3}}{-1} = -\sqrt{3}$

$\theta = \dfrac{2\pi}{3}$ or $\theta = \dfrac{5\pi}{3}$

Polar: $\left(-2, \dfrac{2\pi}{3}\right)$ or $\left(2, \dfrac{5\pi}{3}\right)$

**13.** Rectangular: $4x - 3y = 12$

Polar: $4r \cos \theta - 3r \sin \theta = 12$

$r(4 \cos \theta - 3 \sin \theta) = 12$

$r = \dfrac{12}{4 \cos \theta - 3 \sin \theta}$

**14.** Polar: $r = 5 \cos \theta$

$r^2 = 5r \cos \theta$

Rectangular: $x^2 + y^2 = 5x$

$x^2 + y^2 - 5x = 0$

**15.** $r = 1 - \cos\theta$

Cardioid

Symmetry: Polar axis

Maximum value of $|r|$: $r = 2$ when $\theta = \pi$.

Zero of $r$: $r = 0$ when $\theta = 0$

| $\theta$ | $0$ | $\dfrac{\pi}{2}$ | $\pi$ | $\dfrac{3\pi}{2}$ |
|---|---|---|---|---|
| $r$ | $0$ | $1$ | $2$ | $1$ |

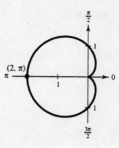

**16.** $r = 5\sin 2\theta$

Rose curve with four petals

Symmetry: Polar axis, $\theta = \dfrac{\pi}{2}$, and pole

Maximum value of $|r|$: $|r| = 5$ when $\theta = \dfrac{\pi}{4}, \dfrac{3\pi}{4}, \dfrac{5\pi}{4}, \dfrac{7\pi}{4}$

Zeros of $r$: $r = 0$ when $\theta = 0, \dfrac{\pi}{2}, \pi, \dfrac{3\pi}{2}$

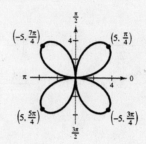

**17.** $r = \dfrac{3}{6 - \cos\theta}$

$r = \dfrac{\frac{1}{2}}{1 - \frac{1}{6}\cos\theta}$

$e = \dfrac{1}{6} < 1$, so the graph is an ellipse.

| $\theta$ | $0$ | $\dfrac{\pi}{2}$ | $\pi$ | $\dfrac{3\pi}{2}$ |
|---|---|---|---|---|
| $r$ | $\dfrac{3}{5}$ | $\dfrac{1}{2}$ | $\dfrac{3}{7}$ | $\dfrac{1}{2}$ |

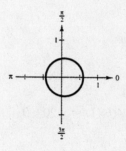

**18.** Parabola

Vertex: $\left(6, \dfrac{\pi}{2}\right)$

Focus: $(0, 0)$

$e = 1$

$r = \dfrac{ep}{1 + e\sin\theta}$

$r = \dfrac{p}{1 + \sin\theta}$

$6 = \dfrac{p}{1 + \sin(\pi/2)}$

$6 = \dfrac{p}{2}$

$12 = p$

$r = \dfrac{12}{1 + \sin\theta}$

**19.** $x = 3 - 2\sin\theta, y = 1 + 5\cos\theta$

$\dfrac{x - 3}{-2} = \sin\theta, \dfrac{y - 1}{5} = \cos\theta$

$\left(\dfrac{x - 3}{-2}\right)^2 + \left(\dfrac{y - 1}{5}\right)^2 = 1$

$\dfrac{(x - 3)^2}{4} + \dfrac{(y - 1)^2}{25} = 1$

**20.** $x = e^{2t}, y = e^{4t}$

$x > 0, y > 0$

$y = (e^{2t})^2 = (x)^2 = x^2, x < 0, y > 0$

# PART 2  Solutions to Chapter and Cumulative Tests

## ❏ Chapter Test Solutions for Chapter P

**1.** $-\dfrac{10}{3} = -3\frac{1}{3}$

$-|-4| = -4$

$-\dfrac{10}{3} > -|-4|$

**2.** (a) $\left|\dfrac{3}{8} + \dfrac{1}{6} - 2\right| = \left|\dfrac{9 + 4 - 48}{24}\right|$

$= \left|-\dfrac{35}{24}\right| = \dfrac{35}{24}$

(b) $\left(21 \div \dfrac{3}{4}\right) = \dfrac{21}{1} \cdot \dfrac{4}{3} = 28$

**3.** $\dfrac{2}{3}(x - 1) + \dfrac{1}{4}x = 10$

$12\left[\dfrac{2}{3}(x - 1) + \dfrac{1}{4}x\right] = 12(10)$

$8(x - 1) + 3x = 120$

$8x - 8 + 3x = 120$

$11x = 128$

$x = \dfrac{128}{11}$

**4.** $\dfrac{x - 2}{x + 2} + \dfrac{4}{x + 2} + 4 = 0$

$\dfrac{x + 2}{x + 2} = -4$

$1 \neq -4 \Longrightarrow$ No solution

**5.** $3x^2 + 6x + 2 = 0$

$x = \dfrac{-6 \pm \sqrt{6^2 - 4(3)(2)}}{2(3)}$

$= \dfrac{-6 \pm \sqrt{12}}{6}$

$= \dfrac{-6 \pm 2\sqrt{3}}{6}$

$= \dfrac{-3 \pm \sqrt{3}}{3}$

**6.** $x^4 + x^2 - 6 = 0$

$(x^2 - 2)(x^2 + 3) = 0$

$x^2 = 2 \Longrightarrow x = \pm\sqrt{2}$

$x^2 = -3 \Longrightarrow$ No real solution

**7.** $2\sqrt{3} - \sqrt{2x + 1} = 1$

$-\sqrt{2x + 1} = 1 - 2\sqrt{x}$

$\left(-\sqrt{2x + 1}\right)^2 = \left(1 - 2\sqrt{x}\right)^2$

$2x + 1 = 1 - 4\sqrt{x} + 4x$

$-2x = -4\sqrt{x}$

$x = 2\sqrt{x}$

$x^2 = 4x$

$x^2 - 4x = 0$

$x(x - 4) = 0$

$x = 0 \ \text{ or } \ x = 4$

Only $x = 4$ is a solution to the original equation.
$x = 0$ is extraneous

**8.** $|3x - 1| = 7$

$3x - 1 = 7 \ \text{ or } \ 3x - 1 = -7$

$3x = 8 \qquad\qquad 3x = -6$

$x = \dfrac{8}{3} \qquad\qquad x = -2$

**9.** $y = 4 - \frac{3}{4}x$

No symmetry

$x$-intercept: $\left(\frac{16}{3}, 0\right)$

$y$-intercept: $(0, 4)$

**10.** $y = 4 - \frac{3}{4}|x|$

$y$-axis symmetry

$x$-intercepts: $\left(\pm\frac{16}{3}, 0\right)$

$y$-intercept: $(0, 4)$

**11.** $y = 4 - (x - 2)^2$

Parabola; vertex: $(2, 4)$

No symmetry

$x$-intercepts:

$(0, 0)$ and $(4, 0)$

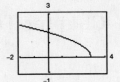

$$0 = 4 - (x - 2)^2$$

$$(x - 2)^2 = 4$$

$$x - 2 = \pm 2$$

$$x = 2 \pm 2$$

$$x = 4 \quad \text{or} \quad x = 0$$

$y$-intercept: $(0, 0)$

**12.** $y = \sqrt{3 - x}$

Domain: $x \le 3$

No symmetry

$x$-intercept: $(3, 0)$

$y$-intercept: $\left(0, \sqrt{3}\right)$

**13.** (a) $(-3, 6), (3, 2)$

$$m = \frac{2 - 6}{3 - (-3)} = -\frac{4}{6} = -\frac{2}{3}$$

Since the slope is negative, the line falls from left to right.

$$y - 6 = -\frac{2}{3}[x - (-3)]$$

$$3y - 18 = -2x - 6$$

$$2x + 3y - 12 = 0$$

(b) $(4, -2), (7, -2)$

$$m = \frac{-2 - (-2)}{7 - 4} = 0 \Rightarrow \text{The line is horizontal.}$$

$$y - (-2) = 0(x - 4)$$

$$y + 2 = 0$$

**14.** $y^2(4 - x) = x^3$. The graph does not represent $y$ as a function of $x$. Some $x$ values correspond to two $y$-values.

**15.** $f(x) = 10 - \sqrt{3 - x}$

$$f(t - 3) = 10 - \sqrt{3 - (t - 3)}$$

$$= 10 - \sqrt{6 - t}$$

Domain: $(-\infty, 6]$

**16.** (a) $\frac{1}{2}g(x - 2)$

Horizontal shift two units to the right and a vertical shrink by $\frac{1}{2}$.

| $x$ | $-1$ | 1 | 2 | 3 |
|---|---|---|---|---|
| $g(x)$ | 2 | 2 | $-2$ | $-2$ |

| $x$ | 1 | 3 | 4 | 5 |
|---|---|---|---|---|
| $\frac{1}{2}g(x - 2)$ | 1 | 1 | $-1$ | $-1$ |

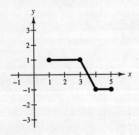

(b) $g\left(\frac{1}{2}x\right) - 1$

Change in $x$ by a factor of $\frac{1}{2}$.

Vertical shift one unit downward.

| $x$ | $-2$ | 2 | 4 | 6 |
|---|---|---|---|---|
| $g\left(\frac{1}{2}x\right) - 1$ | 1 | 1 | $-3$ | $-3$ |

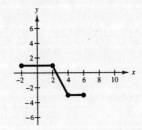

**17.** $(f - g)(x) = f(x) - g(x) = x^2 - \sqrt{2 - x}$

Domain: $(-\infty, 2]$

**18.** $\left(\dfrac{f}{g}\right)(x) = \dfrac{f(x)}{g(x)} = \dfrac{x^2}{\sqrt{2 - x}}$

Domain: $(-\infty, 2)$

**19.** $(f \circ g)(x) = f(g(x)) = f\left(\sqrt{2 - x}\right) = \left(\sqrt{2 - x}\right)^2 = 2 - x$

Domain: $(-\infty, 2]$

**20.** $g(x) = \sqrt{2 - x}, x \leq 2, y \geq 0$

The domain of $g^{-1}(x)$ is the range of $g(x)$.

$$y = \sqrt{2 - x}$$
$$x = \sqrt{2 - y}$$
$$x^2 = 2 - y$$
$$y = 2 - x^2, x \geq 0$$

$g^{-1}(x) = 2 - x^2, x \geq 0$

Domain: $[0, \infty)$

**21.** $(100 \text{ km/hr})\left(2\frac{1}{4} \text{ hr}\right) + (x \text{ km/hr})\left(1\frac{1}{3} \text{ hr}\right) = 350 \text{ km}$

$$225 + \tfrac{4}{3}x = 350$$
$$\tfrac{4}{3}x = 125$$
$$x = \tfrac{375}{4} = 93\tfrac{3}{4} \text{ km/hr}$$

# ❑ Chapter Test Solutions for Chapter 1

**1.** $\theta = \dfrac{5\pi}{4}$

(a)

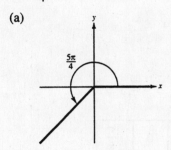

(b) $\dfrac{5\pi}{4} + 2\pi = \dfrac{13\pi}{4}$

$\dfrac{5\pi}{4} - 2\pi = -\dfrac{3\pi}{4}$

(c) $\dfrac{5\pi}{4}\left(\dfrac{180°}{\pi}\right) = 225°$

**2.** $90\,\dfrac{\text{km}}{\text{hr}} \times \dfrac{1 \text{ hr}}{60 \text{ min}} \times \dfrac{1000 \text{ m}}{1 \text{ km}} = 1500 \text{ meters per minute}$

$\text{Circumference} = 2\pi\left(\dfrac{1}{2}\right) = \pi = \pi \text{ meters}$

$\dfrac{\text{Revolutions}}{\text{minute}} = \dfrac{1500}{\pi}$

$\text{Angular speed} = \dfrac{1500}{\pi} \cdot \pi = 1500 \text{ radians per minute}$

**3.** $x = -2, y = 6$

$r = \sqrt{(-2)^2 + (6)^2} = 2\sqrt{10}$

$\sin\theta = \dfrac{y}{r} = \dfrac{6}{2\sqrt{10}} = \dfrac{3}{\sqrt{10}}$

$\cos\theta = \dfrac{x}{r} = \dfrac{-2}{2\sqrt{10}} = -\dfrac{1}{\sqrt{10}}$

$\tan\theta = \dfrac{y}{x} = \dfrac{6}{-2} = -3$

$\csc\theta = \dfrac{r}{y} = \dfrac{2\sqrt{10}}{6} = \dfrac{\sqrt{10}}{3}$

$\sec\theta = \dfrac{r}{x} = \dfrac{2\sqrt{10}}{-2} = -\sqrt{10}$

$\cot\theta = \dfrac{x}{y} = \dfrac{-2}{6} = -\dfrac{1}{3}$

**4.** $\sin\theta = \dfrac{\text{opp}}{\text{hyp}} = \dfrac{3}{\sqrt{13}}$

$\cos\theta = \dfrac{\text{adj}}{\text{hyp}} = \dfrac{2}{\sqrt{13}}$

$\csc\theta = \dfrac{\text{hyp}}{\text{opp}} = \dfrac{\sqrt{13}}{3}$

$\sec\theta = \dfrac{\text{hyp}}{\text{adj}} = \dfrac{\sqrt{13}}{2}$

$\cot\theta = \dfrac{\text{adj}}{\text{opp}} = \dfrac{2}{3}$

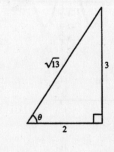

**5.** $\theta = 290°$

$\theta' = 360° - 290° = 70°$

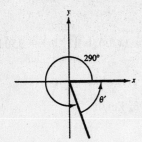

**6.** $\sec \theta < 0$ and $\tan \theta > 0$

$\dfrac{r}{x} < 0$ and $\dfrac{y}{x} > 0$

Quadrant III

**7.** $\cos \theta = -\dfrac{\sqrt{3}}{2}$

Reference angle is 30° and $\theta$ is in Quadrant II or III.

$\theta = 150°$ or $210°$

**8.** $\csc \theta = 1.030$

$\dfrac{1}{\sin \theta} = 1.030$

$\sin \theta = \dfrac{1}{1.030}$

$\theta = \arcsin \dfrac{1}{1.030}$

$\theta \approx 1.33$ and $\pi - 1.33 \approx 1.81$

**9.** $g(x) = -2 \sin\left(x - \dfrac{\pi}{4}\right)$

Period: $2\pi$

Amplitude: $|-2| = 2$

Shifted to the right by $(\pi/4)$ units and reflected in the $x$-axis.

| $x$ | 0 | $\dfrac{\pi}{4}$ | $\dfrac{\pi}{2}$ | $\dfrac{3\pi}{4}$ | $\pi$ |
|---|---|---|---|---|---|
| $y$ | $\sqrt{2}$ | 0 | $-\sqrt{2}$ | $-2$ | $-\sqrt{2}$ |

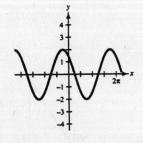

**10.** $f(\alpha) = \dfrac{1}{2} \tan 2\alpha$

Period: $\dfrac{\pi}{2}$

Asymptotes: $x = -\dfrac{\pi}{4}, x = \dfrac{\pi}{4}$

| $\alpha$ | $-\dfrac{\pi}{8}$ | 0 | $\dfrac{\pi}{8}$ |
|---|---|---|---|
| $f(\alpha)$ | $-\dfrac{1}{2}$ | 0 | $\dfrac{1}{2}$ |

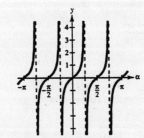

**11.** $y = \sin 2\pi x + 2 \cos \pi x$

Periodic: period $= 2$

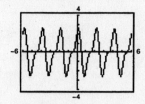

**12.** $y = 6e^{-0.12t} \cos(0.25t),\ 0 \le t \le 32$

Not periodic

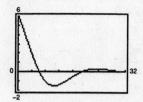

**13.** $f(x) = a \sin(bx + c)$

Amplitude: $2 \Longrightarrow |a| = 2$

Reflected in the $x$-axis: $a = -2$

Period: $4\pi = \dfrac{2\pi}{b} \Longrightarrow b = \dfrac{1}{2}$

Phase shift: $\dfrac{c}{b} = -\dfrac{\pi}{2} \Longrightarrow c = -\dfrac{\pi}{4}$

$f(x) = -2 \sin\left(\dfrac{x}{2} - \dfrac{\pi}{4}\right)$

**14.** Let $u = \arccos \dfrac{2}{3}$,

$\cos u = \dfrac{2}{3}$.

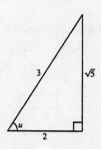

$\tan\left(\arccos \dfrac{2}{3}\right) = \tan u = \dfrac{\sqrt{5}}{2}$

**15.** $f(x) = 2 \arcsin\left(\dfrac{1}{2}x\right)$

Domain: $[-2, 2]$

Range: $[-\pi, \pi]$

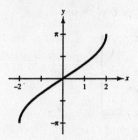

**16.**

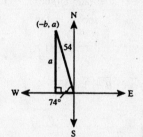

$\sin 74° = \dfrac{a}{54} \Longrightarrow a \approx 51.91$ nautical miles north

$\cos 74° = \dfrac{b}{54} \Longrightarrow b \approx 14.88$ nautical miles west

Coordinates: $(-14.88, 51.91)$

# ❏ Chapter Test Solutions for Chapter 2

**1.** $\tan \theta = \dfrac{3}{2}$ and $\cos \theta < 0$

$\theta$ is in Quadrant III.

$$\sec \theta = -\sqrt{1 + \tan^2 \theta} = -\sqrt{1 + \left(\dfrac{3}{2}\right)^2} = -\dfrac{\sqrt{13}}{2}$$

$$\cos \theta = \dfrac{1}{\sec \theta} = -\dfrac{2}{\sqrt{13}}$$

$$\sin \theta = \tan \theta \cos \theta = \left(\dfrac{3}{2}\right)\left(-\dfrac{2}{\sqrt{13}}\right) = -\dfrac{3}{\sqrt{13}}$$

$$\csc \theta = \dfrac{1}{\sin \theta} = -\dfrac{\sqrt{13}}{3}$$

$$\cot \theta = \dfrac{1}{\tan \theta} = \dfrac{2}{3}$$

**2.** $\csc^2 \beta (1 - \cos^2 \beta) = \dfrac{1}{\sin^2 \beta}(\sin^2 \beta) = 1$

**3.** $\dfrac{\sec^4 x - \tan^4 x}{\sec^2 x + \tan^2 x} = \dfrac{(\sec^2 x + \tan^2 x)(\sec^2 x - \tan^2 x)}{\sec^2 x + \tan^2 x}$

$$= \sec^2 x - \tan^2 x = 1$$

**4.** $\dfrac{\cos \theta}{\sin \theta} + \dfrac{\sin \theta}{\cos \theta} = \dfrac{\cos^2 \theta + \sin^2 \theta}{\sin \theta \cos \theta} = \dfrac{1}{\sin \theta \cos \theta}$

**5.** $y = \tan \theta,\ y = -\sqrt{\sec^2 \theta - 1}$

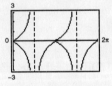

$\tan \theta = -\sqrt{\sec^2 \theta - 1}$ on

$$\dfrac{\pi}{2} < \theta \le \pi,\ \dfrac{3\pi}{2} < \theta < 2\pi.$$

**6.** $y_1 = \cos x + \sin x \tan x,\ y_2 = \sec x$

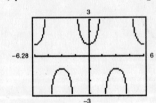

It appears that $y_1 = y_2$.

$$\cos x + \sin x \tan x = \cos + \sin x \dfrac{\sin x}{\cos x}$$

$$= \cos + \dfrac{\sin^2 x}{\cos x}$$

$$= \dfrac{\cos^2 x + \sin^2 x}{\cos x}$$

$$= \dfrac{1}{\cos x} = \sec x$$

**7.** $\sin\theta\sec\theta = \sin\dfrac{1}{\cos\theta} = \dfrac{\sin\theta}{\cos\theta} = \tan\theta$

**8.** $\sec^2 x\tan^2 x + \sec^2 x = \sec^2 x(\sec^2 x - 1) + \sec^2 x$
$$= \sec^4 x - \sec^2 x + \sec^2 x$$
$$= \sec^4 x$$

**9.** $\dfrac{\csc\alpha + \sec\alpha}{\sin\alpha + \cos\alpha} = \dfrac{\dfrac{1}{\sin\alpha} + \dfrac{1}{\cos\alpha}}{\sin\alpha + \cos\alpha} = \dfrac{\dfrac{\cos\alpha + \sin\alpha}{\sin\alpha\cos\alpha}}{\sin\alpha + \cos\alpha} = \dfrac{1}{\sin\alpha\cos\alpha}$

$$= \dfrac{\cos^2\alpha + \sin^2\alpha}{\sin\alpha\cos\alpha} = \dfrac{\cos^2\alpha}{\sin\alpha\cos\alpha} + \dfrac{\sin^2\alpha}{\sin\alpha\cos\alpha}$$

$$= \dfrac{\cos\alpha}{\sin\alpha} + \dfrac{\sin\alpha}{\cos\alpha} = \cot\alpha + \tan\alpha$$

**10.** $\cos\left(x + \dfrac{\pi}{2}\right) = \cos\left(\dfrac{\pi}{2} - (-x)\right) = \sin(-x) = -\sin x$

**11.** $\sin(n\pi + \theta) = (-1)^n\sin\theta$, $n$ is an integer.

For $n$ odd: $\sin(n\pi + \theta) = \sin n\pi\cos\theta + \cos n\pi\sin\theta$
$$= (0)\cos\theta + (-1)\sin\theta = -\sin\theta$$
For $n$ even: $\sin(n\pi + \theta) = \sin n\pi\cos\theta + \cos n\pi\sin\theta$
$$= (0)\cos\theta + (1)\sin\theta = \sin\theta$$
When $n$ is odd, $(-1)^n = -1$. When $n$ is even $(-1)^n = 1$. Thus, $\sin(n\pi + \theta) = (-1)^n\sin\theta$ for $n$ is and integer.

**12.** $(\sin x + \cos x)^2 = \sin^2 x + 2\sin x\cos x + \cos^2 x$
$$= 1 + 2\sin x\cos x$$
$$= 1 + \sin^2 2x$$

**13.** $\tan^2 x + \tan x = 0$
$\tan x(1 + \tan x) = 0$
$\tan x = 0 \quad\text{or}\quad \tan x + 1 = 0$
$x = 0, \pi \qquad\qquad \tan x = -1$
$$x = \dfrac{3\pi}{4}, \dfrac{7\pi}{4}$$

**14.** $\sin 2\alpha - \cos\alpha = 0$
$2\sin\alpha\cos\alpha - \cos\alpha = 0$
$\cos\alpha(2\sin\alpha - 1) = 0$
$\cos\alpha = 0 \quad\text{or}\quad 2\sin\alpha - 1 = 0$
$\alpha = \dfrac{\pi}{2}, \dfrac{3\pi}{2} \qquad\qquad \sin\alpha = \dfrac{1}{2}$
$$\alpha = \dfrac{2\pi}{6}, \dfrac{5\pi}{6}$$

**15.** $4\cos^2 x - 3 = 0$
$$\cos^2 x = \dfrac{3}{4}$$
$$\cos x = \pm\sqrt{\dfrac{3}{4}} \pm\dfrac{\sqrt{3}}{2}$$
$$x = \dfrac{\pi}{6}, \dfrac{5\pi}{6}, \dfrac{7\pi}{6}, \dfrac{11\pi}{6}$$

**16.**    $\csc^2 x - \csc x - 2 = 0$

$(\csc x - 2)(\csc x + 1) = 0$

$\csc x - 2 = 0$     or     $\csc x + 1 = 0$

$\csc x = 2$               $\csc x = -1$

$\dfrac{1}{\sin x} = 2$           $\dfrac{1}{\sin x} = -1$

$\sin x = \dfrac{1}{2}$          $\sin x = -1$

$x = \dfrac{\pi}{6}, \dfrac{5\pi}{6}$         $x = \dfrac{3\pi}{2}$

**17.** $3 \cos x - x = 0$

$x \approx -2.9381, -2.6632, 1.1701$

**18.** $\cos^2 x + \cos x - 6 = 0$

$\cos^2 x + \cos x = 6$

But the maximum value of $\cos^2 x$ is 1 and the maximum value of $\cos x$ is 1. Thus, $|\cos^2 x + \cos x| \leq 2$ for all $x$ and $\cos^2 x + \cos x$ can never equal 6.

**19.** $105° = 135° - 30°$

$\cos 105° = \cos(135° - 30°)$

$= \cos 135° \cos 30° + \sin 45° \sin 30°$

$= -\cos 45° \cos 30° + \sin 45° \sin 30°$

$= \left(-\dfrac{\sqrt{2}}{2}\right)\left(\dfrac{\sqrt{3}}{2}\right) + \left(\dfrac{\sqrt{2}}{2}\right)\left(\dfrac{1}{2}\right)$

$= \dfrac{-\sqrt{6} + \sqrt{2}}{4} = \dfrac{\sqrt{2} - \sqrt{6}}{4}$

**20.** $\sin 2u = 2 \sin u \cos u$

$= 2\left(\dfrac{2}{\sqrt{5}}\right)\left(\dfrac{1}{\sqrt{5}}\right) = \dfrac{4}{5}$

$\tan 2u = \dfrac{2 \tan u}{1 - \tan^2 u} = \dfrac{2(2)}{1 - (2)^2} = \dfrac{4}{-3} = -\dfrac{4}{3}$

# ❑ Cumulative Test Solutions for Chapters 1–3

**1.** (a)

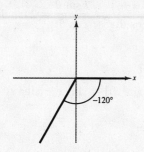

(e) $\sin(-120°) = -\sin 60° = -\dfrac{\sqrt{3}}{2}$

$\cos(-120°) = -\cos 60° = -\dfrac{1}{2}$

$\tan(-120°) = \tan 60° = \sqrt{3}$

$\csc(-120°) = \dfrac{1}{-\sin 60°} = -\dfrac{2\sqrt{3}}{3}$

(b) $-120° + 360° = 240°$

(c) $-120°\left(\dfrac{\pi}{180°}\right) = -\dfrac{2\pi}{3}$

$\sec(-120°) = \dfrac{1}{-\cos 60°} = -2$

(d) $-120°$ is located in Quadrant III.

$240° - 180° = 60°$

$\cot(-120°) = \dfrac{1}{\tan 60°} = \dfrac{\sqrt{3}}{3}$

**2.** $2.35\left(\dfrac{180°}{\pi}\right) \approx 134.6°$

**3.** $\tan \theta = \dfrac{y}{x} = -\dfrac{4}{3} \Rightarrow r = 5$

$\theta$ is in Quadrant IV $\Rightarrow x = 3$.

$\cos \theta = \dfrac{x}{r} = \dfrac{3}{5}$

**4.** (a) $f(x) = 3 - 2 \sin \pi x$

Period: $\dfrac{2\pi}{\pi} = 2$

Amplitude: $|a| = |-2| = 2$

Upward shift of 3 units (reflected in $x$-axis prior to shift)

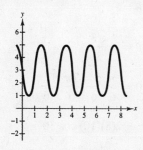

(b) $g(x) = \dfrac{1}{2} \tan\left(x - \dfrac{\pi}{2}\right)$

Period: $\pi$

Asymptotes: $x = 0, x = \pi$

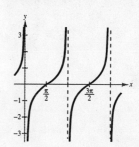

**5.** $h(x) = a\cos(bx + c)$

Graph is reflected in $x$-axis.

Amplitude: $a = -3$

Period: $2 = \dfrac{2\pi}{\pi} \Rightarrow b = \pi$

No phase shift: $c = 0$

$h(x) = 3\cos(\pi x)$

**6.** $y = \arccos(2x)$

$\sin y = \sin(\arccos(2x)) = \sqrt{1 - 4x^2}$

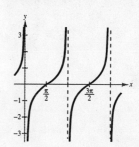

7. $\dfrac{\sin\theta - 1}{\cos\theta} - \dfrac{\cos\theta}{\sin\theta - 1} = \dfrac{\sin\theta - 1}{\cos\theta} - \dfrac{\cos\theta(\sin\theta + 1)}{\sin^2 - 1}$

$\qquad\qquad = \dfrac{\sin\theta - 1}{\cos\theta} - \dfrac{\cos\theta(\sin\theta + 1)}{-\cos^2\theta} = \dfrac{\sin\theta - 1}{\cos\theta} + \dfrac{\sin\theta + 1}{\cos\theta} = \dfrac{2\sin\theta}{\cos\theta} = 2\tan\theta$

8. (a) $\cot^2\alpha(\sec^2\alpha - 1) = \cot^2\alpha\,\tan^2\alpha = 1$

   (b) $\sin(x + y)\sin(x - y) = \frac{1}{2}[\cos(x + y - (x - y)) - \cos(x + y + x - y)]$

$\qquad\qquad\qquad\qquad = \frac{1}{2}[\cos 2y - \cos 2x] = \frac{1}{2}[1 - 2\sin^2 y - (1 - 2\sin^2 x)] = \sin^2 x - \sin^2 y$

   (c) $\sin^2 x \cos^2 x = \left(\dfrac{1 - \cos 2x}{2}\right)\left(\dfrac{1 + \cos 2x}{2}\right)$

$\qquad\qquad\qquad = \dfrac{1}{4}(1 - \cos 2x)(1 + \cos 2x)$

$\qquad\qquad\qquad = \dfrac{1}{4}(1 - \cos^2 2x)$

$\qquad\qquad\qquad = \dfrac{1}{4}\left(1 - \dfrac{1 + \cos 4x}{2}\right)$

$\qquad\qquad\qquad = \dfrac{1}{8}(2 - (1 + \cos 4x))$

$\qquad\qquad\qquad = \dfrac{1}{8}(1 - \cos 4x)$

9. (a) $2\cos^2\beta - \cos\beta = 0$

$\qquad \cos\beta(2\cos\beta - 1) = 0$

$\qquad \cos\beta = 0 \qquad\qquad 2\cos\beta - 1 = 0$

$\qquad\qquad \beta = \dfrac{\pi}{2}, \dfrac{3\pi}{2} \qquad\qquad \cos\beta = \dfrac{1}{2}$

$\qquad\qquad\qquad\qquad\qquad\qquad \beta = \dfrac{\pi}{3}, \dfrac{5\pi}{3}$

$\qquad$ Answer: $\dfrac{\pi}{3}, \dfrac{\pi}{2}, \dfrac{3\pi}{2}, \dfrac{5\pi}{3}$

   (b) $3\tan\theta - \cot\theta = 0$

$\qquad 3\tan\theta - \dfrac{1}{\tan\theta} = 0$

$\qquad \dfrac{3\tan^2\theta - 1}{\tan\theta} = 0$

$\qquad 3\tan^2\theta - 1 = 0$

$\qquad \tan^2\theta = \dfrac{1}{3}$

$\qquad \tan\theta = \pm\dfrac{\sqrt{3}}{3}$

$\qquad \theta = \dfrac{\pi}{6}, \dfrac{5\pi}{6}, \dfrac{7\pi}{6}, \dfrac{11\pi}{6}$

10. (a) $\dfrac{\sin B}{8} = \dfrac{\sin 30°}{9}$

$\qquad \sin B = \dfrac{8}{9}\left(\dfrac{1}{2}\right)$

$\qquad B = \arcsin\left(\dfrac{4}{9}\right)$

$\qquad B \approx 26.4°$

$\qquad C = 180° - A - B \approx 123.6°$

$\qquad \dfrac{c}{\sin 123.6°} = \dfrac{9}{\sin 30°}$

$\qquad\qquad c \approx 15.0$

   (b) $a^2 = 8^2 + 10^2 - 2(8)(10)\cos 30°$

$\qquad a^2 \approx 25.4$

$\qquad a \approx 5.0$

$\qquad \cos B = \dfrac{5.0^2 + 10^2 - 8^2}{2(5.0)(10)}$

$\qquad \cos B = 0.61$

$\qquad B \approx 52.4°$

$\qquad C = 180° - A - B \approx 97.6°$

11. Height of smaller triangle:

$$\tan 16°\,45' = \frac{h_1}{200}$$

$$h_1 = 200 \tan 16.75° \approx 60.2 \text{ feet}$$

Height of larger triangle:

$$\tan 18° = \frac{h_2}{200}$$

$$h = 200 \tan 18° \approx 65.0 \text{ feet}$$

Height of flag:

$$h_2 - h_1 = 65.0 - 60.2 \approx 5 \text{ feet}$$

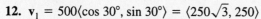

Not to scale.

12. $\mathbf{v}_1 = 500\langle \cos 30°, \sin 30° \rangle = \langle 250\sqrt{3}, 250 \rangle$

$\mathbf{v}_2 = 50\langle \cos 60°, \sin 60° \rangle = \langle 25, 25\sqrt{3} \rangle$

$\mathbf{v} = \mathbf{v}_1 + \mathbf{v}_2 = \langle 250\sqrt{3} + 25, 250 + 25\sqrt{3} \rangle \approx \langle 458.0, 293.3 \rangle$

$\|\mathbf{v}\| = \sqrt{(458.0)^2 + (293.3)^2} \approx 543.9$

$\tan \theta = \dfrac{293.3}{458.0} \approx 0.6404 \implies \theta \approx 32.6°$

The plane is traveling N 32.6° E at 543.9 kilometers per hour.

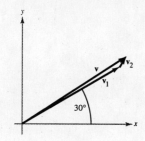

13. $\mathbf{u} = \langle -4, 3 \rangle$, $\mathbf{v} = \langle -1, 5 \rangle$

$$\text{proj}_{\mathbf{v}}\mathbf{u} = \left( \frac{\mathbf{u} \cdot \mathbf{v}}{\|\mathbf{v}\|^2} \right)\mathbf{v}$$

$$= \frac{19}{26}\langle -1, 5 \rangle$$

# ❑ **Chapter Test Solutions for Chapter 4**

**1.** $-2 + \sqrt{-64} = -2 + 8i$

**2.** (a) $10i - \left(3 + \sqrt{-25}\right) = 10i - (3 + 5i)$
$$= -3 + 5i$$

(b) $(3 + 5i)^2 = 9 + 30i + 25i^2$
$$= -16 + 30i$$

(c) $\left(2 + \sqrt{3}i\right)\left(2 - \sqrt{3}i\right) = 4 - 3i^2 = 4 + 3 = 7$

(d) $\dfrac{5}{2 + i} = \dfrac{5}{2 + i} \cdot \dfrac{2 - i}{2 - i} = \dfrac{5(2 - i)}{4 + 1} = 2 - i$

**3.** $f(x) = x^4 - x^3 - 1$
$x \approx 1.380$ and $x \approx -0.819$

**4.** $f(x) = 3x^5 + 2x^4 - 12x - 8$
$x \approx \pm 1.414$ and $x \approx -0.667$

**5.** $h(x) = x^4 + 3x^2 - 4$

Zeros: $-1, 1 \implies (x + 1)(x - 1) = x^2 - 1$ is a factor of $h(x)$

$$
\begin{array}{r}
x^2 + 4 \phantom{)} \\
x^2 - 1 \overline{\smash{)}\, x^4 + 3x^2 - 4} \\
\underline{x^4 - \phantom{3}x^2} \phantom{-4} \\
4x^2 - 4 \\
\underline{4x^2 - 4} \\
0
\end{array}
$$

Thus, $h(x) = (x^2 - 1)(x^2 + 4)$
$$= (x - 1)(x + 1)(x - 2i)(x + 2i)$$

The zeros of $h(x)$ are: $\pm 1, \pm 2i$

**6.** $g(v) = 2v^3 - 11v^2 + 22v - 15$

Zero: $\frac{3}{2} \implies 2v - 3$ is a factor of $g(v)$

$$
\begin{array}{r}
v^2 - \phantom{0}4v + \phantom{0}5 \phantom{)} \\
2v - 3 \overline{\smash{)}\, 2v^3 - 11v^2 + 22v - 15} \\
\underline{2v^3 - \phantom{0}3v^2} \phantom{+ 22v - 15} \\
-8v^2 + 22v \phantom{- 15} \\
\underline{-8v^2 + 12v} \phantom{- 15} \\
10v - 15 \\
\underline{10v - 15} \\
0
\end{array}
$$

Thus, $g(v) = (2v - 3)(v^2 - 4v + 5)$. By the Quadratic Formula, the zeros of $v^2 - 4v + 5$ are $2 \pm i$.

The zeros of $g(v)$ are: $\frac{3}{2}, 2 \pm i$

**7.** $f(x) = x(x - 3)[x - (3 + i)][x - (3 - i)]$
$$= (x^2 - 3x)[(x - 3) - i][(x - 3) + i]$$
$$= (x^2 - 3x)[(x - 3)^2 - i^2]$$
$$= (x^2 - 3x)(x^2 - 6x + 10)$$
$$= x^4 - 9x^3 + 28x^2 - 30x$$

**8.** $f(x) = \left[x - \left(1 + \sqrt{3}i\right)\right]\left[x - \left(1 - \sqrt{3}i\right)\right](x - 2)(x - 2)$
$$= \left[(x - 1) - \sqrt{3}i\right]\left[(x - 1) + \sqrt{3}i\right](x^2 - 4x + 4)$$
$$= [(x - 1)^2 - 3i^2](x^2 - 4x + 4)$$
$$= (x^2 - 2x + 4)(x^2 - 4x + 4)$$
$$= x^4 - 6x^3 + 16x^2 - 24x + 16$$

**9.** No, complex zeros occur in conjugate pairs for polynomial functions with *real* coefficients. If $a + bi$ is a zero, so is $a - bi$.

**10.** $-2 + 2i$

$r = \sqrt{(-2)^2 + 2^2} = 2\sqrt{2}$

$\tan \theta = -\dfrac{2}{2}$, $\theta$ is in Quadrant II

$\theta = 135°$

$-2 + 2i = 2\sqrt{2}(\cos 135° + i \sin 135°)$

**11.** $[4(\cos 30° + i \sin 30°)][6(\cos 120° + i \sin 120°)]$

$= (4)(6)[\cos (30° + 120°) + i \sin(30° + 120°)]$

$= 24(\cos 150° + i \sin 150°)$

$= 24\left(-\dfrac{\sqrt{3}}{2} + \dfrac{1}{2}i\right)$

$= -12\sqrt{3} + 12i$

**12.** $1 = \cos 0° + i \sin 0°$

The cube roots of 1 are : $\cos\left(\dfrac{0° + 360°k}{3}\right) + i \sin \left(\dfrac{0° + 360°k}{3}\right)$; $k = 0, 1, 2$

For $k = 0$: $\cos 0° + i \sin 0° = 1$

For $k = 1$: $\cos 120° + i \sin 120° = -\dfrac{1}{2} + \dfrac{\sqrt{3}}{2}i$

For $k = 2$: $\cos 240° + i \sin 240° = -\dfrac{1}{2} - \dfrac{\sqrt{3}}{2}i$

## ❑ Chapter Test Solutions for Chapter 5

**1.** $f(x) = 2^{-x/3}$

| $x$ | $-3$ | $0$ | $3$ | $9$ |
|---|---|---|---|---|
| $y$ | $2$ | $1$ | $\frac{1}{2}$ | $\frac{1}{8}$ |

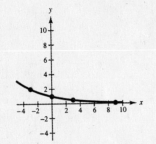

**2.** $f(x) = \dfrac{1000}{1 + 4e^{-0.2x}}$

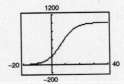

Horizontal asymptotes: $y = 0$ and $y = 1000$

**3.** (a) $A = 5000\left(1 + \dfrac{0.065}{4}\right)^{(4)(30)} \approx \$34{,}596.89$

   (b) $A = 5000e^{(0.065)(30)} \approx \$35{,}143.44$

**4.** $200{,}000 = P\left(1 + \dfrac{0.08}{365}\right)^{(365)(20)}$

   $P \approx \$40{,}386.38$

**5.** $\log_4 64 = 3 \implies 4^3 = 64$

**6.** $5^{-2} = \frac{1}{25} \implies \log_5 \frac{1}{25} = -2$

**7.** $g(x) = \log_3(x - 2)$

Vertical asymptote: $x = 2$

| $x$ | $3$ | $5$ | $2\frac{1}{3}$ |
|---|---|---|---|
| $y$ | $0$ | $1$ | $-1$ |

**8.** $\ln\left(\dfrac{6x^2}{\sqrt{x^2 + 1}}\right) = \ln 6x^2 - \ln\sqrt{x^2 + 1}$

$= \ln 6 + \ln x^2 - \ln(x^2 + 1)^{1/2}$

$= \ln 6 + 2\ln x - \dfrac{1}{2}\ln(x^2 + 1)$

**9.** $3\ln z - [\ln(z + 1) + \ln(z - 1)] = \ln z^3 - \ln(z + 1)(z - 1)$

$= \ln\left(\dfrac{z^3}{z^2 - 1}\right)$

**10.** $\log_6 \sqrt{360} = \dfrac{1}{2}\log_6(36 \cdot 10)$

$= \dfrac{1}{2}[\log_6 6^2 + \log_6 10]$

$= \dfrac{1}{2}[2 + \log_6 10]$

$= 1 + \dfrac{1}{2}\log_6 10$

**11.** $e^{x/2} = 450$

$\dfrac{x}{2} = \ln 450$

$x = 2\ln 450 \approx 12.218$

**12.** $\left(1 + \dfrac{0.06}{4}\right)^{4t} = 3$

$\ln(1.015)^{4t} = \ln 3$

$4t\ln 1.015 = \ln 3$

$t = \dfrac{\ln 3}{4\ln 1.015}$

$t \approx 18.447$

**13.** $5\ln(x + 4) = 22$

$\ln(x + 4) = 4.4$

$x + 4 = e^{4.4}$

$x = e^{4.4} - 4 \approx 77.451$

**14.** $(0, 28{,}000)$, $(1, 20{,}000)$

$y = Ce^{kt}$

$28{,}000 = Ce^{k(0)} \implies C = 28{,}000$

$20{,}000 = 28{,}000e^{k(1)}$

$\frac{5}{7} = e^{k}$

$\ln\left(\frac{5}{7}\right) = k$

$y = 28{,}000e^{t \ln(5/7)}$

When $t = 3$: $y = 28{,}000e^{3 \ln(5/7)} \approx \$10{,}204$

**15–17 Use** $p(t) = \dfrac{1200}{1 + 3e^{-t/5}}$

**15.** $p(0) = \dfrac{1200}{1 + 3e^{0}} = 300$

**16.** $p(5) = \dfrac{1200}{1 + 3e^{-1}} \approx 570$

**17.** $800 = \dfrac{1200}{1 + 3e^{-t/5}}$

$1 + 3e^{-t/5} = \dfrac{1200}{800}$

$3e^{-t/5} = 0.5$

$e^{-t/5} = \dfrac{0.5}{3}$

$-\dfrac{t}{5} = \ln\left(\dfrac{0.5}{3}\right)$

$t = -5 \ln\left(\dfrac{0.5}{3}\right) \approx 9 \text{ years}$

**18.** We can eliminate (a) because (a) has a $y$-intercept of $(0, 6)$. We can eliminate (b) because the graph shown has $y$-axis symmetry and a $y$-intercept of $(0, 0)$. Choice (b) does not have $y$-axis symmetry (you cannot replace $x$ with $-x$ and get an equivalent equation) and it has a $y$-intercept of $(0, 3)$. The graph matches equation (c).

# ❑ Cumulative Test Solutions for Chapters 4–6

**1.** $3 - \sqrt{-25} = 3 - 5i$

**2.** $\dfrac{2i}{3 - 4i} = \dfrac{2i}{3 - 4i} \cdot \dfrac{3 + 4i}{3 + 4i} = \dfrac{6i + 8i^2}{9 + 16}$

$$= \dfrac{-8 + 6i}{25} = -\dfrac{8}{25} + \dfrac{6}{26}i$$

**3.** $0 = x^3 + 4x^2 + 5x$

$0 = x(x^2 + 4x + 5)$

$x = 0 \quad \text{or} \quad x = \dfrac{-4 \pm \sqrt{4^2 - 4(1)(5)}}{2(1)} = \dfrac{-4 \pm \sqrt{-4}}{2} = -2 \pm i$

The zeros are: $x = 0, x = -2 \pm i$

**4.** $-12 + 5i$

$r = \sqrt{(-12)^2 + 5^2} = \sqrt{169} = 13$

$\tan \theta = -\dfrac{5}{12}$

Since $\theta$ is in Quadrant II, $\theta \approx 2.7468$ radians and $z \approx 13(\cos 2.7468 + i \sin 2.7468)$.

**5.** $[3(\cos 30° + i \sin 30°)]^4 = 3^4(\cos 120° + i \sin 120°)$

$$= 81\left(-\dfrac{1}{2} + \dfrac{\sqrt{3}}{2}i\right)$$

$$= -\dfrac{81}{2} + \dfrac{81\sqrt{3}}{2}i$$

**6.** (a) $f(x) = 6(2^{-x}) = 6\left(\dfrac{1}{2}\right)^x$

Exponential decay

Horizontal asymptote: $y = 0$

Intercept: $(0, 6)$

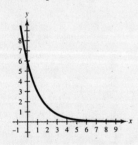

(b) $g(x) = \log_3 x \Longrightarrow 3^y = x$

Vertical asymptote: $x = 0$

Intercept: $(1, 0)$

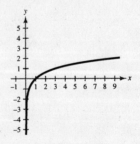

**7.** (a) $6e^{2x} = 72$

$e^{2x} = 12$

$2x = \ln 12$

$x = \frac{1}{2} \ln 12 \approx 1.2425$

(b) $\log_2 x + \log_2 5 = 6$

$\log_2 5x = 6$

$5x = 2^6$

$x = \dfrac{64}{5}$

**8.** $y^2 - 4x + 4 = 0$

$y^2 = 4(x - 1)$

Parabola, opens to the right.

Vertex: $(1, 0)$

Focus: $(2, 0)$

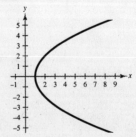

**9.** $\dfrac{(x-2)^2}{4} + \dfrac{(y+1)^2}{9} = 1$

Ellipse

Center: $(2, -1)$

Vertices: $(2, -4)$ and $(2, 2)$

Foci: $\left(2, -1 \pm \sqrt{5}\right)$

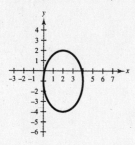

**10.** $\dfrac{x^2}{1} - \dfrac{y^2}{4} = 1$

Hyperbola

Center: $(0, 0)$

$a = 1, b = 2, c = \sqrt{5}$

Horizontal transverse axis

Vertices: $(\pm 1, 0)$

Foci: $\left(\pm\sqrt{5}, 0\right)$

Asymptotes: $y = \pm 2x$

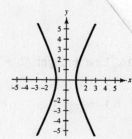

**11.** $x^2 - 4y^2 - 4x = 0$

$(x^2 - 4x + 4) - 4y^2 = 0 + 4$

$\dfrac{(x-2)^2}{4} - \dfrac{y^2}{1} = 1$

Vertices: $(0, 0), (4, 0)$

Foci: $\left(2 \pm \sqrt{5}, 0\right)$

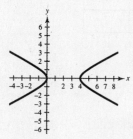

**12.** Parabola

Vertex: $(3, -2) \Longrightarrow y = a(x-3)^2 - 2$

Point: $(0, 4) \Longrightarrow 4 = a(0-3)^2 - 2$

$$6 = 9a \Longrightarrow a = \tfrac{2}{3}$$

Equation: $y = \tfrac{2}{3}(x-3)^2 - 2$

$$3y = 2(x^2 - 6x + 9) - 6$$

$$0 = 2x^2 - 12x - 3y + 12$$

**13.** Hyperbola

Foci: $(0, 0)$ and $(0, 4) \Longrightarrow$ Center: $(0, 2)$ and vertical transverse axis

Asymptotes: $y = \pm\dfrac{1}{2}x + 2 \Longrightarrow \dfrac{a}{b} = \dfrac{1}{2} \Longrightarrow 2a = b$

$c^2 = a^2 + b^2 \Longrightarrow 4 = a^2 + 4a^2 \Longrightarrow a^2 = \dfrac{4}{5}$ and $b^2 = \dfrac{16}{5}$

Equation: $\dfrac{(y-2)^2}{4/5} - \dfrac{x^2}{16/5} = 1$

$$20(y-2)^2 - 5x^2 = 16$$

$$20y^2 - 80y - 5x^2 + 80 = 16$$

$$0 = 5x^2 - 20y^2 + 80y - 64$$

**14.** $x^2 + 6xy + y^2 - 6 = 0$

(a) $\cot 2\theta = \dfrac{1-1}{6} = 0 \Longrightarrow 2\theta = \dfrac{\pi}{2} \Longrightarrow \theta = \dfrac{\pi}{4} = 45°$

(b) To use a graphing utility, solve for $y$ in terms of $x$.

$$y^2 + 6xy = 6 - x^2$$

$$y^2 + 6xy + 9x^2 = 6 - x^2 + 9x^2$$

$$(y + 3x)^2 = 8x^2 + 6$$

$$y = -3x \pm \sqrt{8x^2 + 6}$$

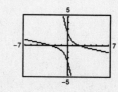

**15.** $x = 2 + 3\cos\theta \implies \dfrac{(x-2)}{3} = \cos\theta$

$y = 2\sin\theta \implies \dfrac{y}{2} = \sin\theta$

$\dfrac{(x-2)^2}{9} + \dfrac{y^2}{4} = 1$ Ellipse

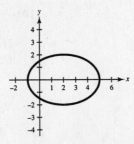

**16.** Line through: $(2, -3)$ and $(6, 4)$

$x = x_1 + t(x_2 - x_1) = 2 + 4t$

$y = y_1 + t(y_2 - y_1) = -3 + 7t$

**17.** $x^2 + y^2 - 6y = 0$

$r^2 - 6r\sin\theta = 0$

$r(r - 6\sin\theta) = 0$

$r = 6\sin\theta$

**18.** (a) $r = \dfrac{4}{1 + \cos\theta}$

$e = 1$

Parabola with a vertical directrix to the right of the pole.

| $\theta$ | 2 | $\pi/2$ | $3\pi/2$ |
|---|---|---|---|
| $r$ | 2 | 4 | 4 |

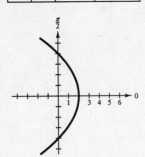

(b) $r = \dfrac{4}{2 + \cos\theta} = \dfrac{2}{1 + (1/2)\cos\theta}$

$e = \dfrac{1}{2}$

Ellipse with a vertical directrix to the right of the pole.

| $\theta$ | 0 | $\pi/2$ | $\pi$ | $3\pi/2$ |
|---|---|---|---|---|
| $r$ | 4/3 | 2 | 4 | 2 |

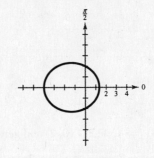

**19.** (a) $r = 2 + 3\sin\theta$ is a limaçon. Matches (iii).

(b) $r = 3\sin\theta$ is a circle. Matches (i).

(c) $r = 3\sin 2\theta$ is a rose curve. Matches (ii).